Environmental Change in the Pacific Basin

Other books by Patrick D. Nunn

Oceanic Islands (1994, Blackwell)
Pacific Island Landscapes (1998, The University of the South Pacific)

More details available at web site –

http://www.usp.ac.fj/~geography/

Environmental Change in the Pacific Basin

Chronologies, Causes, Consequences

Patrick D. Nunn
Professor of Oceanic Geoscience, The University of the South Pacific

JOHN WILEY & SONS
Chichester · Weinheim · New York · Brisbane · Singapore · Toronto

National   01243 779777
International   (+44) 1243 779777
e-mail (for orders and customer service enquiries): cs-books@wiley.co.uk
Visit our Home Page on http://www.wiley.co.uk
                   or http://www.wiley.com

Reprinted August 2000

Other Wiley Editorial Offices

John Wiley & Sons, Inc., 605 Third Avenue,
New York, NY 10158-0012, USA

WILEY-VCH Verlag GmbH, Pappelallee 3,
D-69469 Weinheim, Germany

Jacaranda Wiley Ltd, 33 Park Road, Milton,
Queensland 4064, Australia

John Wiley & Sons (Asia) Pte Ltd, 2 Clementi Loop #02-01,
Jin Xing Distripark, Singapore 129809

John Wiley & Sons (Canada) Ltd, 22 Worcester Road,
Rexdale, Ontario M9W 1L1, Canada

Library of Congress Cataloging-in-Publication Data

Nunn, Patrick D., 1955–
    Environmental change in the Pacific Basin: chronologies, causes, consequences / Patrick D. Nunn.
        p.   cm.
    Includes bibliographical references and index.
    ISBN 0-471-94945-0
    1. Oceanography—Pacific Ocean Region.   2. Climatic changes—Pacific Ocean Region.   3. Paleoecology—Pacific Ocean Region.
    I. Title
    GC771.N86   1999
    551.46′5—dc21   98-47323
                                                   CIP

British Library Cataloguing in Publication Data

A catalogue record for this book is available from the British Library

ISBN 0 471 94945 0

Typeset in 10/12 pt Times by MHL Typesetting Ltd, Coventry
Printed and bound in Great Britain by Bookcraft (Bath) Ltd, Midsomer Norton
This book is printed on acid-free paper responsibly manufactured from sustainable forestry, in which at least two trees are planted for each one used for paper production.

for Fipe

Contents

PART F   CONCLUSIONS

Preface and Acknowledgements

The subject of environmental change has become highly topical in recent years for two main reasons. Firstly, observations have been made increasingly of deteriorating environments, commonly perceived to be the direct result of human mismanagement. Secondly, the spectre of human-enhanced climate change – the ubiquitous 'greenhouse effect' – looms ever larger and with it the realization that the 21st century may see changes to environments at rates greater than many in the recorded past.

To be properly understood, the issue of contemporary and future environmental change needs to be placed in its historical context. Our world has been evolving continuously and will do so irrespective of what humanity does in the future. The study of what changed in the past and why these changes occurred is instructive in its own right but particularly so for persons concerned with planning our collective future. This is the fundamental rationale for this book. Such a rationale has motivated the writing of other fine books about environmental change, but this one strives to be distinctive in two ways.

Firstly it focuses on the Pacific Basin, a region which has suffered much from being marginalized in thematic books about the natural sciences and has had very few such books devoted to it. By writing such a book, the intention is to add to the number of reference texts to which the inhabitants of the Pacific Islands and Pacific Rim can relate directly while at the same time demonstrating that the subject is of interest and relevance to a wider geographical audience.

The second way in which this book tries to be distinct from others written on the subject is by not overplaying the role of humans in environmental change. One of the greatest current misconceptions in the popular psyche is that environmental change is the result largely of human impact and that, were environmentally detrimental human impacts to be stopped, environments would revert to an unchanging, pristine condition, the way they allegedly were before humans arrived in them. Yet the history of environmental change is not merely the history of human interaction with the environment; environments have been changing ever since they were created. Humans are a comparatively recent cause of change who, because they have developed the potential to alter environments more profoundly than any other species, have acquired a moral obligation to understand how to manage those environments effectively. This book was written with this purpose in mind.

The final draft of this book was completed during a six-month sabbatical at the Quaternary Research Center at the University of Washington in Seattle, USA. The staff of the Center could not have been more hospitable nor more supportive of my work. I thank the University of the South Pacific for permitting me sabbatical leave.

It has been my good fortune to discuss environmental change with a number of stimulating people in recent years. Among this number are Bill Aalbersberg, Atholl Anderson, Russell Blong, John Chappell, William Clarke, Keith Crook, Bill Dickinson, Geoff Hope, Roger McLean, Nobuo Mimura, Cliff Ollier, Yoko Ota, Paolo Pirazzoli, Steve Porter, David Stoddart, James Terry, Randy Thaman, Michael Tooley, Eric Waddell and Paul Williams. These people do not necessarily endorse any views I express in this book.

For specific help, I thank Pat Anderson, Stephen Athens, Brian Atwater, Susan Bartsch-Winkler, Kelvin Berryman, Ted Bryant, John Clague, Chalmers Clapperton, Darrel Cowan, George Denton, John Flenley, Paul Geraghty, Jack Golson, Rick Grigg, Bernard Hallet, Keiji Imamura, Thea Johanos, Toshio Kawana, James Kennett, Patrick Kirch, Yasuo Kitagamae, Larry Lawver, Wesley LeMasurier, Karl Lilliquist, Peter Newell, Akio Omura, Luc Ortlieb, Judith Totman Parrish, Brad Pillans, Charles Repenning, Jim Specht, Minze Stuiver, Bob Tingey, Lisa Wells, Peter White and Jack Wolfe. Particular thanks are due to Bill Dickinson for reading and commenting on the entire manuscript.

Many of the drawings were produced by Turenga Christopher, Faustino Yarofaisug, Finau Ratuvili, Naz Bibi and Savitri Karunairetnam under the direction of Bruce Davis, James Britton and George Saemane in the Geography Department's GIS Laboratory at the University of the South Pacific. Special thanks go to Sharon McGowan for efficient and cheerful secretarial support. The reconstructions of the Pacific shown in Figure 3.2 were prepared specially by Lisa Gahagan at the PLATES Project, Austin Institute for Geophysics, The University of Texas. Phil Woodward, formerly of SOPAC, redrew this figure and others with his usual flair. I have been fortunate to work with a positive and well-organised team at Wiley. In particular, I thank Sally Wilkinson, Louise Portsmouth, Isabelle Strafford and Teresa MacLeod.

As ever, I thank my family – Fipe, Rachel, Warwick and Mary Nunn – for their unfailing support and encouragement without which it would have been impossible to write this book.

Patrick D. Nunn

Laucala Bay, Fiji, and
Seattle, USA

June 1998

Organization of this Book

This book is divided into six unequal parts.

Part A establishes a context for the rest of the book; it defines the Pacific Basin, and describes the important attributes of its environments and the processes which control their current development (Chapter 1). There follows an account of the history of scientific investigation of the Pacific Basin, and the range of techniques employed to reconstruct its long-term history (Chapter 2). Ideas concerning the origin and physical development of the Pacific Basin are presented next (Chapter 3).

Part B reviews the history of the Pacific Basin before the Mesozoic–Cenozoic boundary ~66 Ma. Chapter 4 deals with the earliest Pacific, during the Precambrian and Palaeozoic eras. Chapter 5 focuses on the Mesozoic Era.

Part C looks at the history of the Pacific Basin during the Tertiary Period. Chapter 6 describes the changes which took place during the Palaeogene. Neogene changes are recounted in Chapter 7.

Part D focuses on the Pleistocene, Chapter 8 being a brief discussion of its chronology. Chapter 9 looks at Pacific Basin environments during the early and middle Pleistocene, Chapter 10 examines those of the Last Interglacial in detail and Chapter 11 considers those of the various stages of the Last Glacial.

Part E covers the Holocene, the last 12 000 calendar years of Earth history, and that part for which most information is available. This was also the epoch during which human impact on the environment first became widespread, in deference to which Chapter 12 outlines the history of human arrival in the Pacific Basin. Chapter 13 is devoted to the early Holocene, Chapter 14 to the middle Holocene and Chapter 15 to the late Holocene, the time leading up to the present during which human impact in many parts of the region increased rapidly in both magnitude and extent.

Part F concludes the book with a single chapter (Chapter 16) highlighting key areas for future research.

Conventions for Expressing Age

The abbreviation 'Ma' is used for millions of years ago. By itself, the abbreviation 'ka' is used for thousands of calendar years ago. The abbreviation 'ka BP' is used for thousands of radiocarbon years before AD 1950. The abbreviation 'BP' is used for radiocarbon years before AD 1950. As the present is approached, it is deemed preferable to use BC and AD to identify calendar years. No attempt has been made to convert radiocarbon ages to calendar ages although conversion tables were given by Stuiver and Pearson (1993) and other articles in the same issue of the journal *Radiocarbon*.

PART A

Context

CHAPTER 1

The Pacific Basin

Let us go forth, the tellers of tales, and seize whatever prey the heart longs for, and have no fear.
Everything exists, everything is true, and the earth is only a little dust under our feet

W.B. Yeats

1.1 DEFINITION

The Pacific Basin is that part of the Earth's surface covered by the Pacific Ocean, the islands within it and those parts of the continents around its edge, the origins of which are linked to those of the Pacific Ocean, and/or the environments of which closely affect or are influenced by the Pacific Ocean (Figure 1.1). For ease of discussion, the Pacific Basin is divided into two parts: (i) the *Pacific Rim*, and (ii) the *Pacific Ocean and Islands*. The former is mostly land, the latter mostly ocean with a little land.

The Pacific Basin has integrity in many senses. In a physical sense, it includes the modern Pacific Ocean floor – formed as much as 174 Ma – along with disaggregated fragments of ancestral Pacific Ocean floors which have become accreted onto the surrounding continents. In another sense, the interaction between different parts of the Rim, the Ocean and the Islands is fundamental to the global economy and likely to become more so if the forthcoming 'Pacific Century' progresses as predicted (Linder, 1986).

Similar comments apply to environmental change. Both literally and figuratively, winds of change blow across the Pacific Basin in all directions. Cyclonic storms which form in the west Pacific 'warm pool' near Solomon Islands may range across its island neighbours to the southeast, or move south along the east coast of Australia, often gathering strength and destructive force as they proceed. Air from Siberia sweeps across Japan in winter producing cold monsoonal

rains in sharp contrast to the summer monsoon associated with warm water driven northwards from the Equator. In common with the Americas, most Pacific islands were colonized initially by people from the central western part of the Rim. The cultural and behavioural baggage of the earliest settlers transformed the environments they encountered. Yet today, the numbers of Pacific Islanders in the metropolitan centres of the Pacific Rim greatly exceed numbers on their islands of origin – a reverse colonization with less evident environmental effects.

The diversity of physical environments, of human history and culture, within the Pacific Basin accounts largely for the varying degree of information about environmental change across this vast region. The Pacific Rim in North America, occupied by just two nations, yielded the secrets of its past earlier and more readily than the less hospitable terrain of Pacific South America, which has long been divided between a greater number of nations. The geotectonic and environmental history of East Asia is still unclear to many English-speaking scientists in contrast to the Australasia region, where understanding of such topics is well in advance of that in many other parts of the world. Owing largely to the harshness of its terrain, most of Beringia in the far north Pacific remains less well known than other parts of the Pacific Rim with the sole exception of Antarctica, where ice cover restricts observation and understanding. Information is comparatively sparse for most island groups. Some like Hawaii[1] have had human populations for only a fraction of

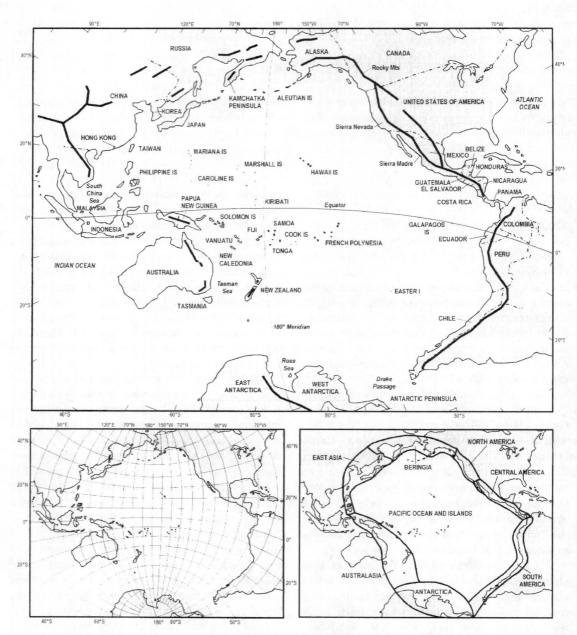

Figure 1.1   The Pacific Basin. Major mountain ranges are shown in black where this would not obscure other detail. Insets show the geographic grid and the constituent regions used in the text

the time that the Pacific Rim has been settled. Others like Japan have been settled for much longer and have an uncommonly long written record of environmental change.

Many parts of the Pacific Rim are mountainous and records of environmental change are linked closely to an understanding of mountain origin and landscape development. Other parts are lower with landscapes ranging from polar to equatorial. An understanding of how these environments have changed can therefore be viewed as analogous to the ways in which environments worldwide have changed. With the inclusion in the region of a huge ocean, it is clear that the Pacific Basin can be considered a microcosm of the entire surface of the Earth.

1.1.1 The Pacific Rim

The Pacific Rim is an area of considerable genetic variability. Most is composed of detached pieces (named terranes) of oceanic or continental landmasses which have become accreted onto older continental cores (cratons). Primarily to facilitate discussion, the Pacific Rim is divided into regions, illustrated in Figure 1.1, and referred to throughout this book.

The *Antarctica* region includes the Antarctic periphery between 140°E and 70°W together with adjoining islands and ocean floor. This includes the Ross Sea and extends as far inland as the Pacific flanks of the Transantarctic Mountains. The western coasts of the Antarctic Peninsula are also included within this region.

The *South America* region extends northwards from Drake Passage along the western seaboard of South America to the Colombia–Panama border. The westernmost divide of the Andes forms the eastern border of most of this region. The Galápagos Islands are included in the Pacific Ocean and Islands, not this region.

The *Central America* region extends from Panama to the Mexico–United States border near

the head of the Gulf of California, the eastern boundary of the Pacific Rim approximately bisecting the region along its long axis. The Yucatan Peninsula, and the Caribbean halves of Honduras and Mexico are not therefore included in this region.

The *North America* region extends north to the Alaska–Canada border. In the United States, the eastern border of the Pacific Rim includes the coastal ranges and extends to the main divide of the Sierra Nevada in California, and that of the Cascade Range farther north, but does not reach as far inland as the main chain of the Rocky Mountains. A comparable situation applies in Canada.

The *Beringia* region covers the northernmost part of the Pacific Rim, extending from southern Alaska in the east to easternmost Russia including the Kamchatka Peninsula in the west.

The *East Asia* region extends from northeast China through Hong Kong to the southernmost islands of the Philippines. It includes the Kuril Islands, the Korean Peninsula, Japan, the Ryukyu Islands and Taiwan.

The *Australasia* region extends from Halmahera island in Indonesia through New Guinea (Irian Jaya – mainland Papua New Guinea) and along the eastern seaboard of Australia through Tasmania to Macquarie island. In Australia, the eastern slopes of the Great Dividing Range form the western boundary of this region which includes New Zealand and New Caledonia.

The constituent regions of the Pacific Rim are merely a vehicle for discussion; they have geographic integrity but the degree to which they are physically integrated is highly variable.

The disparities between the *Pacific Rim* and the *Pacific Ocean and Islands* (see below) in terms of an understanding of environmental change and the amount of effort which has been put into this are worthy of note at this point. Both now and in the past, the number of people investigating environmental change in the Rim has been far greater than those employed in similar work in the Ocean and Islands. Most major cities along the Pacific Rim have institutions which have played major roles in enhancing understanding of

---

[1] The glottal stop is used to distinguish the island Hawai'i from the Hawaii island group in this book

environmental change within the region yet most researchers in such institutions have comparatively parochial interests. As a result of this, there are gaps in the understanding of environmental change in the Pacific Rim, which generally broaden as one goes farther back in time. Examples include Central America, western Beringia, parts of mainland East Asia, and the Philippines.

1.1.2 The Pacific Ocean and Islands

The Pacific Ocean and Islands covers approximately 45% of the Earth's surface yet contains less than 0.01% of both its total land area and human population. Most islands are located in the southwest quadrant of the Pacific Ocean. Of the remainder, fewest islands are found in the northeast quadrant. To many outsiders, the Pacific Ocean is truly 'Earth's empty quarter', a place which most people cross by air or by sea, often blithely unaware of the complexity of the environments on the islands over the horizon or the ocean floor below (Ward, 1989).

Pacific islands vary immensely in physiography and history. Large islands, many once regarded (incorrectly) as continental fragments, dominate in the southwest quadrant particularly along the main line of lithospheric plate convergence between Papua New Guinea and Tonga. Elsewhere islands are generally smaller and often solely the product of one or two volcanoes in the case of volcanic islands, or the upgrowth of coral reef and an associated accumulation of detritus in the case of atolls. High limestone islands also exist, commonly in areas of tectonic disruption (Nunn, 1994a).

The principal island groups can be identified by dividing the Pacific Ocean and Islands into four quadrants using the Equator and the 180° meridian (Figure 1.1). Kiribati is a nation spanning all four quadrants; it is represented by the Line Islands in the northeast, the Phoenix Islands in the southeast, and the Gilbert group in the west.

The northeast quadrant is dominated by the Hawaiian islands, part of the Aleutian group, the Islas Revillagigedo, Clipperton and Guadalupe. It

is an area of mostly Pacific Plate although there is a complex area of convergence and slip in its northeast part adjoining the North American coast where smaller plates (the Gorda and Juan de Fuca plates) exist.

The southeast quadrant is dominated by the East Pacific Rise, a divergent plate boundary not breaking the ocean surface. Nearby islands include Easter Island (Rapanui), Sala y Gomez and the equatorial Galápagos group. Farther east, on the Nazca Plate, are the Juan Fernandez and San Felix–San Ambrosio groups. Most other islands in this quadrant fall politically within French Polynesia and include the Marquesas, Tuamotu, Society and Austral (Tubuai) groups. Others are the Cook Islands, Niue, Samoa and Tokelau on the Pacific Plate and the Tonga, Kermadec, Wallis and Futuna groups along the plate boundary to the west. The Lau group of eastern Fiji also falls within this quadrant.

Most islands in the southwest quadrant formed as a result of plate convergence; these include the main groups of Fiji, Vanuatu (formerly New Hebrides), Solomon Islands and the outer islands of Papua New Guinea. The atoll nation Tuvalu and the isolated island Nauru also lie within this quadrant. On account of their continental origins, New Caledonia and New Zealand are included under the Pacific Rim in this classification.

The northwest quadrant includes the Marshall Islands, the Federated States of Micronesia – Chuuk (Truk), Pohnpei, Kosrae – the Belau (Palau) group, the Marianas, Volcano and Nampo groups together with nearby isolated islands and part of the Aleutian group. The island groups on the Asian side of the westernmost ocean trenches are treated as part of the Pacific Rim.

Ever since the 1950s the study of the Pacific Ocean floor has burgeoned, and many useful data relating to environmental change have been acquired. It is yet worth considering that even such a huge volume of studies has hardly scratched the surface of most parts of this vast region. The geography of the Pacific Islands is one reason why the study of environmental change therein has been so uneven. Some islands can be reached only after days by boat; boats may call for just a few hours, perhaps only every few months. The political complexity of the Pacific Islands and Ocean has also played a role in frustrating research; some island nations welcome outside researchers, others are more circumspect. The present writer has been fortunate to work at the University of the South Pacific, an international university serving 12 island nations across whose territories he has roamed comparatively freely.

1.2 CHANGING ENVIRONMENTS THROUGH TIME: AN OVERVIEW

The configuration of land and sea in the Pacific Basin has never been fixed. This is something difficult to conceive and to appreciate the importance of when thinking on human timescales, but something which is very real and of critical significance when dealing with the distant past.

It is known, for example, that most continents emerged above sea level around the end of the Precambrian Era. Before that time, perhaps as much as 90% of the Earth's surface was under water. Continental freeboard – the amount of coastline in existence – was much less, and it is consequently misleading to try to describe ocean basins at that time in the way we can today. Closer to the present, we know that late Cenozoic ice-volume changes caused sea-level changes which periodically uncovered islands and vast areas of continental shelf in the Pacific Basin which are submerged today. These sea-level changes brought about environmental changes of many kinds. They altered ocean circulation and climate, which caused variations in rates and types of geomorphic processes. They led to speciation and adaptation of living things. At different times, they both stimulated and inhibited movements of the earliest peoples in the Pacific.

In the past the Pacific Basin looked different – in general, more diffuse. More than that though, the factors which controlled the nature of environments of the Pacific Basin were also different, both in kind and in magnitude. For example, unlike today, for most of the past 100 million years there was a largely equatorial movement of water across the Pacific and round

the Earth. Only comparatively recently did ocean circulation in the Pacific become driven from high latitudes, notably by a west wind drift named the Antarctic Circumpolar Current[2] in the southern-most Pacific.

An understanding of the formation and development of the modern Pacific Basin requires one to understand a range of Earth-surface processes, to appreciate the origins of a variety of landforms, and to understand the synergy between terrestrial and oceanic evolution. Yet as a preliminary to this understanding, it is necessary to have some idea of what exists today and how it is organized. The main physical elements of the Pacific Basin which control environmental development may be topographic, influencing environments on account of their elevation, location or associated character, or they may be climatic (vegetational), or oceanographic. Each of these groups of elements is considered separately below.

1.3 TOPOGRAPHIC ELEMENTS

Excepting coastal lowlands, which mostly formed comparatively recently, most parts of the Pacific Rim are mountainous. It is uncommon to find lowland parts of the Pacific Rim which are not associated with or bounded by mountain ranges. The few places where this situation occurs – such as the isthmuses of Panama and Tehuantepec in Central America, the Amur valley in far east Russia, or the Yangtze–Huanghe[3] valleys in central China – are all depressions of primarily structural origin.

Interest in the high mountain ranges of the Pacific Rim has been encouraged by their association with earthquakes and volcanic activity, which in turn have provided clues about their long-term development. Most of these mountain

---

[2] The name 'Antarctic Circumpolar Current' is used in preference to 'West Wind Drift' to refer to the broad west–east movement of ocean surface water in the southernmost Pacific
[3] Yangtze and Huanghe are synonymous with the names Changjiang and Yellow (River) respectively

ranges run parallel to the continental border of the Pacific suggesting that they formed and grew upwards as the result of processes which affected large stretches of the Pacific Rim simultaneously. Such processes include collisions of continental blocks, the overriding of an oceanic plate by a continental one and *vice versa*.

The alignments of islands and island chains within the Pacific Ocean commonly bear little relationship to trends of its continental margins. This is a reflection of those islands' oceanic origins. Most island chains in the Pacific mark the direction of former plate movements and thus often occur either at right angles or parallel to the trends of underwater mountain ranges or trenches marking divergent or convergent plate boundaries respectively. Thus the Easter–Sala y Gomez trend is at right angles to that of the divergent East Pacific Rise, from which it may have originated; the Tonga–Kermadec islands occur in lines parallel to the axis of the adjacent trench to the east. Many west Pacific island groups are arcuate because they parallel the curved axes of convergent plate boundaries. Many others are arranged in lines or near-linear clusters while a few appear truly isolated.

Some islands included as part of the Pacific Rim in this book are detached pieces of continents and exhibit similar trends to those of adjacent continental margins; Sakhalin and Vancouver islands are good examples. Other elongate island groups like Japan are largely the products of plate convergence in places where the plate boundary runs parallel to the continental margin, but this need not happen. Although formed by similar processes, the Philippines area is more irregular and displays no clear geometrical relationships with nearby continental margins.

The break-up of the eastern Gondwana continental amalgamation some 90 Ma in what is now the southwest Pacific Basin saw New Zealand and New Caledonia becoming detached from Australia–Antarctica and moving into an ancestral Pacific Ocean. The similarity in trend between the Norfolk Ridge, which links New Caledonia and New Zealand, and the shelf of the Australian continent, belies their similar history.

1.4 CLIMATIC AND VEGETATIONAL ELEMENTS

In a crude sense, zonal (north–south) variations of Earth-surface temperature account for the distribution of climates and natural vegetation types in the Pacific Basin. Within latitudinal climate zones, there are also intrazonal variations attributable to other influences, particularly elevation (Plate 1.1). Temperature varies markedly with elevation as does precipitation, particularly where rain-bearing winds blow directly onto mountain ranges (Figure 1.2). Since temperature and precipitation are the principal controls of vegetation type, this often varies with elevation in a similar way to that in which it varies with latitude.

Interannual variability of precipitation is most conspicuous in areas affected by the El Niño–Southern Oscillation (ENSO) phenomenon, discussed below. Similarly when tropical cyclones[4] affect particular parts of the Pacific Basin, the amount of precipitation within a few days may exceed average monthly, sometimes even mean annual, totals. The amount of environmental change accomplished at such times may vastly exceed that associated with 'normal' conditions.

To help understand the details of the climate and vegetation of the constituent regions of the Pacific Basin given below, the ground-surface pressure for both January and July is shown in Figure 1.3. Data characterizing named localities are listed in Tables 1.1–1.5 and 1.7–1.13.

1.4.1 Climate and Vegetation of Antarctica

Antarctica is the highest, coldest and driest continent. In its interior, mean annual temperatures vary from −64°C to −32°C. Along coasts and in lower parts of West Antarctica, temperatures typically range from −32°C to 0°C within a single year. The low temperatures are a product of both the low latitude of the continent, which means that

---

[4] The term 'tropical cyclones' is used in preference to its synonyms 'hurricanes' and 'typhoons' in this book

Plate 1.1   Typical view of *Nothofagus* forest and Magellanic moorland, Peninsula Brunswick, Strait of Magellan,
southernmost Chile
Stephen C. Porter

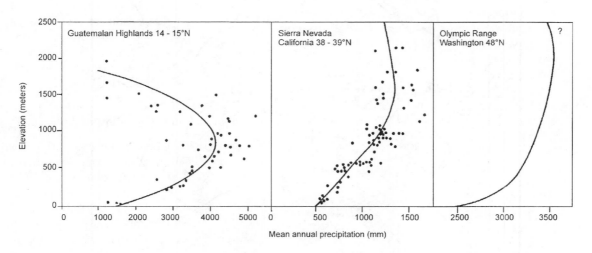

Figure 1.2   Curves showing the relationship between elevation and mean annual precipitation for west-facing
slopes in Central and North America. Note how precipitation begins to decrease in the highest parts
After Barry and Chorley (1982) with permission of Methuen and Co.

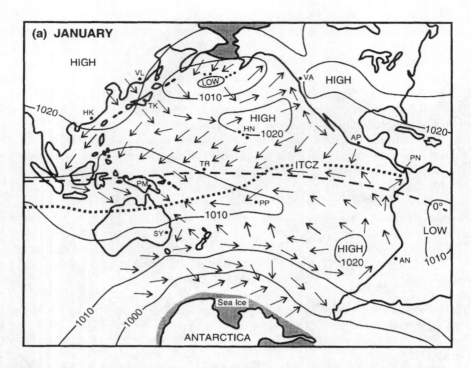

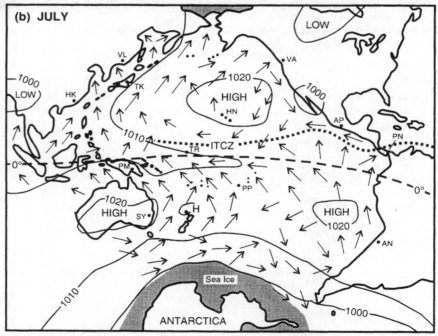

less solar radiation is received than in most other places on Earth, and the enveloping cover of snow and ice – more than 4.5 km thick in places – which causes most solar radiation to be reflected back into the atmosphere.

Most of inland Antarctica receives less than 50 mm of precipitation (water equivalent) annually. This aridity is explained by the moderate high-pressure cell centred on high eastern Antarctica which prevents moisture-bearing winds from penetrating far inland. Along the coast 200–800 mm of precipitation may occur.

Antarctica is a cold desert. Its greatest floral diversity is provided by lichens and mosses which live mostly around its periphery.

1.4.2 Climate and Vegetation of South America

Southern Chile lies within the zones of westerlies and is one of the wettest temperate locations in the world: San Pedro (47°43′S) receives more than 4480 mm of precipitation annually. Most of this zone is covered with dense forest. At higher elevations, zones of mixed grassland and trees gradually give way to little but fescue grass.

Farther north the climate falls under the influence of the permanent high-pressure cell in the southeast Pacific, close to Easter Island, and becomes drier. This effect is also accounted for by the presence of unexpectedly cold water, associated with upwelling (see below), close to the shore and the proximity of the Andes mountains which cast a rain shadow across much of Pacific South America. This area is one of the driest parts of the Pacific Rim. It includes the Atacama Desert – the only hot desert within the Pacific Basin – and stretches for more than 3000 km of the Rim between Valparaiso and the westernmost point of South America. The coastal city Antofagasta in Chile is representative of this area (Table 1.1). Its annual 14 mm of rain falls mostly during the winter months (June–August) when southerly winds occasionally blow onshore. Intra-annual temperature changes reflect those in solar radiation.

Those parts of the Pacific Rim of South America around and north of the Equator are generally characterized by higher temperatures and higher precipitation than the area to the south. In the northernmost part of this region, winds blow onshore from both the Pacific and Atlantic bringing plentiful precipitation – as much as 2000 mm annually – to both sides of the main divides. Topography and altitude are the principal reasons for variations in temperature and precipitation within tropical Pacific South America. Adequate precipitation falls along much of the coast yet little falls in most higher parts, especially between May and October, because onshore winds are too weak to carry moisture far up the mountain flanks. Moisture carried by winds blowing from the Atlantic across the South American continent falls almost exclusively on the eastern sides of the mountains. Snow-capped peaks marking the crests of these mountain ranges separate dry windswept plateaux called *altiplanos*.

Table 1.1   Monthly precipitation and temperature for Antofagasta, Chile

	Jan	Feb	Mar	Apr	May	Jun	Jul	Aug	Sep	Oct	Nov	Dec	Year or mean	Range
Precipitation (mm)	0	0	0	0	0	3	5	3	0	3	0	0	14	
Temperature (°C)	20.5	20.5	19.5	17.5	16	14.5	14	14	15	16	17.5	19	17	6.5

Figure 1.3 (opposite)   Major pressure systems and wind directions in the Pacific Basin in (a) January and (b) July. The approximate position of the ITCZ (Inter-Tropical Convergence Zone) is shown, together with the extent of sea ice in Arctic and Antarctic waters. Abbreviations refer to cities for which climate data are presented in Tables 1.1–1.5 and 1.7–1.13, as follows: AN, Antofagasta; PN, Panama; AP, Acapulco; VA, Vancouver; VL, Vladivostok; HK, Hong Kong; TK, Tokyo; PM, Port Moresby; SY, Sydney; PP, Papeete; TR, Tarawa; HN, Honolulu

Forests dominate the wetter areas of Pacific South America. In the north of the region, from the Gulf of Guayaquil to Panama, the tropical rainforest is known as *selva*, characterized by a floor more open than its counterparts elsewhere in the Pacific Basin. On the western flanks of the Andes above the treeline (3000–3500 m) to about 4500 m lie grassland-covered areas. In the south, where it is dry, these are known as *puna*: farther north, where annual precipitation is higher, as *páramo*.

1.4.3 Climate and Vegetation of Central America

The principal controls on the climate of the Central American part of the Pacific Rim are its tropical location and the proximity of its entire area to the sea.

Winds across the southern part of the region come mostly from the Atlantic. Where they encounter mountains, a rain shadow is created along their Pacific flanks although south-westerlies, particularly in winter, reduce the consequent aridity. Gaps in the mountain spine of Central America tend to be wetter than other parts of the Pacific Rim in this region. Close to the narrowest point of Central America, Panama City (Table 1.2) receives most rain during summer months (May–October), largely on account of the broad high-pressure cell centred over the Azores

in the North Atlantic. During late winter (January–April), the Inter-Tropical Convergence Zone (ITCZ) lies close to Panama; little air is driven onland from either side of the isthmus, so drier conditions prevail. The constantly high temperatures typify tropical locations.

In winter (November–May) northeast trade-winds affect the area farther north in Mexico and produce little precipitation since they have crossed only the North American continent. The disappearance of the high-pressure centre from this continent and the strengthening of that in the northeast Pacific during summer (June–October) generates onland winds which bring 500 mm or more of monsoonal rainfall to most of the northern part of this region. The coastal city Acapulco is typical (Table 1.3). Drier areas farther north along the Pacific Rim in Central America generally occur in rain shadows of the mountain ranges. For example, the Sonoran Desert fringing the east coast of the Gulf of California may lie in the rain shadow of either the Sierra Madre Occidental or the central range of Baja California at certain times of year.

Lowland tropical forests cover most of the tropical part of this region, giving way to grasslands and savannas, such as the *páramo*, in higher areas. The drier subtropical parts are covered with scrub and steppe but this is replaced by chaparral vegetation where annual rainfall increases. Higher areas are covered by pine–oak woodland or grassland.

Table 1.2   Monthly precipitation and temperature for Panama City, Panama

	Jan	Feb	Mar	Apr	May	Jun	Jul	Aug	Sep	Oct	Nov	Dec	Year or mean	Range
Precipitation (mm)	25	10	18	74	203	213	180	201	208	257	259	122	1770	
Temperature (°C)	26.5	27	27	27	26.5	27	27	26.5	26	26	26	27	26.6	1

Table 1.3   Monthly precipitation and temperature for Acapulco, Mexico

	Jan	Feb	Mar	Apr	May	Jun	Jul	Aug	Sep	Oct	Nov	Dec	Year or mean	Range
Precipitation (mm)	5	0	0	0	36	282	257	251	351	160	28	8	1378	
Temperature (°C)	26.5	26.5	26.5	27.5	28.5	29	28.5	29	28	28	27.5	26.5	27.7	2.5

1.4.4 Climate and Vegetation of North America

Temperatures in this sector of the Pacific Rim are governed largely by the amount of solar radiation received although the cool California current reduces coastal temperatures. Most precipitation in the south (the United States) falls in winter (October–March), largely as a result of migratory low-pressure systems from the west. Summers tend to be dry. Although precipitation levels commonly fall as one moves inland, marked spatial variations are produced by mountain ranges which create barriers for westerly and southerly air flows.

The climate of the northwest United States has more in common with that of the Pacific Rim in Canada. High winter precipitation is due largely to the relative proximity of the Hawaiian high-pressure centre which has been displaced from its summer position by the development of the Aleutian low. In winter, moisture-bearing winds are thus driven onland. In summer, this high-pressure centre broadens and moves west. Since its locus is farther offshore than in winter it produces winds which blow generally along this part of the North American coast. In some places, summer winds blow offshore allowing upwelling of deep water.

Typical of the climates of the northern part of this region is Vancouver, the principal urban centre on the west coast of Canada (Table 1.4). The low summer (May–August) rainfall here is the result of winds diverging away from the coast around the broad high-pressure cell over Hawaii. In winter (November–February), this cell is weaker, smaller and closer to this part of the North American coast, and produces westerlies which bring more rain than in summer.

Various types of grassland or woodland, dominated by pine, oak and juniper, are found in the southern part of the coastal zone and on lower mountain slopes in southern California. The Great Basin is a high (~1500 m) cool desert owing its aridity to the rain-shadow effects of the Sierra Nevada and the Rocky Mountains between which it lies. Coniferous forests dominate mountain vegetation farther north; pine, spruce and fir are the most common components of native forests.

In response to the rainfall all year round, the lowland part of the northern Pacific coast of North America supports a broad-leaved forest, dominated by oak, beech, chestnut and maple. Evergreen, largely pine, forests occur at higher elevations. In the Yukon, in the far north of this region, boreal forests (*taiga*) are found.

1.4.5 Climate and Vegetation of Beringia

Mean annual temperatures in eastern Beringia (Alaska) may average 3°C in parts of the coastal zone warmed by the Alaska Current, but are more often below freezing inland. The high-pressure cell in the northeast Pacific drives wet winds onto Alaska during summer which bring more than 500 mm of precipitation annually in the southern third of the peninsula. With the development of the Aleutian low-pressure centre in winter, the coastal zone draws some moist air from the south which keeps it wet, but the interior is largely dry.

Boreal forest dominated by spruce and hemlock is found in southern Alaska. The Yukon–Kuskokwim lowlands in the interior of Alaska are also forested; spruce and birch dominate. The remainder is covered mostly by treeless tundra.

The Pacific part of western Beringia receives an average 400 mm of precipitation during

Table 1.4 Monthly precipitation and temperature for Vancouver, Canada

	Jan	Feb	Mar	Apr	May	Jun	Jul	Aug	Sep	Oct	Nov	Dec	Year or mean	Range
Precipitation (mm)	218	147	127	84	71	64	30	43	91	147	211	224	1457	
Temperature (°C)	2.5	4	6.5	9	13	16	17.5	17.5	13.5	10.5	6.5	4	10	15

Table 1.5   Monthly precipitation and temperature for Vladivostok, Russia

	Jan	Feb	Mar	Apr	May	Jun	Jul	Aug	Sep	Oct	Nov	Dec	Year or mean	Range
Precipitation (mm)	8	10	18	30	53	74	84	119	109	48	30	15	598	
Temperature (°C)	−13.5	−9	−2.5	4.5	9.5	14	19	21	16.5	9	−0.5	−9	4.9	32.5

summer from wet winds driven onland from the high-pressure cell in the northeast Pacific. In winter, the situation is completely altered because of the strong central Asian high-pressure cell which moves dry air offshore; only the southeast part of the Kamchatka Peninsula receives significant rainfall at this time.

The high latitude and proximity of sea ice ensure that winter (November–March) temperatures in the coastal city Vladivostok in Russia are below freezing (Table 1.5). Precipitation at this time is also low because dry winds are driven seawards across the area out from the Asian high-pressure centre. In summer (June–September), temperatures rise significantly as warm air reaches Vladivostok from the south. Rainfall also reaches a maximum around this time because of the dominance of onshore winds.

Most of western Beringia is covered by tundra. Some river valleys and much of the Kamchatka Peninsula support boreal forests dominated by fir, larch and pine.

1.4.6   Climate and Vegetation of East Asia

Although temperature in East Asia is largely a function of latitude, it is the Asian monsoon which is the principal control on intra-annual air movements, precipitation and vegetation. Its influence reaches even northern China (Table 1.6).

In winter, the strong Siberian high-pressure cell in central Asia pushes cold dry air outwards. Little rain falls except in areas where the air is blown across water, as in the southern part of the Korean Peninsula, China south of the Yangtze River, and offshore islands such as Taiwan and those in Japan. When the land warms in summer, a broad band of low pressure covers continental East Asia and winds blow onland bringing moisture in often great quantities. Coastal areas and offshore islands generally receive most precipitation, particularly where there is a range of coast-parallel mountains as in southern China or on Kyushu island in southern Japan to produce an orographic effect.

Hong Kong is representative of this climate type (Table 1.7). In summer (May–September), warm wet winds blow onshore and considerable orographic rain falls on the city which lies seaward of high mountains. Since this air comes from lower latitudes, summer temperatures are more characteristic of a location well within the tropics, even though Hong Kong lies on their periphery. In winter (November–March), the winds influencing Hong Kong come from central

Table 1.6   Dominant wind directions in summer and winter in China
After Barry and Chorley (1982)

Location	Summer (July)	Winter (January)
North China	57% of winds from SE, S and SW	60% of winds from W, NW and N
Southeast China	56% of winds from SE, S and SW	88% of winds from N, NE and E

Table 1.7   Monthly precipitation and temperature for Victoria, Hong Hong

	Jan	Feb	Mar	Apr	May	Jun	Jul	Aug	Sep	Oct	Nov	Dec	Year or mean	Range
Precipitation (mm)	33	46	74	137	292	394	381	361	257	114	43	30	2162	
Temperature (°C)	15.5	15	17.5	21.5	25.5	27.5	28.5	28.5	27	25	20.5	17.5	22.5	13.5

Table 1.8   Monthly precipitation and temperature for Tokyo, Japan

	Jan	Feb	Mar	Apr	May	Jun	Jul	Aug	Sep	Oct	Nov	Dec	Year or mean	Range
Precipitation (mm)	48	74	107	135	147	165	142	152	234	208	97	56	1565	
Temperature (°C)	3.5	4.5	7	12.5	17	20.5	24.5	26	22.5	17	11	6	14.3	22.5

Asia and are dry; winter temperatures are more typical of the latitude.

Winter in the coastal zones of North Korea and northern Japan is marked by an influx of polar air bringing snow and low temperatures to most parts. In summer, high rainfall is associated with the southeast monsoon. Summer rainfall in Japan commonly exhibits a double maximum. The *baiu* season occurs at the monsoon's peak. A second wet season, the *shurin*, occurs when the Hawaiian high-pressure centre in the subtropical Pacific contracts sufficiently to allow Pacific cyclones to reach Japan.

The climate of Tokyo, on the Pacific coast of Honshu island in Japan, exemplifies this climate (Table 1.8). Rain occurs throughout the year but is least in winter (November–February) when winds moving off the Asian continent drop most of their rain on the Asian side of Honshu. Winds in Tokyo are generally offshore at this time. In summer (May–September), rainfall is generally higher because the winds blow onshore; note the maxima in June (*baiu*) and September (*shurin*). Summer temperatures are higher than expected for a place with this latitude because of the warming effect of winds and ocean currents from the tropical Pacific.

Most of the East Asia region within the Pacific Rim receives sufficient precipitation to sustain forest[5]. In the north, temperate forest (broad-leaved deciduous) dominated by oak, beech and maple is typical. Farther south, once

covering most of southern China, rainforest is found but tropical rainforest is not encountered until the southernmost China coast, Hainan island and southern Taiwan. Tropical rainforest dominates the islands of the Philippines to the south.

A similar division can be made of the vegetation in Japan. The northern parts and the mountains are characterized by mainly coniferous forest dominated by fir, pine, spruce and larch, the centre by temperate forest, and the southern parts by diverse subtropical rainforests.

1.4.7   Climate and Vegetation of Australasia

In summer (January) air is swept outwards across the Pacific Rim in Australasia from the high-pressure cell in central Asia and, where it crosses the ocean, it picks up moisture which it drops on the windward sides of the next landmass it encounters. In this way, Papua New Guinea and northern Australia receive considerable summer rainfall. By comparison, winters may be quite dry owing to the reversal and weakening of the monsoon.

---

[5] Much of this has now been cleared (see Chapter 15) and replaced by crops. The wheat–millet growing areas of North China and the wetter rice-growing areas of South China are defined largely by climate, the dividing line running along the 33rd parallel between the Huanghe and Yangtze valleys

Port Moresby on the southwest coast of New Guinea (Table 1.9) has a climate typical of places influenced by the wet northwest monsoon during late summer (December–April). The much lower rainfall in late winter (June–October) is attributable to the weaker southeast tradewinds reaching Port Moresby at this time. Temperatures reflect the city's tropical location; annual range is characteristically low.

In contrast, on account of its temperate location farther south, Sydney on Australia's east coast has a much greater intra-annual temperature range (Table 1.10). A subtropical high-pressure belt passes through central and southern Australia in winter (May–September) but is farther south in summer (November–March). More rain falls in Sydney in autumn as the high pressure in this band strengthens and it moves northwards driving the tradewinds before it. Conversely, spring is drier as it weakens and moves south.

Similar reasons explain the climate in other parts of southern Australasia. The same subtropical band of high pressure just clips the north of New Zealand in winter allowing wet westerly winds passage across most of the island and bringing abundant precipitation to its windward (western) slopes. In summer, this band of high pressure covers most of New Zealand which experiences generally drier conditions as a result.

New Guinea is covered with tropical rainforest, except where this is limited by altitude and replaced by less dense forests and grasslands. Much of the coastal fringe of eastern Australia is also rainforest-covered, tropical in the north, temperate (dominated by eucalypts) farther south. Moving inland, precipitation declines rapidly and forests become less dense, often savanna-like, eventually giving way to grassland. Subtropical and temperate rainforests dominate the wetter parts of New Zealand and New Caledonia; grassland is typical of rain-shadow areas.

1.4.8 Climate and Vegetation of the Pacific Ocean and Islands

Weak high-pressure centres over the North and South Poles cause an outflow of air from both which gives rise to easterly winds in both hemispheres to about 60°N and S. These are generally weaker in the north Pacific where the high-pressure centre is more subdued than in the south. Although the same basic pattern of zonal climate pertains in both the north and south Pacific, the configuration of land and sea in the former constrains the movement of air (and ocean) more than in the latter.

Moving north from the periphery of Antarctica in the south Pacific, an area of low pressure marking the convergence of polar easterlies and the westerlies to the north is encountered. These westerlies derive from the subtropical high-pressure belt found between about 25°S and 35°S

Table 1.9   Monthly precipitation and temperature for Port Moresby, Papua New Guinea

	Jan	Feb	Mar	Apr	May	Jun	Jul	Aug	Sep	Oct	Nov	Dec	Year or mean	Range
Precipitation (mm)	178	193	170	107	64	33	28	18	25	36	48	112	1012	
Temperature (°C)	28	27.5	27.5	27.5	27	26	25.5	25.5	26	27	27.5	28	26.9	2.5

Table 1.10   Monthly precipitation and temperature for Sydney, Australia

	Jan	Feb	Mar	Apr	May	Jun	Jul	Aug	Sep	Oct	Nov	Dec	Year or mean	Range
Precipitation (mm)	89	102	127	135	127	117	117	76	74	71	74	74	1183	
Temperature (°C)	22	22	20.5	18	15	12.5	12	13	15	17.5	19.5	21	17.3	10

within a single year. The most persistent feature of this high-pressure belt is the anticyclone in the southeast Pacific, close to Easter Island.

The region of the south Pacific north of this high-pressure belt is affected by the southeast tradewinds, which blow across a much larger area in winter (May–September) than in summer. They blow north to converge with their northern counterparts (the northeast tradewinds) and rise at the ITCZ. In summer (November–February), the ITCZ bends southwards in the central Pacific so tradewinds generated in the eastern Pacific are forced to move due west in this area; in other words, they become easterlies.

Papeete, the capital city of French Polynesia, is located on the northwest side of the high island Tahiti (Table 1.11). Summer (November–April) rainfall is explained by the easterlies which affect this coast at this time; the lower rainfall in winter (June–September) is explicable by the dominance of southeast tradewinds which have dropped most of their moisture on the windward (southeast) side of Tahiti before reaching Papeete. The high temperatures are characteristic of tropical maritime locations in the region.

The position of the ITCZ in the eastern Pacific changes little in the course of an average year. Yet in the western Pacific, the ITCZ moves a considerable distance within a single year, largely because of the development of the massive Siberian high-pressure centre by January. At this time, the ITCZ usually passes through northern Fiji, Vanuatu and northern Australia. The area to the south is subject to easterlies (in the east) and southeasterlies (in the west). The area to the north, including Solomon Islands and the outer islands of Papua New Guinea, is affected by the northwest monsoon. In contrast, by July the ITCZ has moved north and the whole area is under the influence of southeast tradewinds.

Movements of the ITCZ in the central Pacific often produce significant intra-annual variations in wind direction and rainfall. During summer, Bairiki on low-lying Tarawa Atoll in western Kiribati is within the region affected by the southeast tradewinds; note the rainfall maximum in June–September (Table 1.12). In winter (December–February), Bairiki lies within the band of northeast tradewinds so experiences a second rainfall maximum at this time. In the intervening periods, the ITCZ passes across this part of Kiribati and winds are less constant and bring generally less rainfall. Like Papeete, temperatures are typical of central Pacific locations.

The pattern of winds in the tropical and subtropical North Pacific varies depending on seasonal pressure variations. The high-pressure cell close to Hawaii is the most enduring feature. In summer, it forms part of a much broader band of high pressure which produces westerlies to the north and comparatively subdued northeast tradewinds to the south from the Tropic of Cancer

Table 1.11 Monthly precipitation and temperature for Papeete, Tahiti

	Jan	Feb	Mar	Apr	May	Jun	Jul	Aug	Sep	Oct	Nov	Dec	Year or mean	Range
Precipitation (mm)	251	244	429	142	102	76	53	43	53	89	150	249	1881	
Temperature (°C)	27	27	27	27	26	25.5	25	25	25.5	26	26.5	26.5	26.2	2

Table 1.12 Monthly precipitation and temperature for Bairiki, Tarawa Atoll, Kiribati

	Jan	Feb	Mar	Apr	May	Jun	Jul	Aug	Sep	Oct	Nov	Dec	Year or mean	Range
Precipitation (mm)	315	206	180	94	53	99	155	193	122	99	152	239	1592	
Temperature (°C)	27	27.5	28	28	28	27.5	27.5	27.5	28	27.5	27.5	27.5	27.6	1

to around 10–15°N where the ITCZ lies. Owing to the growth of the central Asian high in summer, the Hawaiian high-pressure cell becomes smaller and shifts southeast. Yet owing to the southward displacement of the ITCZ, the northeast trade-winds influence a greater area at this time than in winter.

The effects of these changes are well illustrated by Honolulu on the south coast of Oahu island in the Hawaii group (Table 1.13). Little rain falls during summer (June–September), when the high-pressure centre often straddles the islands, although tropical cyclones occasionally reach them at this time. In winter (November–March), when the high-pressure cell has shrunk and moved southeast, Oahu receives westerly and south-westerly winds which bring most of the islands' annual rainfall. Compared to Papeete and Bairiki, the temperature range is slightly greater at Honolulu because this site receives air from cooler latitudes during winter and is in the subtropics, where temperatures commonly vary more than in the tropical Pacific.

In the northernmost Pacific, the Hawaii anticyclone dominates air movement during summer, but in winter the Aleutian low-pressure centre is established and drives westerlies across the region. Polar easterlies enter the north Pacific in winter.

The natural vegetation of wet tropical Pacific islands is rainforest. On drier tropical islands, or the drier (windward) sides of large islands, grassland–savannas exist. On account of their poor soils and small water reserves, most atolls have natural vegetation consisting of only a very few species. The comparatively few Pacific islands outside the tropics support vegetation similar to that of the continent(s) from which it derived. Owing to the often great distances from continental source areas, Pacific island natural vegetation tends to be less diverse than that on source continents yet often has a greater number of endemic species, reflecting the reduced competition and low diversity; the extraordinary number of endemic lobelia (*Rollandia* spp.) on the islands of Hawaii provides an example.

1.5 OCEANOGRAPHIC ELEMENTS

The Pacific is the largest ocean basin in the world, for which reason it has the lowest salinity of any. Some comparative data are given in Table 1.14. Compared to other oceans, the Pacific also receives little terrigenous sediment because of the existence of sediment traps – island arcs, trenches and marginal basins – around much of its periphery. For this reason, past changes in the nature of sedimentation in the central Pacific have been considered good indicators of past changes in the geochemical character of oceans *per se*. Owing to the presence around most parts of the Pacific Rim of mountain belts which inhibit development of large rivers entering the Pacific Ocean, it also receives little fresh water compared to the other oceans of the world.

In contrast to the other ocean basins which have been growing for the past few hundred million years or so, the Pacific has been decreasing in size. The sedimentary record of the ancient (pre-early Mesozoic) Pacific has been mostly lost, or crumpled up and stored in sediment prisms adjoining the convergent plate boundaries along its margins.

1.5.1 Surface-water Circulation

For more than 200 million years, the system of surface-water circulation in the Pacific Basin has

Table 1.13   Monthly precipitation and temperature for Honolulu, Hawaii

	Jan	Feb	Mar	Apr	May	Jun	Jul	Aug	Sep	Oct	Nov	Dec	Year or mean	Range
Precipitation (mm)	97	69	89	38	30	13	13	15	15	48	81	86	594	
Temperature (°C)	22.5	22.5	23	24.5	25	26	27	27	27	26.5	25	23	21.2	4.5

Table 1.14   Characteristics of the world's oceans
After Gross (1972) and sources therein

Ocean	Area $(10^6 \, \text{km}^2)$	Land area $(10^6 \, \text{km}^2)$	Average depth (m)	Average salinity (‰)	Average temperature (°C)
Pacific	180	19	3940	34.62	3.36
Atlantic	107	69	3310	34.90	3.73
Indian	74	13	3840	34.76	3.72
World	361	101	3730	34.72	3.52

been dominated by two gyres in which surface water circulates. Water in the north Pacific gyre circulates clockwise around the subtropical high-pressure cell(s) close to Hawaii around 30°N. The south Pacific gyre circulates anticlockwise around the large Easter Island high-pressure cell in the southeast Pacific (Figure 1.4).

In the north Pacific, the easterly North Equatorial Current forms the southern part of the gyre. The Kuroshio Current forms its western limb bringing warm water from the equatorial Pacific to keep Japan's Pacific coasts warmer than would otherwise be expected for their latitudes. The cold Oyashio Current which enters the Pacific from the Arctic Ocean through the Bering Strait meets the Kuroshio around the latitude of the Kuril Islands. They merge to flow east as the North Pacific Current which meets the coast of North America, and flows south as the cool California Current. The west coast of central North America is thus kept cooler than expected given its latitude. The California Current runs into the North Equatorial Current farther south to complete the north Pacific gyre. An offshoot of the North Pacific Current flows anticlockwise within the northeasternmost Pacific keeping most of the Pacific coasts of northern Canada and Alaska ice-free.

In the southernmost Pacific, surface-water flow is dominated by the cold Antarctic Circumpolar Current which insulates Antarctica from the influence of warmer water by carrying cold water around the Earth from west to east at a high southern latitude. Where the Antarctic Circumpolar Current meets South America, part of it flows north as the Peru Current. The Pacific coasts of South America are thus cooler than they might

otherwise be. When the Peru Current reaches the Equator, it flows west as the South Equatorial Current, a minor part of which exits the Pacific Basin around Papua New Guinea and Indonesia along a number of comparatively shallow seaways. The major part of the South Equatorial Current flows south along the east coasts of Australia which are thus much warmer than expected at this latitude. This explains why coral reefs are found well outside the tropics in this region – on Lord Howe island for instance (31°33'S). Where the East Australian Current meets the Antarctic Circumpolar Current, the gyre is completed.

Between the two west-flowing equatorial currents in the Pacific runs the east-flowing Equatorial Counter Current which varies markedly in strength and continuity from year to year. An unexpectedly strong manifestation of the Equatorial Counter Current – perhaps associated with an El Niño event – may explain the earliest human colonization of the Marquesas Islands east from Samoa around 2000 years ago.

1.5.2   Surface-water Convergence and Divergence

In the South Pacific, the Antarctic Convergence (latitude 50–60°S) marks the place where cold water, cooled in high latitudes, converges with warmer water from low latitudes. Being denser, the cold water is thrust downwards eventually to become bottom water. Sea-surface temperatures drop rapidly as the Antarctic Convergence is crossed from north to south. The convergence between cold and warm surface waters in the

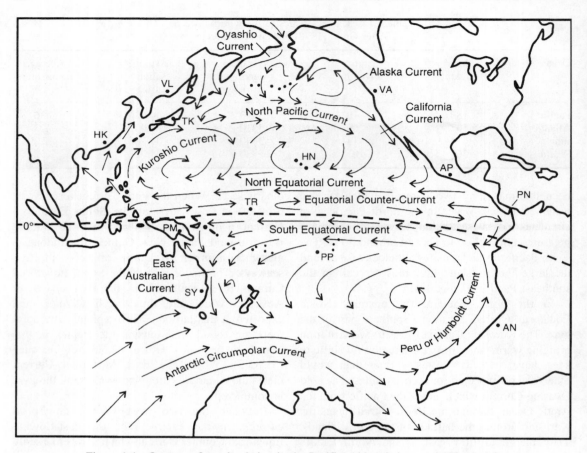

Figure 1.4   Ocean surface circulation in the Pacific. Abbreviations as in Figure 1.3

North Pacific is less defined because Arctic waters are able to enter the Pacific Basin only at shallow depth through the Bering Strait.

Winds blowing off the land along some coasts of the Pacific Rim carry surface waters away from the coast. This allows cold deep water rich in nutrients, particularly phosphates, silicates and nitrates, to reach the surface. The economic importance of such upwelling off the west coasts of both North and South America is immense since the nutrient-rich waters attract huge numbers of fish; some 90% of the world's finfish are caught in regions of upwelling (Plate 1.2). Particularly off the Pacific coast of South America, vast numbers of seabirds are also attracted to upwelling waters. Through guano deposition, these birds have given rise to an important fertilizer industry.

1.5.3   The El Niño–Southern Oscillation (ENSO)

Every few years, the circulation of the Walker Cell in the tropical Pacific reverses (Figure 1.5). At such times, winds along the Pacific coasts of North and South America, which normally blow offshore allowing upwelling, blow onshore increasing precipitation, preventing upwelling and thereby depleting fish stocks. This effect, known as El Niño, has been linked to the Walker Circulation reversal (the Southern Oscillation); the combined phenomenon is referred to as ENSO[6].

---

[6] Recently Philander (1998: 170) argued that the popular term 'El Niño' should replace the 'uninspired acronym ENSO'

Plate 1.2   Nutrient-rich upwelling waters of the Pacific Ocean off the coasts of mid-latitude North and South America sustain uncommonly high concentrations of marine organisms. Humans have not been slow to exploit these resources, as shown here by a Chinese fishing–shrimping camp on San Francisco Bay in 1889 established to supply the burgeoning urban population of the region
National Archives (NWDNS-22-FA-145)

ENSO is measured by the Southern Oscillation Index (SOI). When the SOI is positive, the situation is normal; the east Pacific is cool and the low equatorial Pacific islands of Kiribati, for example, experience mostly easterly winds and are dry. When the SOI is negative, then an El Niño condition prevails. The east Pacific is hot. Kiribati receives more rain than normal while islands close to the outer limits of the tropics, like most of those in Fiji for example, commonly experience prolonged droughts. The situation is summarized in Table 1.15.

El Niño has been linked to extreme weather conditions in most parts of the Pacific Basin including heavy rain and flooding in South America, storms and associated mass movements in California, drought and famine in Indonesia and the Philippines, and droughts and destructive bush fires in Australia. The magnitude of these events is often severe: at Guayaquil on the coast of Ecuador, annual precipitation averages 843 mm, but in 1983, during a strong El Niño, it was 3950 mm (Iriondo, 1994). The potential for accelerated environmental change at such times

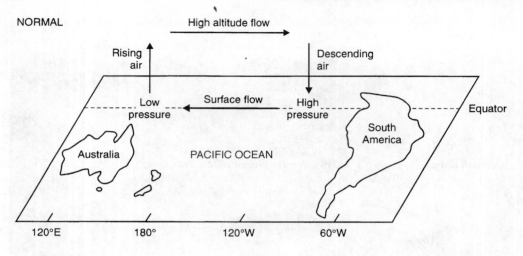

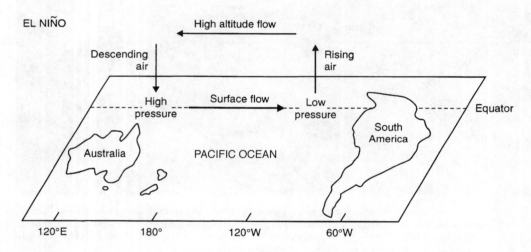

Figure 1.5   Contrasting nature of 'normal' and El Niño conditions in the Pacific Ocean as related to the Walker Circulation

is manifest. One of the challenges of recon-structing the history of environmental change, during the Holocene in particular, is to successfully marry the effects of such short-lived yet extreme changes with the effects of long-term changes of lesser magnitude.

Table 1.15   Abnormal conditions associated with El Niño (negative-ENSO) events in the
South Pacific Ocean

Region	Conditions
West Pacific	Above-normal pressure Below-average precipitation Weakened South Asian summer (July) monsoon Longer dry season Sea-level fall
Central Pacific	Above-average precipitation South Pacific Convergence Zone shifts east Tropical cyclones occur further east Tradewinds weaken/reverse
East Pacific	Below-normal pressure (Easter Island high weakens) Above-average precipitation Sea-surface temperatures rise Upwelling is suppressed, thermocline sinks Reduced nutrients at ocean surface Sea-level rise

CHAPTER 2

Investigating the Pacific Basin

We seem to exist in a hazardous time,
Driftin' along here through space;
Nobody knows just when we begun
Or how fur we've gone in the race.
Scientists argy we're shot from the sun,
While others we're goin' right back,
An' some say we've allers been here more or less,
An' seem to establish the fact.

Ben King, *Evolution*

2.1 HISTORY OF SYSTEMATIC INVESTIGATION

It is impossible to mark the point at which the systematic investigation of the Pacific began. We know that human ancestors may have reached the Pacific Rim as much as 1.7 Ma. Modern humans[1] were probably present there several tens of thousands of years ago and settled all parts of the Rim except Antarctica within the last 12 000 years or so. The earliest systematic investigations of most Pacific islands are assumed to have begun around 4 ka. Using celestial navigation techniques, the descendants of these first settlers crossed the Pacific from west to east, long before Europeans first saw this vast ocean. Many discoveries made in the Pacific Basin since it was first colonized have been forgotten as 'western' scientific methods and information have come to dominate it (Turnbull, 1994; Clarke, 1995).

2.2 THE EARLIEST HUMANS IN THE PACIFIC BASIN

The pattern of human colonization of the Pacific Basin is explored in more detail in Chapter 12. This account simply outlines the chronology of human appearance in the region.

Homo erectus may have been present 1.7 Ma in southern China. The later occupation of inland eastern China by these and other pre-modern foraging and fire-using humans may have compelled modern humans, who appeared there within the last 100 000 years, to occupy the coastline (and offshore islands) and develop an appropriate lifestyle. It is plausible to suppose that earlier humans were eventually extirpated by the inland spread of modern humans with their more advanced tools and strategies for obtaining food. Humans from East Asia were the first to colonize North America. Most commentators aver that they crossed through Beringia at the height of the Last Glacial (Last Ice Age) when the area was dry (because the sea level was low), then spread south through the Americas.

Early humans (*Homo erectus*) reached southeast Asia around 1 Ma where they were joined by modern humans perhaps 80–50 ka. As in eastern China, long-established inland populations of *Homo erectus* may have forced these modern

---

[1] When comparing modern humans with their earliest ancestors in this book, the term 'modern human' is reserved for *Homo sapiens*. 'Human' is the term used for *Homo erectus* and descendants. Earlier human ancestors are referred to as hominids

arrivals to settle island shorelines – environments unfamiliar to them – entailing lifestyle changes which eventually led them to traverse an ocean gap several tens of kilometres broad, and colonize Papua New Guinea and Australia more than 40 ka, taking advantage of low glacial sea levels exposing land connections which are submerged today. Most parts of Australia were occupied by humans 30 ka.

The earliest human colonists of the Pacific Basin probably did not explore the lands they occupied solely out of curiosity. Any exploration they did would have been largely in response to particular environmental factors – for instance, a need to find water, or to follow herds of migratory animals, a need to escape increasing population pressure or rising sea levels or, in the case of those accustomed to a nomadic lifestyle, as an unquestioned part of their existence. Many such people retained a memory of where they had been – the mental maps or songlines of native Australians (wonderfully elucidated by Chatwin, 1987) are an example – so that they might tell of their past or even revisit former 'homes'. Later, when many Pacific people began to occupy large settlements permanently and to depend on each other and on nearby agriculture to survive, pictures began to develop in people's minds about the geography of large areas. People would eventually have become cognizant of adjacent such areas, endowed with different qualities to the home area, leading to the establishment of trading networks which help explain the modern settlement pattern of the Pacific Basin.

Much writing on the subject of early human (say, pre-16th century) understanding of the natural environment is permeated, albeit largely unthinkingly, by eurocentric or other nationalistic ideas to such a degree that a realistic treatment of the subject in the Pacific Basin is impossible in the present writer's view; in the blunt assessment of Jared Diamond, 'the whole subject of human differences in level of civilization still reeks of racism' (Diamond, 1992: 236). For this reason among others, no attempt is made here at synthesis. Rather four case studies – from Australasia, China, the Pacific islands and South America – are given to demonstrate the inroads which certain groups of early Pacific people made into the understanding of the environments they inhabited.

2.2.1 Early Human–Environment Interactions in Australasia

In a controversial though largely persuasive book, Tim Flannery (1994) argued that the crossing by humans of the Wallace Line (perhaps initially between Bali and Lombok islands), marking the first human incursion into the Australasian fragment of Gondwana from southeast Asia, led to a 'great leap forward' in the evolution of *Homo sapiens*. Flannery reasons that humans had hitherto been merely an unremarkable part of ecosystems, compelled in southeast Asia to compete for the same prey as tigers and leopards. Yet once humans crossed the Wallace Line, they entered an ecosystem which had evolved in their absence, indeed in the absence of similar predators, so they found food comparatively easy to obtain. This may have led them to experience more 'leisure', more time to develop an intelligence higher than other animals.

Whether or not this is true, it is clear that the earliest inhabitants of Australasia produced innovations which were in advance of humans elsewhere in the world at the same time. One of the earliest of these were the waisted stone blades (Plate 2.1) being made and hafted in New Guinea at least 40 000 years ago (Groube et al., 1986). Such innovations undoubtedly produced undesirable environmental pressures, and almost certainly led to the extinction of some species of indigenous biota, but eventually also led to a degree of understanding about environmental processes which allowed the earliest modern humans of Australia and New Guinea to live in a state of benign interaction with the environments they inhabited. A good example is the traditional land system classification of the Gidjingarli people of Arnhem Land in northern Australia, strikingly similar to the independently formulated and much later classification of the modern Australian government (Jones, 1985).

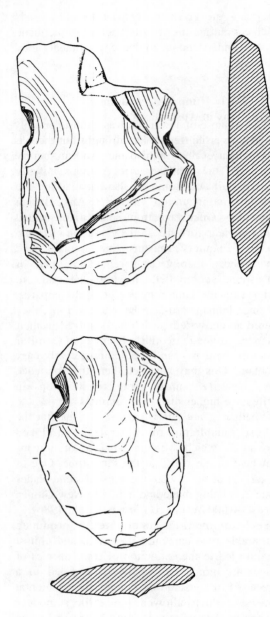

Plate 2.1   Examples of waisted blades from the Kosipe
site in the Papua New Guinea highlands. Such blades
were hafted to make axes used to clear forest around this
site perhaps 40 000 years ago
J. Peter White

2.2.2   Early Environmental Understanding in China

The comparatively sophisticated socio-political
system which evolved several thousand years
ago in China – indeed in many parts of East Asia
– before anywhere else in the Pacific Basin, was
marked by an unusual degree of pragmatism
concerning the natural environment. For example,
the comparatively high incidence of earthquakes
led to a comprehensive recording of both their
precursors and their effects with a view to
prediction and mitigation respectively. The
development and intensification of agriculture in
coastal areas of China also led to the recording and
prediction of tides and various types of weather,
and an understanding of flooding of large lowland
rivers (Yang et al., 1989).

Chinese scientists were also quick to learn
some of the fundamental truths of geology, long
before their western counterparts had much
inkling of these. Of particular note was the
Chinese recognition of the true character and
significance of fossils, and some apprehension of
the true antiquity of the Earth. It is likely that early
Chinese 'scientists' derived this advantage from
both a greater appreciation of the role of
catastrophic events in Earth history and from their
Buddhist beliefs which espoused an uncreated
universe, without beginning or end. Such views
left Chinese thinkers unrestrained compared to
European scientists who, for much of the last
thousand years have had to be careful not to make
statements which conflicted with the prevailing
Judeo-Christian belief system.

2.2.3   The Initial Human Colonization of the Pacific Islands

Modern humans colonized Papua New Guinea and
Solomon Islands more than 40 ka but apparently
did not move further east until around 3.5 ka when
an almost instantaneous colonization of Vanuatu,
New Caledonia, Fiji, Tonga and Samoa took place
within perhaps just 500 years. The remainder of
the south and east Pacific islands were not
colonized until later: many, including New

Zealand, only within the last millennium. Considered from even a modern standpoint let alone from that of other seafaring nations 3.5 ka, the long-distance voyaging which the initial human colonization of the island Pacific involved was awesome. At the time (and for a long time afterwards) it was unprecedented (Turnbull, 1994).

Initial exploration may have been driven by false expectations arising from the perception of the Pacific by the earliest islanders as a 'sea of islands'. Continental people often envisage the Pacific as an empty ocean (Levison et al., 1973: 62):

> It was the expanses of the empty Atlantic and eastern Pacific which formed the ocean images of Europeans [as continental people], but people entering the Pacific from its western margins [as the earliest settlers did] might well expect the seas to the east to be island-studded as the margins they knew. If one believes that island-studded seas are the norm and has not had experience of the empty ocean wastes, then one may well sail forth with confidence. And failure to find new land or survive landing is unlikely to be reported back to alter the images of the home community.

Later, when sailing techniques became more sophisticated and the chances of returning to a home base from a distant island were increased, more deliberate exploration of the Pacific Ocean took place. Pragmatic reasons for exploration are testified to by the comprehensive cargo of crops and food animals which many early voyagers took with them (Kirch, 1997).

2.2.4 Environmental Determinism in Pacific South America

The prehispanic[2] civilizations of the Pacific Rim in South America were notable for their ingenuity in developing or adapting techniques of irrigated agriculture which could sustain large numbers of people in inhospitable environments. One example

is the dry *altiplano* around Lake Titicaca, site of the pre-Inca Tiwanaku civilization (300 BC–AD 1100) centred around 190 km^2 of raised fields, which might have sustained 760 000 people in its heyday yet today – inoperational and degraded – sustains only about 40 000 (Binford et al., 1997; Plate 2.2).

The collapse of Tiwanaku was almost certainly linked to a prolonged severe drought but there is also evidence that prehispanic Andean people were able to predict rain and drought cycles through the ritual exchange with coastal people of *Spondylus princeps*, a bivalve mollusc whose abundance varied according to the variable effects of El Niño warm water incursions offshore (Lumbreras, 1988).

Along the dry coastal plains of Pacific South America, year-round agriculture is possible only with irrigation from some of the large rivers passing through the area. The single most ambitious project of this kind in the prehispanic era was that of the Chicama–Moche Canal in northern Peru constructed about AD 1200 to connect the rivers of the same names. The hydraulic and surveying expertise employed was not matched in Europe and North America until the late 19th century (Ortloff et al., 1982).

2.3 SYSTEMATIC INVESTIGATIONS (POST-15TH CENTURY) OF THE PACIFIC BASIN

This account is divided between the Pacific Rim, and the Ocean and Islands. Both the geographical (spatial) and environmental (ecological) understanding of the earliest post-15th century human inhabitants of the Pacific Basin were confined to particular portions of the Pacific Rim, and/or to particular groups of islands. It does not belittle these achievements to acknowledge that the later history of this region has been dominated by science of European origin. For better or worse, at the time of their first 'discovery' of the Pacific Basin, Europeans had the rudiments of a geographically valid world vision which was being crystallized through exploration at a much faster rate than that of any

---

[2] Used synonymously with 'precolumbian' in this book

Plate 2.2   The centre of the Andean state of Tiwanaku, 3870 m above sea level, southeast of Lake Titicaca in Bolivia. This urban centre was once home to 20 000–40 000 people supported by carefully managed agricultural systems in the surrounding district but collapsed around AD 1100 as a result of a period of prolonged extreme aridity Marilyn Bridges

human group elsewhere. This view is used to explain but not necessarily justify the uncritical imposition of European science (particularly scientific methods) on both Pacific peoples and the Pacific environment. It is heartening to see 'traditional' science reasserting itself in many parts of the Pacific Basin (e.g. Morrison et al., 1994) demonstrating that true science[3] is a common legacy of human reason.

2.3.1   The Roots of Modern Scientific Exploration of the Pacific Rim

Owing to the continuing unfamiliarity of Pacific islands to most continental people today, the early accounts of these areas by European explorers are often considered to be their major achievements. Yet many such people also gathered considerable information about parts of the Pacific Rim. Indeed, many observations made in the Rim influenced those which were later made in the islands. Witness Charles Darwin who, upon being confronted with the blatant evidence of land uplift

---

[3] The word 'science' means 'knowledge' in Latin

and seismicity along the Pacific fringe of South America (Plate 2.3), was led to consider a movement in the opposing direction to explain the formation of atolls before he ever saw one.

The political and commercial interests in Europe which validated Pacific 'exploration' focused more on the Pacific Rim than on the Pacific islands, whether recognizing their potential as penal settlements, whaling stations, sources of mineral riches, or as places over which to establish

Plate 2.3   The ruins of Concepçion, Chile, devastated in the earthquake of 20 February 1835: detail from a drawing by J.C. Wickham, first lieutenant on HMS *Beagle*. When Charles Darwin walked around the area a few days later, he found fresh mussel beds above high tide – the result of the land having risen a metre or so. He deduced that this was the way in which once-submerged mountains had been lifted above the sea. Later Darwin realized that land could also move downwards, giving rise in the coral seas to atolls
From the narrative of the *Voyage of the Beagle*

sovereignty (or to counter the extension of a rival sovereignty). By AD 1790, most parts of the Pacific Rim had been divided up, at least notionally, among European powers, except for eastern Asia, which was mostly under the control of China and Russia.

The systematic exploration and inventorizing of the resources of these 'new' lands accompanied European settlement and led ultimately to the establishment of branches of various learned societies in these lands. It is clear that the dominance of European thinking, particularly a tendency to see non-European environments as 'alien' and needing to be 'tamed', inhibited their understanding and contributed to their subsequent mismanagement (Stafford, 1988).

By the mid-19th century, many countries (colonial or independent) which extended into the Pacific Rim had developed sufficient scientific expertise to undertake systematic surveys of their own lands; the Geological Survey of Canada, for example, was established in 1842. Yet scientific interest was far from uniform throughout these nations, particularly as they are delineated today. In North America, for example, almost all scientific interest was in the 'tamed' lands of the east and midwest. In 1846, 'from the very centre of Mexico, all the way to the Columbia [River in the northwest USA], is a terra incognita, rich, it would seem, in geological phenomena' (Rogers, 1896: 265). Later developments are typified by California where the discovery of gold in 1848 catalysed environmental interest and understanding, manifested by the appointment of the first state geologist, Josiah Whitney, in 1860.

2.3.2   Modern Understanding of the Pacific Rim

Until the 1960s, the science practised in many parts of the Pacific Basin could still reasonably be described as 'colonial'; for the colonial investigator, 'natural history was the ideal refuge from the more perilous enterprise of embarking upon theoretical constructions by which he would be pitched into naked competition with the best scholars of all countries' (Fleming, 1964: 182).

While this suggestion may be controversial, it is clear that the degree of understanding of local environments by colonial officials was inadequate for many purposes. This point is exemplified by the deliberate introduction into alien ecosystems throughout the 'colonial' Pacific Basin of particular plants and animals, which then went on to occupy niches hitherto unrecognized and to cause untold disruption (Burt and Williams, 1988). The European colonization of Australasia was premised on the naive assumption that its ecosystems 'worked pretty much like the European ones' (Flannery, 1994: 306), a view directly responsible for the looming ecological and environmental crisis in many parts of this region today.

In contrast to most other parts of the Pacific Rim, the countries of East Asia retained traditional science and methods of investigation for a long time, even after adopting global frameworks of environmental investigation and analysis. Until very recently in China, for example, both archaeology and earthquake prediction depended on methods established largely independently of those regarded as standard by 'western' scientists (Hashimoto et al., 1995).

The situation today has changed for, while culture may still influence the approaches to scientific research in parts of the Pacific Basin, the methodological tools are largely standard, the end results directly comparable, and transglobal networks of researchers increasingly commonplace.

One of the prime movers in drawing together Pacific researchers has been the Pacific Science Association. The first Pan-Pacific Science Conference was held in 1920 in Hawaii and attracted 103 participants from throughout the Pacific Basin and beyond. The last Congress in Beijing in 1995 brought 1296 scientists together for a week of interaction.

2.3.3   The First European Explorers in the Pacific Ocean and Islands

Until his death in 1506, Christopher Columbus believed he had reached Asia when he captained the first recorded voyage from Europe across the Atlantic. The sighting of the Pacific by Balboa in 1513 was followed by the circumnavigation by Magellan in 1519–1522 which made the existence and breadth of the Pacific widely known.

Spurred on by the promise of lucrative trade with Pacific Basin peoples, European explorers made numerous ocean journeys into and across the Pacific Ocean in the following few hundred years – voyages made increasingly easy by the development of more sophisticated sailing technology and navigational equipment, particularly the chronometer in 1735 which allowed the accurate determination of longitude. Tahiti was the first mid-Pacific island base of importance, largely because of its suitable location as a revictualling and watering post for vessels traversing the ocean. Captain James Cook's expeditions introduced cattle, goats, horses, pigs and poultry to Tahiti, with the principal intention of ensuring that the island would be able to supply European vessels with what they wanted in the future.

After Cook's first visit in 1778, the Hawaii group also grew in strategic importance owing to its suitability as a place for breaking journeys across the Pacific Ocean. By 1794, cattle, goats and geese had been landed there by Captain George Vancouver. As the numbers of foreigners coming into contact with Hawaii increased, so too did the foreign population, whose influence with the native Hawaiians grew. Interest turned to the islands themselves, and sandalwood became an important export. Similar patterns of European interaction with Pacific people and islands occurred elsewhere in the region.

European trade and cultural exchange with East Asia, particularly China and Japan, became an important stimulus to systematic exploration of the Pacific Ocean and Islands. European colonization of Australasia led to increased knowledge of the southwest Pacific islands.

Increasing Russian involvement in the north part of the Pacific was stimulated initially by the fur trade and thence by sovereign intentions. An ambitious expression of these was the Russian claim to Hawaii in 1816, eventually countered by

King Kamehameha and his British allies. Hawaii provides a good case study of the rush by colonial powers to claim Pacific lands. The annexation of the island group by the United States in 1898 was partly to pre-empt the growing Japanese influence.

The earliest European explorers in the Pacific were trained navigators and surveyors, concerned especially with locating islands and mapping their outlines and anchorages; few had particular interests in natural science and consequently did not take a rigorous approach to their other discoveries. Much early Pacific exploration led to a romanticized view of the reality. The crux of Bernard Smith's (1985) history of the portrayals of Pacific Islands by pre-20th century artists is the contrast between the economical, often spartan, sketches of shipboard artists with those of Europe-based artists who reinterpreted these sketches in the spirit and lavish style of the times, giving rise to grossly misleading impressions of the region among those Europeans who had never visited it.

2.3.4 The Roots of Modern Scientific Exploration of the Pacific Ocean and Islands

A new approach was ushered in with Captain James Cook's first voyage in 1768 – a voyage sponsored by the Royal Society (United Kingdom) and including several scientists, among them Joseph Banks, many of whose observations on Pacific fauna and flora still provide a descriptive yardstick for scientists. 'No people ever went to sea better fitted out for the purpose of Natural History, nor more elegantly' wrote a contemporary (quoted by Frost, 1988: 32).

Later Pacific voyages included naturalists such as Charles Darwin, Joseph Hooker and T.H. Huxley, for all of whom the experience of seeing parts of the Pacific world helped them clarify ideas which subsequently proved highly influential in the development of various facets of science. Some of the earliest systematic geoscientific information collected about the Pacific Islands was by Darwin in the *Beagle* (1831–1836). Some results of that trip – particularly Darwin's 1842 book *Structure and*

Distribution of Coral Reefs – were notable in drawing geoscientists' attention to Pacific island environments.

Around this time, interest from Pacific Rim countries in the environmental history of the Pacific Ocean and Islands burgeoned. The United States Exploring Expedition (1838–1842) for example made many useful observations but, as an example of a nascent 'colonial' scientific corpus straining to shake off its European straitjacket, the Funafuti expeditions are hard to match. It was these expeditions, particularly the second in 1897, which demonstrated to London-based scientists that their Australian counterparts, led by Edgeworth David, were equal to the task of mounting scientific enterprises of global significance: in this instance, boring through the coral cap of Funafuti Atoll in Tuvalu to see whether it covered a submerged volcanic foundation (Cantrell and Rodgers, 1989).

The most impressive body of information collected during the 19th century about the Pacific Ocean was that associated with the *Challenger* expedition (1872–1876; Plate 2.4). This represented the first time that a global circumnavigation had been dedicated wholly to science rather than devoted, at least in part, to more pragmatic purposes. The *Challenger* traversed the Pacific from northwest to southeast in 1874–1875, made extensive collections of deep-water material, living and non-living, and surveyed key areas; a depth of nearly 7000 m recorded near Japan proved the ocean to be much deeper than hitherto realized. The 40 volumes of scientific results concerned mostly oceanography although those dealing with ocean-floor geology

Plate 2.4   The HMS *Challenger* at Tahiti in 1874. Sponsored by the Royal Society of London and the British Admiralty, the *Challenger* Expedition gathered huge quantities of oceanographic and ocean-floor data and also reported on the character and composition of many of the lands in the Pacific Basin
Edinburgh University Library

attracted considerable interest from geoscientists. The *Challenger* expedition proved a model on which much 20th century scientific exploration of the Pacific was based. At the same time, it marked the end of the era of largely unsystematic and piecemeal exploration. From this time onwards, most scientific 'expeditions' in the Pacific addressed precisely formulated questions in specified areas.

It is also worth noting that, from this age to the mid-20th century, there was progressively more contact between the peoples of the Pacific Basin and outsiders: contact that – for all its attendant shortcomings – facilitated the subsequent rational and systematic exploration of the Pacific.

2.3.5 Modern Exploration of the Pacific Ocean and Islands

Voyages seeking to gather geological and biological information in the Pacific grew in number in the first part of the 20th century. The *Dana* and the *Carnegie* both sailed in 1929, the *Albatross* and *Galathea* in 1947 and 1952 respectively. Many ships were engaged in mapping the ocean floor, a task made easier with the development of automatic echo sounders in the 1920s. The systematic collection of oceanographic data by ships crossing the Pacific and the archiving of these data in the United States, United Kingdom and Japan led to a rapid increase in the understanding of Pacific oceanography in the first part of the 20th century. This has been supplemented by Australian, New Zealand and Canadian data in the second half.

The Second World War (1939–1945) not only saw a great mobilization of people and ships in the Pacific but also stimulated a degree of international cooperation without which the Pacific might never have been systematically explored as it has been. Utilizing technological innovations (like sonic profiling) and many of the resources which had been mobilized during the war, the United States began systematic studies of the Pacific Ocean floor in earnest in the 1950s and 1960s. Cruises by vessels from the Scripps Institution of Oceanography, Woods Hole

Oceanographic Institution and the United States Navy Electronics Laboratory were notable; later, data were gathered and exchanged by most Pacific Rim countries. Mapping of the ocean floor revealed for the first time the full extent of the underwater mountain range (mid-ocean ridge) named the East Pacific Rise, and the ocean trenches, together with the numerous seamounts and guyots.

Understanding was aided by global co-operation. For example, during the International Geophysical Year 1957–1958 and International Geophysical Cooperation (1959), 12 specially designed and equipped vessels from France, Japan, the former USSR and the USA each made at least one cruise in the tropical Pacific Ocean with the express purposes of gathering oceanographic data. Investigations of ocean-floor sediments and the underlying basement were carried out on most of these cruises.

Coral reefs were a particular target of a smaller group, such as the Pacific-wide investigations of reefs by the *Albatross* in 1899–1900 and the British Great Barrier Reef Expedition 1928–1929. The decision by the United States to test atomic weapons at Bikini Atoll in 1945 (and later, elsewhere in the region) led to a flurry of investigations of the structure and origin of Bikini and surrounding atolls as part of Operation Crossroads (Plate 2.5). Similar work preceded and accompanied the use of Mururoa and other atolls in French Polynesia for nuclear testing from the 1960s to the 1990s.

Geophysical studies in the Pacific began early in the 20th century and came to focus on magnetic and gravity measurements, seismic and heat-flow data. Ideas about palaeomagnetism were tested in the Pacific. From studies of ocean-floor sediments and lavas, it was demonstrated that the Earth's magnetic field had reversed periodically in the past. Magnetic stripes of variable polarity were arranged symmetrically around mid-ocean ridges, which were also areas of high heat flow and conspicuous seismicity compared to surrounding areas. From such lines of evidence, the sea-floor spreading hypothesis was proposed in 1961; from there it took only a few years for the more rounded

Plate 2.5   Atomic cloud during 'Able' Day blast at Bikini Atoll, northwest Pacific, Monday 1 July 1946. This was
the first of 23 nuclear blasts on Bikini between 1946 and 1958, the effects of which have been manifest in humans
and island ecosystems across a large region
United States Navy National Archives 80-G-396226 and Greenpeace Pacific

model of plate tectonics to be formulated and become widely accepted (see Chapter 3).

The age of the ocean basins was targeted by the JOIDES (Joint Oceanographic Institutions for Deep Earth Sampling) programme. The results demonstrated that the age of the ocean floor (generally from the ages of microfossils within overlying basal sediments) increased away from the mid-ocean ridges and that the rate of age increase at particular times was the same on both sides of a ridge – fully supporting plate tectonics. The success of JOIDES spawned continuing initiatives involving the drilling of the ocean floor, notably DSDP (Deep Sea Drilling Project) and ODP (Ocean Drilling Program). Through these, ocean cores have become available for a variety of purposes, including palaeoclimatic analysis. The systematic understanding of palaeoclimate and palaeoceanography in the Pacific can be linked largely to the rise of deep-sea drilling.

In recent decades, the Pacific Ocean has received more scientific interest than any other, particularly from the United States, Australia and, more recently, Japan and Russia. Particular foci of interest include: (i) the complex plate interactions associated with the Gorda and Juan de Fuca plates in the northeast Pacific, which are linked to shallow earthquakes in the northwest United States and southwest Canada; (ii) the plate boundaries of the southwest Pacific, where recent volcanism may have created traps for hydrocarbons migrating upwards through the associated zones of lithospheric weakness; (iii) the economic mineral potential of the ocean floor, particularly that associated with hydrothermal vents; and (iv) the hazard potential of active ocean-floor volcanoes, especially in Hawaii.

In contrast, work on the Pacific Islands themselves has grown at a far slower pace, partly because of their few economic resources and partly because of political constraints. Most investigations are applied and focus on resource management issues although research less applied in character continues and is yielding data of relevance to questions of environmental change. Insights into the management and evolution of Pacific island environments have been hindered by the cultural and scientific baggage of outside investigators, specifically their uncritical application to the Pacific Islands of models developed for other places (Lasaqa, 1973; Cox and Elmqvist, 1997; Nunn, 1999b).

2.4 REACHING THE PAST

Imagination was an important component of analysis and understanding when people first began asking how environments had changed, long before recorded accounts were made. Gradually, as observations were amassed, imagination became tempered with inference and deduction, and thenceforth by the development of specific empirical and analytical techniques.

Although many such techniques were developed for and applied to terrestrial environments in the Pacific Basin, many detailed and informative data have come from ocean-floor sediment sequences, which are generally far more complete than their onland counterparts. Indeed, the quest for terrestrial records of environmental change of similar detail and duration to those obtained from the ocean floor has largely been given up. The longest and least disturbed terrestrial records are those from lake bottoms, such as Wien Lake in Alaska, cored by Hu et al. (1993). Although core length was only 2.55 m, it yielded a 12 000 year record of vegetation change and soil development. Other sequences derived from sediments are commonly of shorter duration and have numerous discontinuities (Plate 2.6). The most conspicuous exception comes from areas of the Earth's surface which have had a long, unbroken history of ice accumulation. Coring of 2546 m of ice at Vostok in East Antarctica, for example, has yielded information about temperature change over the past 200 000 years (Jouzel et al., 1993).

Early efforts to understand the origin and character of the ocean floors through sampling of ocean-bottom materials and bathymetric mapping was pioneered by HMS *Challenger* (see above). By the 1950s and 1960s, the need for a systematic survey of the ocean basins was realized. It had become clear that in order for their history to be properly understood, it would be necessary to drill 500–1000 m through sediments and rocks lying beneath several kilometres of

Plate 2.6   One way to gain detailed insights into terrestrial environmental change is to analyse cores through sediments which have accumulated slowly in swamps and lakes. This is one example from research coordinated by the writer on Yacata island in northeast Fiji. What were thought initially to be shallow swamps turned out to be deep karstic depressions filled with sediments, the analysis of which has focused on pollen as a record of changing vegetation assemblages in the catchment, and on the incidence of charcoal indicating burning, possibly instigated by humans

water. This represented a technological challenge which is routinely overcome today on specially designed research vessels like the *JOIDES Resolution* (Plate 2.7). For example, an 88.5 m core drilled at DSDP site 594 in the southwest Pacific was made beneath 1204 m of water.

Plate 2.7   The *JOIDES Resolution*, a specially converted research vessel designed to drill into the ocean floor beneath as much as 8230 m of water
Ocean Drilling Program

Analysis of this core provided a high-resolution record of the changing climate throughout much of the late Quaternary (Nelson et al., 1993).

The potential for extracting complete records of past environmental change from the ocean floor stimulated the development of new analytical techniques, including oxygen-isotope analysis as a proxy for ocean-surface palaeotemperatures. Much pioneering research into the use of deep-sea cores for palaeotemperature records used Pacific equatorial cores (Shackleton and Opdyke, 1973, 1976). Although this technique has been used most often to illuminate later Quaternary environmental changes, it has also been successfully applied to parts of the earlier Cenozoic despite the comparative weakness of the $\delta^{18}O$ signals.

Ocean-floor sediments have been analysed in a number of ways. For example, particularly during the 1970s, there were a number of studies in the Pacific Basin which linked spatial maxima of ice-rafted debris with late Cenozoic glacial periods (Kent et al., 1971; Keany et al., 1976). More recently, work has been carried out on the proportions of biogenic calcium carbonate ($CaCO_3$) in marine sediments. The increase observed by Farrell and Prell (1991) since 3.9 Ma is probably due to falling sea level associated with a long-term increase in global ice volume. The value of bringing several techniques to bear on the same question is exemplified by the recent study of Ortiz et al. (1997) who discovered hitherto unsuspected subtleties in their reconstruction of the California Current around the Last Glacial Maximum.

The difficult task of creating accurate maps of the Pacific has been revolutionized over the past 20 years with the increasingly widespread availability of satellite imagery. The use of Global Positioning Systems is likely to revolutionize ground-based mapmaking in the next 20 years, and is also being used to measure rates of slow environmental changes. More difficult proved the problem of how to make accurate maps of the ocean floor. Most success has been gained with echo-sounding techniques which have evolved only comparatively recently into sonic imaging techniques providing three-dimensional views of the ocean floor (Schlee et al., 1995).

2.5 DATING THE PAST

So many different techniques of dating the past are now available that a comprehensive review of these is inappropriate in this book. Only the main groups are outlined below.

Methods of ascertaining ages for particular environmental changes in the distant past may be divided into (i) those which date the past absolutely, (ii) those which allow past events to be correlated over wide areas, and (iii) those involving counting from the present backwards.

The most powerful dating techniques to have been developed over the past few decades are those based upon the long-term decay at known rates of radioactive isotopes in various materials. Such techniques have revolutionized the understanding of past environmental changes. Table 2.1 gives a list of some common dating techniques and the effective age ranges over which they can be used.

Large numbers of dates, particularly of ocean floor samples, have allowed other timescales to be developed. Subsequent samples can be correlated using these timescales without recourse to radiometric dates. The principal methods of correlation which have given rise to their own timescales are listed in Table 2.2; the most important for the study of Quaternary environmental change in the Pacific Basin are discussed in more detail in Chapter 8. In the Pacific Ocean, one of the most commonly used is the palaeomagnetic timescale, developed by dating ocean-floor rocks exhibiting known palaeomagnetic polarity. This involves the recognition of distinct time zones called chrons (and subchrons) during which the Earth's magnetic polarity was the same (Table 2.3).

Other timescales dependent on radiometric dating have been developed, including ones based on marine micropalaeontology (Vincent and Berger, 1981), pollen (Webb, T., 1985), and oxygen isotopes (Pisias et al., 1984). Oxygen-isotope analysis has proved one of the most

Table 2.1   Common dating methods used for understanding environmental change in the Pacific Basin

Method	Age range (years ago)	Materials/landforms
Amino acid racemization	50 000 to 1 000 000	Organic material
Fission-track (^{238}U decay)	150 000+	Volcanic ash, metamorphic rocks
Potassium–argon (K/Ar)	100 000 to 4.5×10^9	Igneous rocks (particularly micas, hornblende)
Radiocarbon (^{14}C)	100 to 70 000	Organic material such as wood, bone, shell, charcoal and coral
Luminescence	0 to 150 000+	Loess, volcanic ash
Uranium–thorium (U/Th)	0 to 350 000	Shells, corals, speleothems

Table 2.2   Selected correlation methods used for understanding environmental change in the Pacific Basin

Method	Age range (years ago)	Materials/landforms
Loess stratigraphy	10–500 000	Loess
Marine fossils	Various	Marine sediments
Oxygen isotope analysis	Various	Deep-sea cores, cave deposits
Palaeomagnetism	Various	Sediments, volcanic rocks, loess
Pollen analysis	Various	Organic sediments
Tephrochronology	Variable	Volcanic ash

Table 2.3   Average ages of the four geomagnetic polarity chrons of the Pliocene and Quaternary Compiled from Cande and Kent (1992), Shackleton et al. (1990) and Berggren et al. (1995)

Chron (subchron)	Polarity	Age of base (Ma)	Age of top (Ma)
Brunhes	Normal	0.78	–
Matuyama	Reversed	2.58	0.78
(Jaramillo)	Normal	1.07	0.99
(Olduvai)	Normal	1.95	1.77
(Réunion II)	Normal	2.15	2.14
(Réunion I)	Normal	2.23	2.20
Gauss	Normal	3.58	2.58
Gilbert	Reversed	5.89	3.58

Table 2.4   Major controls on oxygen-isotope composition of open-ocean planktonic and benthic foraminifera. Note that this table assumes that foraminiferal species do not change their depth preferences
After Gasperi and Kennett (1993)

No net change in $\delta^{18}O$
- Warming of surface waters is equal to increase in ice-volume effect
- Cooling of surface waters is equal to decrease in ice-volume effect

Increase in $\delta^{18}O$
- Cooling of surface waters
- Increase in ice volume
- Cooling of surface waters and an increase in ice volume
- Cooling of surface waters has a greater influence than the decrease in ice volume
- Increase in ice volume has a greater influence than the warming of surface waters

Decrease in $\delta^{18}O$
- Warming of surface waters
- Decrease in ice volume
- Warming of surface waters and a decrease in ice volume
- Warming of surface waters has a greater influence than the increase in ice volume
- Decrease in ice volume has a greater influence than the cooling of surface waters

---

powerful tools for determining the course of Cenozoic climate change; the critical indicator is $\delta^{18}O$, the significance of which is summarized in Table 2.4.

A final group of methods are those dependent on counting individual records which are known to have been produced at regular intervals (Table 2.5). These include varves and tree rings.

Laminated sediments in which the deposition for single years can be distinguished are termed varves. These occur commonly in places with marked seasonal contrasts in the rate and/or the nature of sedimentation. Counting the number of varves is a method of directly dating a particular deposit; noting the changing characteristics of varves within a sequence allows insights into environmental changes. The alternations between diatom-rich and clay-rich sediments in the Gulf of California are a good example (Calvert, 1964).

Seasonal changes in wood growth in trees produce annual rings. Tree rings can be counted to determine the age of a particular tree. Variations in the width of rings can be used to infer details of contemporary climate. Tree-ring studies throughout North America allowed Fritts and Lough (1985) to estimate annual average temperatures for AD 1602–1961.

Bracketed with counting methods are trapped-electron dating techniques, including luminescence (TL or OSL) and electron spin resonance, and those such as amino acid racemization which are based on slow chemical reactions.

Table 2.5   Selected counting methods used for understanding environmental change in the Pacific Basin

Method	Age range (years ago)	Materials/landforms
Dendrochronology	0 to 4000	Wood
Lichenometry	0 to 5000	Lichens
Varve analysis	0 to 2000	Rhythmically banded sediments

2.6   ANALYSING THE PAST

The ability to reach the past and date it does not guarantee its understanding. As for many other facets of science, the history of environmental change research is littered with the debris of forsaken and forgotten models invalidated by subsequent discoveries. Many others were conceived in such simple terms that they now fail to explain the observed complexity satisfactorily.

A case in point, particularly relevant to the Holocene history of the Pacific Basin (see Chapters 12–15), concerns the interpretation of charcoal in sedimentary sections. There is an excellent correlation between massive charcoal accumulation and the earliest human colonization of particular parts of the region – the study from Tinian Island in the northwest Pacific (Athens and Ward, 1998) is a recent example – implying beyond reasonable doubt that the first human colonizers of the island used fire as a management tool. Yet the relationship is far less clear for other parts of the Pacific, and attempts to reconcile the record of forest burning with that of human settlement history have proved controversial.

The history of environmental change in many areas of the Earth's surface remains comparatively unknown because certain techniques do not work as effectively under certain conditions as others or because the past cannot be reached so easily as it can elsewhere. A good example is the impotence of pollen analysis in palaeoenvironmental reconstruction on many tropical Pacific islands owing to a lack of diversity of pollen-producing source plants, both today and in the past, and the paucity of environments suitable for long-term pollen preservation.

Scale is another cause of difficulty. The most complete and detailed chronologies usually refer to the last few millennia (or centuries or years) for particular places but, on account of an absence of data for earlier times, it is difficult to determine whether or not such recent changes conform to a longer-term trend. The observed increase in tropical-cyclone frequency in most parts of the Pacific Ocean over the past few decades is a good example (Radford et al., 1995).

It is becoming increasingly apparent that many assumptions underlying particular techniques of absolute dating are incorrect, and that the validity of dates obtained in the past must be re-evaluated. The most widely used technique for the late Quaternary – radiocarbon dating – has attracted considerable criticism. For example, the assumption made initially that the production (and therefore the amount) of ^{14}C in the atmosphere was constant in the past is now recognized as invalid, and all radiocarbon dates need to be calibrated in order to determine age accurately (Edwards et al., 1993; Kitagawa and van der Plicht, 1998).

Despite these drawbacks, understanding of past environments has progressed greatly in recent decades. This understanding has helped humankind reach a better understanding of their role in ecosystems. It is creating opportunities for learning from the past in order to better understand and manage the future – the search for analogues of current/future 'global warming' in both the Last Interglacial (~125 ka) and the Little Climatic Optimum (~700 BP) exemplifies the point (Lorius and Oeschger, 1994). The imperative of examining ancient sediment cycles for evidence of abrupt changes in the thermohaline circulation in order to determine whether one might be forthcoming is another example (Broecker, 1997).

CHAPTER 3

Ideas Concerning the Origin of the Pacific Basin

... when God said,
'Be gathered now, ye waters under Heaven,
Into one place, and let dry land appear!'
Immediately the mountains huge appear
Emergent, and their broad bare backs upheave
Into the clouds; their tops ascend the sky.
So high as heaved the tumid hills, so low
Down sunk a hollow bottom broad and deep,
Capacious bed of waters: thither they
Hasted with glad precipitance, uprolled,
As drops on dust conglobing, from the dry:
Part rise in crystal wall, or ridge direct,
For haste: such flight the great command impressed
On the swift floods.

John Milton, *Paradise Lost,* VII

3.1 EARLY IDEAS

Over many generations, the history remembered collectively – in the absence of recorded evidence to the contrary – often involves mainly catastrophic events. Mundane everyday occurrences are relegated to a subconscious level. Before the 20th century, when structured forms of enquiry became commonplace amongst scientists, there was a tendency to highlight catastrophic events in Earth history. Many early ideas in 'western' Earth science and cosmogony were influenced by the story of the Deluge in the Book of Genesis in the Christian Bible, culminating in the catastrophism of the early 19th century. The subsequent discovery of ice ages fuelled the catastrophist view although it was eventually overwhelmed by gradualists who emphasized the long-term cumulative effects of processes operating at everyday rates.

Many early ideas concerning the origin of the Pacific Basin were explicitly catastrophist,

reflecting the mood of the times in which they were conceived together with the awe with which most scientific observers regarded the vast ocean with its conspicuous 'ring of fire'.

3.1.1 The Pacific Basin as the Birthplace of the Moon

In the early part of the 20th century, it was widely believed that the Pacific Basin was the scar left behind when the Moon was ejected from the Earth. This would have been a cataclysmic event of unparalleled magnitude and would have to have occurred when the Earth was in a viscous, possibly molten, state (Darwin, 1879). The event was described in fanciful terms by some writers: 'Three quarters of the Earth's surface, to a depth of 35 miles [56 km], was carried away in a trailing mass of ruin. New Zealand itself was just saved to the Earth' (Pickering, 1924: 32). For others, the birth of the Moon from parent Earth initiated continental drift (Evans, 1925).

This theory of the origin of the Moon and – more importantly for our purposes – the Pacific Basin did not long survive the widespread acceptance of mobilist theories about the Earth's outer shell. These treated all ocean floors as having the same origin and, particularly as knowledge about the floor of the Pacific increased, it became abundantly clear that no special case could be made here.

Yet the idea that the Moon came out of the Earth should not be entirely dismissed. Particularly in the light of geochemical evidence of lunar composition gathered by the Apollo missions, the idea that the Moon was derived from the Earth's mesosphere (mantle) after its core formed is considered plausible (Ringwood, 1979).

3.1.2   The Pacific Basin as the Site of a Sunken Continent

The idea that a continental mass once occupied most of the Pacific Basin is one which, with hindsight, could be recognized as an expression of the stupefaction of a continental dweller upon being confronted by something so alien as a vast ocean basin dotted with islands. Keats represented the sensation famously when he described how the first European to see the Pacific – Vasco Nuñez de Balboa on 26 September 1513 – might have felt[1]:

> ... when with eagle eyes
> He star'd at the Pacific – and all his men
> Look'd at each other with a wild surmise –
> Silent, upon a peak in Darien

Yet the idea that a sunken continent underlies the Pacific has never really been fully discredited and, in the light of recent discoveries, may never be so.

Supposed lost continents in the Pacific included Mu, claimed by Churchward (1959) to be a Pacific equivalent of Atlantis and to have been mentioned by Plato in that context (which it was not). Also claimed to have once existed in the Pacific was the continent Rutas – referred to in various Hindu

classics as having once existed only in the Indian Ocean – yet nevertheless considered to have included all the Polynesian islands by some writers. Both Mu and Rutas were used by Theosophists to support various of their ideas concerning human cultural evolution; later development of such ideas led to other continents being postulated to have existed in the Pacific. One of the most elaborate of these was a large triangular continent named Pan in the northern Pacific which, despite the admission that the continent disappeared 24 ka, was described along with a Panic dictionary and alphabet by Newbrough (1932). The idea of Frobenius, between 1908 and 1926, that civilization began on a 'lost' Pacific continent and then spread west across Asia to Africa (Sprague de Camp, 1970) is also worthy of mention in this context. None of these ideas is considered to have any scientific credibility today.

The scientific belief in the disappearance of a huge continent in the Pacific Basin has its roots in the late 19th century preoccupation with 'lost continents' and 'land bridges'. In an age when the level of the ocean surface was thought largely unchanging, 'lost' continents were used to support theories about human exploration and cultural change. In a time when continents and ocean basins were considered to have been fixed relative to each other, land bridges (now disappeared) were called upon to explain the distribution of plants and animals, both extant and fossil, across ocean gaps. In the Pacific, land bridges were routinely invoked to explain floral similarities between South America and Australasia (Arldt, 1918).

There is some valid geological evidence of a 'lost' Pacific continent along the west coast of South America. The nature of middle to late Palaeozoic rocks suggested to Burckhardt (1902) and others that a 'southeast Pacific continent' had once lain oceanwards of the area at the time. Recent work by Bahlburg (1993) has largely affirmed the correctness of this idea although it is clear that the legacy of such a continent in dispersing organisms would be practically indiscernible today.

---

[1] John Keats (*On First looking into Chapman's Homer*). Keats actually wrote that it was Cortéz not Balboa

The possibility of a huge lost continent in the central Pacific persisted into the 1960s before being abruptly marginalized as the models of sea-floor spreading and plate tectonics were successively adopted by most geoscientists. Then in 1977, Nur and Ben-Avraham announced the existence of a former continent in the Pacific – a detached piece of Gondwana which they named Pacifica – pieces of which had drifted across the Pacific becoming accreted onto the edges of the existing continents forming the periphery. Collisions between fragments of Pacifica and the Pacific continental margins created fold mountains, many of which continue to rise as the result of these collisions. More recent work, while acknowledging the former existence of exotic terranes or microcontinents in the central Pacific, has cast doubt on the existence of Pacifica on account of its improbably large size and discrepancies in its proposed life history (see Chapter 5).

3.1.3 The Pacific Basin as a Primordial Earth Structure

Before the ascendancy of mobilist theories of the Earth's surface, most thinking within the Earth sciences was confined within a fixist framework. This entailed the belief that all elements of the Earth's surface were fixed relative to each other: continents were unmoving, ocean basins were primary, unchanging structures.

Even when evidence supporting mobilist views was increasing for other parts of the Earth's surface, little was forthcoming for much of the Pacific Basin and, as late as 1961, a few scientists still favoured its origin as a primordial structure. Their argument was based principally on observations that the thermal gradient beneath the oceans was much less than beneath the continents; in other words, 'the mean depth of heat sources under oceans is greater than under continents' (Magnitsky, 1961: 362). They reasoned that the transition zone from oceanic to continental lithosphere was consequently one of thermal stress, which could explain the belt of seismic and volcanic activity around the Pacific Rim without recourse to a mobile lithosphere.

One way in which such a primordial structure could have been created was by the impact of a huge asteroid, perhaps one-third of the size of the Moon; the additional mass added to the Earth could have been the original Pangaea (Howell, 1959).

It is now clear that the form of the Pacific Basin has changed throughout its history (see Figure 3.2) largely because the lithosphere forming its outermost solid shell is mobile. The oldest ocean floor in the Pacific is of late Jurassic age – very young when compared to the maximum ages of the surrounding continents. The present shape of the Pacific Basin is a residual one, produced by the outward drift of large continental fragments from Pangaea at least 200 Ma and, more recently, by the accretion of exotic terranes onto pre-existing continental margins.

3.2 MODERN IDEAS

The conflict between the fixists, who regarded all the components of the Earth's surface as fixed and unmoving, and the mobilists, who regarded the Earth's surface as dynamic and ever-changing, was not resolved rapidly. Indeed, Alfred Wegener's first enunciation of his theory of continental drift in 1915 received little support. It was not until after the Second World War that discoveries were made which led to the widespread acceptance of a mobilist view. It is worth briefly appraising the role which the Pacific Basin played in the development of mobilist theories.

3.2.1 Continental Drift and the Pacific Basin

Around the beginning of the 20th century, the most globally influential centres of academic endeavour were in Europe and, although most of the world was well known, the Atlantic – particularly the northern part – was, owing to its proximity, better known than the Pacific. Knowledge about the Pacific was often undervalued or sidelined owing to both its greater size and remoteness from the 'centre' which meant that it did not play a great role in the formulation of Earth mobility theories.

The theory (or model) of continental drift was the earliest coherent expression of the idea that the Earth's surface was mobile and dynamic. Much of the argument put forward by Alfred Wegener – its leading proponent – was based on a fit of continental fragments (Wegener, 1929). While it was clear (and had been recognized long before Wegener's time) that continental masses on either side of the Atlantic fitted together well, it seems that the existence of that fit precluded that of any fit across the Pacific. As a result, classic reconstructions of the super-continent Pangaea have proceeded on the assumption that the modern Pacific is the residual of a former world ocean which became reduced in size as the Atlantic Basin opened (see Figure 3.2). Although this conclusion is consistent with what has been discovered since, other reconstructions of comparable cogency would have been possible using Wegener's methods (Dobson, 1992) and could have led to a different orthodoxy for supercontinent reconstruction.

Wegener believed that the fold mountains along the western side of the Americas formed as the result of them drifting westwards and encountering resistance from the Pacific ocean floor. The orientation of fold mountains elsewhere in the Pacific – particularly in Australasia – therefore indicated to Wegener the direction in which these continental fragments had once drifted.

In his reconstruction of Gondwana, Wegener employed many persuasive palaeontological and biogeographical arguments. Notable among these were recognition of past and present floral similarities across the Pacific between South America and Australasia which suggested a former land connection via Antarctica. Pacific islands were regarded – incorrectly in the main – as having been detached from continental blocks as these had drifted across the ocean basin.

A perceived overemphasis on form at the expense of process left most scientists sceptical about continental drift until almost half a century after its first announcement when the idea of sea-floor spreading was first proposed.

3.2.2 Sea-Floor Spreading in the Pacific

The mapping of vast undersea mountain ranges having anomalously high heat flow and being volcanically and seismically active over much of their lengths drew attention to the dynamism of the ocean basins. Subsequent observation that the age progressions and palaeomagnetic polarity of the ocean floors were symmetrical about these mid-ocean ridges led to the view that ocean floor formed along these and gradually moved away on either side of them at similar rates (Dietz, 1961).

Compared to Wegener's time, a more equable amount of evidence from the Pacific and Atlantic in support of sea-floor spreading manifested the increase in importance of Pacific Rim, particularly North American, research centres by this time.

Recognition of ocean-floor creation and spreading along mid-ocean ridges begged the question of where it was destroyed – as it would have to be if the Earth is not expanding. The answer to this question was provided by an understanding of the possible function of ocean trenches and led to the most comprehensively formulated of 20th century mobilist theories – plate tectonics.

3.2.3 Plate Tectonics and the Pacific Basin: an Overview

The relative abundance of ocean trenches outside the northeast quadrant of the Pacific Basin – that part studied most intensively – delayed an appreciation of their importance to theories of Earth-surface mobility. Yet, by the late 1960s, sea-floor spreading was known to be only part of the answer. Zones of ocean-floor creation were matched with ones of ocean-floor destruction – the ocean trenches. This Earth-surface model was complemented by increased knowledge of the structure and composition of the upper Earth, specifically the identification of the fluid layer named the asthenosphere (~70–250 km below sea level) within which molten (destroyed) ocean floor was transported from beneath the ocean trenches to beneath the mid-ocean ridges, thence to be recycled via the Earth's surface.

On account of its comparative plethora of ocean trenches, the Pacific now became the focus of world attention to demonstrate the validity of the plate-tectonics model. The Pacific Basin was found to be divisible into a number of constituent plates (Figure 3.1), most having a single divergent (sea-floor spreading) boundary, a convergent (destructive) boundary and one or more transverse (strike-slip) boundaries. The Pacific is dominated by the vast Pacific Plate, being created along the East Pacific Rise and destroyed along a number of trenches in the western Pacific including the Hikurangi–Tonga Trench, the Marianas Trench, and the Japan, Kuril and Aleutian Trenches. The

most prominent transverse boundary of the Pacific Plate is that running northwards from the Gulf of California and finding surface expression in the active San Andreas Fault.

Later on, intraplate (within-plate) regions in the Pacific attracted more interest from Earth scientists. Many groups of intraplate islands were realized as being parallel to the directions of plate movements and it was found that the rates of these could be calculated accurately from island ages. Intraplate volcanism was explained by 'hotspots' where magma from the asthenosphere reached the Earth's surface at locally elevated intraplate swells.

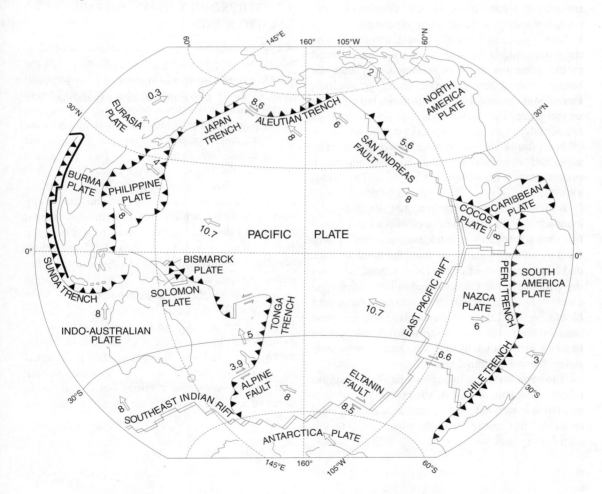

Figure 3.1   Lithospheric plates and associated features in the Pacific Basin

Plate tectonics remains the best explanation for observations made in the Pacific Basin and is consequently used as such throughout this book. Yet there are other tenable explanations, including surge tectonics (Meyerhoff et al., 1992), which involves a contracting Earth, and the expanding-Earth theory. A short account of the latter is given in the following section.

3.2.4   Expanding Earth – Evidence in the Pacific?

The expanding-Earth theory holds that the Earth is expanding and that separation of the continents by the growth of new ocean floor is not significantly compensated by its destruction elsewhere.

The essence of the expanding-Earth theory and its application to the Pacific can be encapsulated in the comment of Sam Warren Carey that the ocean trenches in the Pacific are truly 'pacific'. Far from actively accommodating plate convergence, this view holds that ocean trenches are actually sites of lithospheric extension. Many other features of the Pacific Basin have been explained within an expanding-Earth rather than a plate-tectonic framework (Carey, 1976; King, 1983).

Rather than assuming that the Pacific grew symmetrically, as the plate-tectonics model has it, the expanding-Earth model assumes that it was created asymmetrically. In this, the need to have had large amounts of ocean floor subducted along the circum-Pacific Ocean trenches is avoided. The clearest evidence is that the Asian and Australian blocks, which migrated westwards, left marginal seas in their wake in contrast to the American block which, since it did not move anywhere, never formed marginal seas.

The expanding-Earth explanation for the origin of the Pacific is hard to refute in all its ramifications. Yet the plate-tectonics model remains the more useful and more tested explanatory tool.

3.2.5   Evolution of the Pacific Basin

The plate-tectonics model is the basis for describing and explaining the long-term evolution of the Pacific Basin, summarized in Figure 3.2. This shows how the continental lithosphere of the Earth aggregates periodically under the influence of deep mantle convection to form a 'supercontinent'. The subsequent splitting of a supercontinent into fragments is followed by their disaggregation. The two best-known super-continents, Rodinia and Pangaea, are shown at 750 Ma and 200 Ma respectively.

3.3   THE SINGULARITY OF THE PACIFIC BASIN

An appreciation of Earth-surface mobility, acquired largely through acceptance by the scientific community of the models of continental drift, sea-floor spreading and plate tectonics, led to a revolution in thinking about the origins and development of the ocean basins. It has become clear that rather than being residual features, the ocean basins are dynamic – in fact, the prime movers of the Earth's surface.

The Pacific contains most of the world's convergent (active) plate margins and hardly any passive margins, which dominate the continental margins of the other ocean basins. In addition, the Pacific hosts a divergent (sea-floor spreading) boundary, comparable in size to those in the other ocean basins, about which a well understood system of lithospheric plates has developed. Most of the Pacific Ocean is occupied by the Pacific Plate – the world's largest oceanic plate – which exhibits the classic attributes of the plate-tectonics model. The interior of the Pacific Plate is occupied by islands, the patterns of which demonstrate both past and present plate motions besides allowing areas of current intraplate stress to be precisely delineated.

47

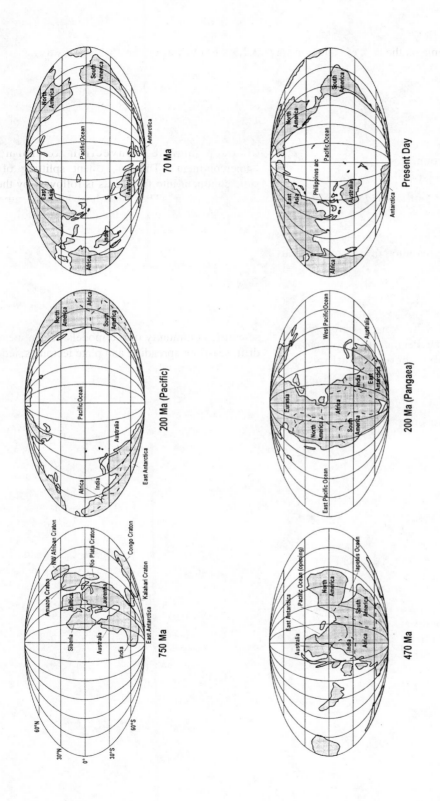

Figure 3.2 The formation of the Pacific Basin by movements of continents. All maps are centred on 0°, 180° except the 200 Ma (Pangaea) one which is centred on 0°, 0°. All maps are based on the Mollweide (equal-area) projection. The 750 Ma map shows the supercontinent Rodinia. The 470 Ma map shows the earliest Pacific Basin forming between Antarctica and North America. Two maps are shown for 200 Ma, the time when the supercontinent Pangaea was aggregated: the upper map is centred on the Pacific, which was at this time a residual ocean; the lower map shows Pangaea. The 70 Ma map shows the larger ancestor of the modern Pacific Basin, which formed only after the equatorial gaps closed

PART B

The Early Pacific

CHAPTER 4

The Earliest Pacific (Before 250 Ma)

I look at the geological record, as a history of the world imperfectly kept, and written in a changing dialect; of this history we possess the last volume alone, relating only to two or three countries. Of this volume, only here and there a short chapter has been preserved; and of each page, only here and there a few lines.

Charles Darwin, *On the Origin of Species* (1859: 310–311)

4.1 INTRODUCTION

When Darwin gave his views on the geological record, he was writing about a history of the Earth far more condensed than that we recognize today. Darwin may have been shocked to learn that life appeared first on Earth more than 3000 million years ago. He might scarcely have credited the sparseness of the record from such distant times yet would doubtless have applauded the efforts of those who used it to reconstruct the history of our planet. The Precambrian and Palaeozoic history of the Pacific Basin are discussed separately in this chapter with emphasis on the environmental changes that took place around their boundary, some 590 Ma.

4.2 THE PACIFIC DURING THE PRECAMBRIAN ERA (BEFORE 590 Ma)

The earliest known Pacific Basin dates from before 750 Ma. Although it probably closed completely before the formation of its modern counterpart, its legacy is clear. The existence of this early Pacific Basin was discovered from the recognition of once-contiguous Precambrian geological units in Antarctica, North America, Australia and India (Moores, 1991; Dalziel, 1991) which had become juxtaposed as the result of the amalgamation of the supercontinent Rodinia about 1000 Ma. Rifting between what is now the southwest United States (Laurentia in Figure 3.2)

and eastern Antarctica led to the opening of the earliest Pacific.

Aside from contributing to the gross form of the modern Pacific Rim, such details could be considered as having little relevance to the history of environmental change in the region. Yet just as these changes set the physical stage for later developments, so too did they have climatic and biogeographical implications that were themselves the product of environmental changes. The nature of these ancient changes can still be discerned, albeit in only a general sense.

Slices of ancient oceanic crust known as ophiolites show an abrupt decrease in thickness around 1000 Ma which has been interpreted as meaning that the surface of the lithosphere was deeper relative to the ocean surface (that is, closer to the Earth's core) before this time. Continental freeboard before 1000 Ma would thus have been much less because of the greater terrestrial submergence; average ocean cover may have been 90–95% rather than the present 70%. This would have meant that the Earth's climates exhibited far less variation, both spatially and temporally, compared to today (Moores, 1993).

The start of production of thinner oceanic crust around 1000 Ma was associated with emergence of the continents – a process lasting perhaps just 180 million years – and an accompanying increase in continental freeboard. By about 800 Ma, it is likely that the Earth's climates displayed far more heterogeneity than at any earlier time. Continental emergence would have increased terrestrial erosion and offshore sedimentation. An increase in atmospheric oxygen from 3% to 20% was also associated with these processes on the emerging continents (Knoll, 1991) and has been held largely accountable for the conspicuous diversification of living organisms during the early Palaeozoic (McMenamin and McMenamin, 1990).

The thinning of oceanic crust 1000–800 Ma may have been caused independently of the contemporary formation and break-up of Rodinia. When continental emergence began about 1000 Ma, Rodinia was already assembled. The process of its disaggregation, 750–500 Ma, may have contributed to the already increasing climatic heterogeneity, and hastened the appearance of new lifeforms and the attendant rise in ecosystem complexity.

The chronology of the Palaeozoic used in this book is illustrated in Figure 4.1.

4.3 ENVIRONMENTAL CHANGE AROUND THE PRECAMBRIAN–PALAEOZOIC BOUNDARY (590 Ma)

Precambrian lifeforms were generally simple; multicellular organisms appeared only in its later part. The diversity of Precambrian life was also very low, in sharp contrast to the early Palaeozoic situation. Life along Precambrian coastlines was dominated by stromatolites: organo-sedimentary structures built by simple, non-nucleated cells (prokaryotes). The end of the Precambrian and beginning of the Palaeozoic (the Cambrian Period) was a time marked by 'a unique radiation of eucoelomate [having a body cavity] animals that permanently changed the nature of ecological and sedimentary systems … truly a period of remarkable evolutionary innovation' (Knoll and Walter, 1992: 673).

This 'Cambrian Explosion' of life is preserved best in only a few places; the world is indebted to

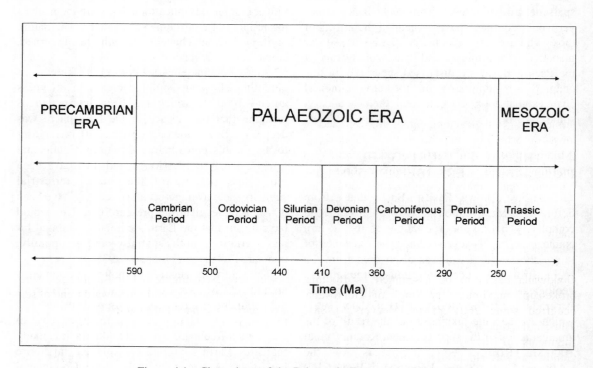

Figure 4.1   Chronology of the Palaeozoic Era used in this book

Stephen Jay Gould for making the story of one of them – the Burgess Shale in the Canadian Cordillera – widely accessible (Gould, 1989). The nature of life along the shores of the early Palaeozoic Pacific is represented by Burgess Shale organisms both here and in Yunnan Province in southern China.

The question as to why the Cambrian Explosion occurred when it did is as yet unanswered to everyone's satisfaction, but environmental change is likely to have been an important cause. It is plausible to suppose that the combined effect of the disaggregation of Rodinia and the emergence of continents increased climatic heterogeneity, ocean-circulation complexity and offshore sedimentation, all of which led to increased oxygen production and the evolution of a variety of lifeforms adapted to the greatly increased possibilities posed by this new environmental complexity.

4.4 THE PACIFIC DURING THE PALAEOZOIC ERA (590–250 Ma)

The movements of the major continental lithospheric fragments during the Palaeozoic are still debated. The majority view to have emerged in the 1980s and early 1990s was that, once Laurentia broke out of Rodinia, as it had by the end of the Precambrian, it no longer had any influence on the development of the remainder, most of which amalgamated into Gondwana (Moores, 1991). An alternative view, which is gaining ground, invokes collisions between Laurentia and Gondwana during the Palaeozoic as the cause of initial formation of mountain ranges, principally the Appalachian and Andean chains of North and South America respectively (Dalziel et al., 1994).

What is generally agreed upon is that Gondwana began to be assembled as Rodinia broke up; maximum 'packing' of Gondwana occurred some 260–250 Ma (Veevers, 1988). Thus, at the time it (re)joined the other continental fragments to form Pangaea in the Mesozoic, Gondwana was already breaking up (see Figure 3.2).

It has been suggested that plate convergence took place continuously along the entire Pacific Rim during the Palaeozoic (Flöttmann et al., 1993). A well documented example is provided by the series of terranes (see Chapter 5) which first began to be accreted onto the Australian and Antarctic cratons perhaps 530 Ma (Figure 4.2), a process explicable most plausibly by contemporary plate convergence. Similar evidence is available for other parts of the Palaeozoic Pacific margin.

The recognition that terranes were being accreted along the Pacific Rim throughout the Palaeozoic requires that they existed and were being created within the Pacific Basin during this era. This cannot be reconciled with the view, once widely accepted in the absence of evidence to the contrary, that the Palaeozoic Pacific was a largely inactive ocean basin in which nothing of consequence to continental evolution was happening. Such an idea is now clearly invalid: 'The great wide blue [Palaeozoic] Pacific has developed spots – terrane fever – with islands and archipelagoes, and little continents' (Waterhouse, 1987: 607).

4.4.1 Sea Levels and Climate

By analogy with the formation of Pangaea some 800 million years later (Vail et al., 1977), very low sea level around the beginning of the Palaeozoic was probably associated with the agglomeration of Rodinia. This is largely because continental collisions reduce the extent of continental lithosphere, allowing the oceans a larger area over which to spread, thus lowering sea level.

The Palaeozoic is marked by a series of sea-level transgressions and regressions which are thought to correlate reasonably well with the seawater strontium isotope ($^{87}Sr/^{86}Sr$) record. As can be seen from Figure 4.3, these changes were superimposed on an overall downward trend of sea level during the later Palaeozoic.

A lack of synchrony between Palaeozoic climate changes and sea-level changes is conspicuous (Frakes et al., 1992). Indeed, owing to the comparative equability of Palaeozoic

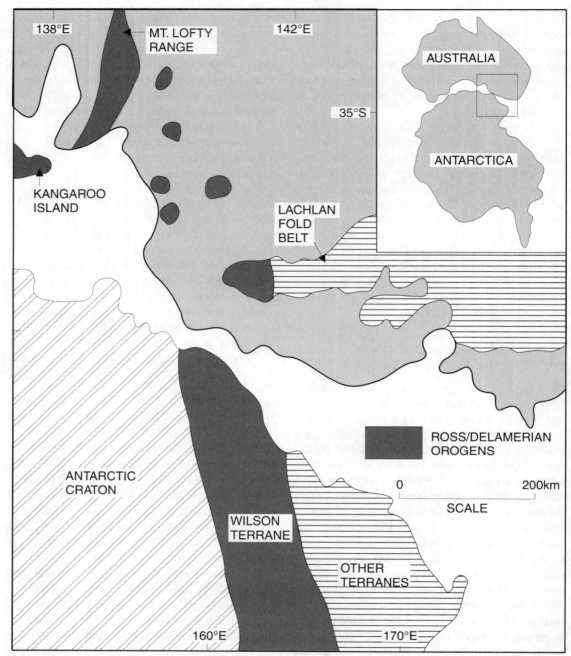

Figure 4.2 Map showing terrane remnants accreted onto the Australian and Antarctic cratons, and how these probably appeared in the Gondwana assembly

Modified, with permission, from Flöttman et al. (1993) in Findlay, R.H., Unrug, R., Banks, M.R. and Veevers, J.J. (eds), *Gondwana Eight – assembly, evolution and dispersal*, Proceedings of the eighth Gondwana symposium, Hobart, Tasmania, Australia, 21–24 June 1991. 638 pp. Hfl. 225/US$130.00. A.A. Balkema, PO Box 1675, Rotterdam, Netherlands

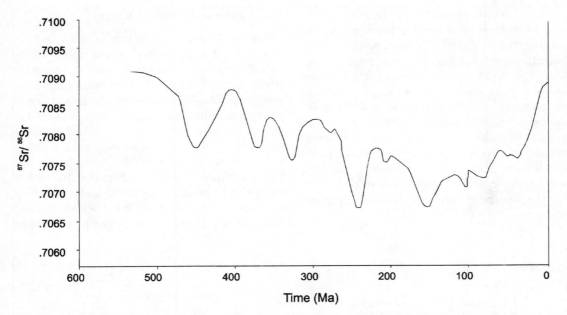

Figure 4.3   The $^{87}Sr/^{86}Sr$ ratio of seawater since the late Cambrian. Since the ratio is lower in seawater (0.7091) than in continental runoff (0.7160), so times of lower $^{87}Sr/^{86}Sr$ values are approximately coincident with times of lower sea level
After Burke et al. (1982), with permission of the authors

climate until the mid-Carboniferous, it has been argued that earlier sea-level changes were of largely tectonic origin (Hallam, 1984), from the successive aggregation and break-up of continental amalgams associated respectively with orogeny and crustal extension. In this scenario, the overall falling sea level in the second half of the Palaeozoic is attributed to the gradual closing of ocean basins which accompanied the agglomeration of Gondwana and Pangaea.

Yet there are some indications that sea-level oscillations of a kind commonly associated with more recent glacial eustatic changes affected the rim of the Palaeozoic Pacific Basin. For example, a Late Ordovician (~440 Ma) submarine fan sequence preserved in the Lachlan Fold Belt of southeast Australia was suggested as having developed under the influence of glacial eustatic oscillations of sea level in this then-equatorial part of the Pacific (Jones et al., 1993).

Slow warming during the early Palaeozoic ended abruptly in the middle Ordovician as polar

regions cooled and ice sheets developed. Subsequent warming lasted until the early Carboniferous (Frakes et al., 1992). The striking mismatch between the timing of this warm period with glaciation in the Andes during the late Devonian and early Carboniferous exemplifies the point that such events cannot have been solely a result of global climate change. In this example, the onset of glaciation has been attributed convincingly to contemporary orogeny (Caputo and Crowell, 1985). Around the same time, the earliest known glaciation in the Pacific half of Australia occurred, possibly as a result of continuing compression producing uplift in the Lachlan Fold Belt. Later glaciations in southern Australia appear to have originated in adjoining Antarctica; direct evidence of these has been found in the Transantarctic Mountains and Ellsworth Mountains. Outside Gondwana, evidence for late Palaeozoic glaciation has been found in western Beringia (Pavlov, 1979).

Around the end of the Palaeozoic, Pangaea was close to its time of maximum packing, and the contemporary Pacific – the world ocean Panthalassa (or EoPacific) – was huge. It was twice the width of the modern Pacific and extended from pole to pole. Ocean circulation may have been simpler than in the modern Pacific. Enormous gyres driven by tradewinds flowed around 85% of the Earth's circumference. Probably the western part of Panthalassa would have been warmer than its eastern part, and there would have been only a modest thermal gradient between the poles and the Equator (Kennett, 1982).

4.4.2 Palaeozoic Biotic Changes

The early Cambrian explosion is marked most clearly by the appearance of the major shelly animal groups (McMenamin and McMenamin, 1990), many of which fed on stromatolites leading to their decline. Reef limestones developed first in association with the Silurian transgression but their diversification was abruptly curtailed during the late Devonian regressions only to be renewed as ocean levels rose once more during the Carboniferous. The development of other categories of biota during the early Palaeozoic is less well known because fewer fossils remain. Vascular land plants had developed by the middle

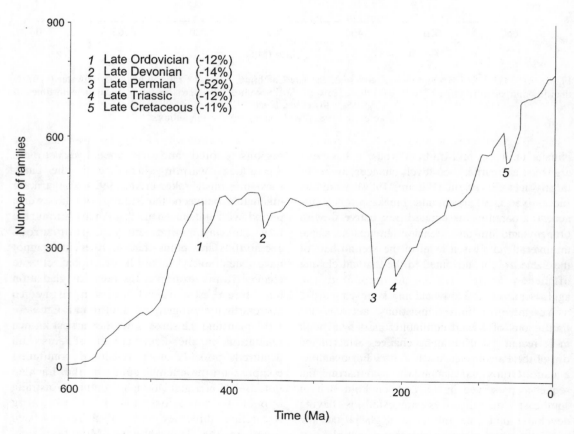

1 Late Ordovician   (-12%)
2 Late Devonian   (-14%)
3 Late Permian   (-52%)
4 Late Triassic   (-12%)
5 Late Cretaceous   (-11%)

Figure 4.4   Variations in the numbers of the best-preserved families of marine vertebrates and invertebrates over the past 600 million years. Five mass extinctions are shown by abrupt drops in the diversity curve
After Raup and Sepkoski (1982), with permission of David Raup and the American Association for the Advancement of Science

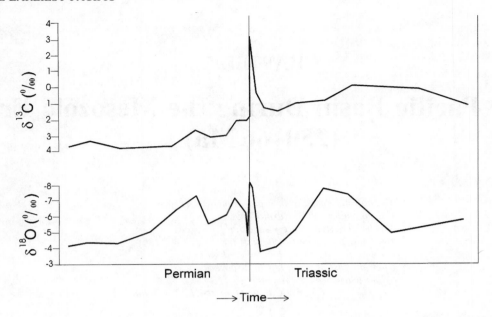

Figure 4.5   Changes in carbon and oxygen isotope values across the Permian–Triassic (P–Tr) boundary at site B, Meishan, near Shanghai, eastern China (after Xu et al., 1989). Note the sharp perturbations at the P–Tr boundary, which may be empirical evidence of a catastrophic event which contributed to the mass extinction around this time.

Palaeozoic, and ferns, gymnosperms and conifers by its end.

Mass extinctions occurred at various times during the Phanerozoic (Figure 4.4). Measured by the numbers of species becoming extinct, the second greatest of these events was that in the late Ordovician (Sepkoski, 1993), the greatest being around the end of the Permian; some 96% of marine species became extinct at this time (Maxwell, 1989).

This mass extinction has been widely regarded as coincident with the boundary between the Palaeozoic and Mesozoic (the Permian–Triassic or P–Tr boundary). Iridium anomalies – interpreted as indicators of catastrophic bolide impacts – occur in some P–Tr boundary sections and are contemporaneous with 'spikes' in other diagnostic parameters (Figure 4.5). Yet largely because these spikes are not as widespread as might be expected, most investigators now conclude that the P–Tr extinctions were not brought about by bolide impacts of the kinds that are now widely accepted as an explanation for the late Devonian (McGhee, 1996) and Cretaceous–Tertiary (see Chapter 5) mass extinctions. As more evidence pertaining to the P–Tr mass extinction has been amassed, so it has become clear that it was a more gradual process than envisaged initially. In this light, the most plausible cause of these extinctions is one relating to global cooling causing sea-level fall leading to a decline in atmospheric oxygen levels (Gruszczynski et al., 1989; Kakuwa, 1996).

CHAPTER 5

The Pacific Basin During the Mesozoic Era (250–66 Ma)

We shall not cease from exploration
And the end of all our exploring
Will be to arrive where we started
And know the place for the first time.
Through the unknown, remembered gate
When the last of earth left to discover
Is that which was the beginning;

T.S. Eliot, *Four Quartets: Little Gidding*

5.1 INTRODUCTION

It is no surprise that evidence for Mesozoic environmental change is more abundant within the modern Pacific Basin than that referring to the Palaeozoic but it is still largely obscured by the overprint of Cenozoic changes. The chronology of the Mesozoic is summarized in Figure 5.1.

The agglomeration and initial break-up of Gondwana preceded the agglomeration of Pangaea. Its subsequent break-up in the latest Palaeozoic and early Mesozoic led to the

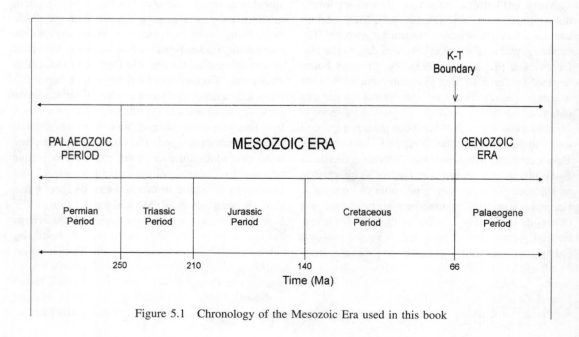

Figure 5.1   Chronology of the Mesozoic Era used in this book

Table 5.1   The increase in exposed land area ($10^6 km^2$), Carboniferous to Cretaceous
data from Parrish (1992)

Time	Northern hemisphere	Southern hemisphere	Total
Early Carboniferous	32.5	79.0	111.5
Late Carboniferous	37.0	91.5	128.5
Late Permian	47.4	84.5	131.9
Early Triassic	63.4	78.3	141.7
Early Jurassic	75.7	71.2	146.9
Late Jurassic	64.9	72.8	137.7
Late Cretaceous	64.7	62.7	127.4

formation of the modern Pacific Basin. Continued collisions between fragments of continental lithosphere brought about an overall global increase in the area of exposed land although the rise of sea level during the late Mesozoic reduced this subsequently (Table 5.1).

About 200 Ma, when Pangaea was assembled most completely, what was to become the Tethys Sea (and later the Indian Ocean) was already opening. The rest of the world ocean occupied most of the western hemisphere, and was larger and less well defined than today. It was the opening of the Atlantic Ocean and the continued break-up of Gondwana which brought about a reduction in size of this early Mesozoic Pacific causing its outline to become recognizable as that of the modern Pacific (see Figure 3.2).

The ocean gap between the two parts of Beringia in the Mesozoic north Pacific had closed by the late Cretaceous, the subsequent movement of terrestrial fauna leading to their conspicuous diversification in North America during the Palaeogene (Webb, S.D., 1985).

The gap existing between North and South America during most of the Mesozoic was responsible for a stronger and warmer Equatorial Current across the Pacific than today (Barron and Peterson, 1989). The closure of this gap for a few million years during the late Cretaceous and Palaeogene allowed mammals to migrate from North America south along the Pacific coast of South America; an important fossil site is near Lake Titicaca in the central Andes (Sigé, 1972). A southward dispersion of hadrosaurs is likewise

testimony to the existence of a late Cretaceous connection between the Americas (Casamiquela, 1980). Biogeographical evidence favours a Central American location rather than one farther east, near the modern Antilles, for this archipelagic 'filter bridge' (Briggs, 1987: 42).

5.2   PLATE EVOLUTION

About one-third of the modern Pacific Ocean floor, particularly the central and western parts, formed during the Mesozoic. For most of this time there were no magnetic reversals so palaeo-magnetism cannot be used to date these 'quiet' zones as it can Cenozoic ocean floor. Renkin and Sclater (1988) got around this problem for the quiet zones of the North Pacific by reconstructing their tectonic history using known Mesozoic plate configurations and plausible rates of sea-floor spreading; they found ocean floor as much as 174 million years old.

This ocean floor dates from a time when three plates dominated the Pacific. In the reconstructions of Cox et al. (1989), a triple junction between the Pacific Plate (in the south), the Izanagi Plate (in the west) and the Farallon Plate was active between 140 Ma and 110 Ma, when the Izanagi–Farallon boundary changed from divergent to convergent, resulting in both the subsumption of the Izanagi by the Pacific Plate and the splitting of the Farallon Plate into two: the Farallon in the eastern Pacific (represented today by Nazca, Cocos and Juan de Fuca plates), and the now-disappeared Kula Plate in the north Pacific.

The growth of the Pacific Plate in the central Pacific was also accommodated by the subduction of the now-disappeared Phoenix Plate along its boundary with the Antarctic Plate.

5.2.1 Terranes

Tectonostratigraphic terranes are large fault-bounded pieces of lithosphere, each having a history distinct from that of surrounding lithosphere. Most terranes which form the modern Pacific Rim originated within the Pacific Ocean. Modern analogues of Mesozoic continental terranes include the Ontong Java and Manihiki plateaux, the Hess and Shatsky rises (Howell, 1985). Most oceanic terranes originated as island arcs dominated by lines of volcanoes and/or emerged, often coral-reef, limestones. The northward movement of the Hess and Shatsky rises – now at the latitude of Japan – through central Pacific low latitudes during the late Mesozoic has been traced from studies of the pelagic limestones which accumulated on their surfaces (Thiede et al., 1981).

Although terranes were accreted onto the Pacific Rim during the Palaeozoic, most terranes mapped along the modern Pacific Rim were accreted during the Mesozoic and Cenozoic. A critical event was the initial opening of the Atlantic Ocean during the early Jurassic which initiated plate convergence along many parts of the Pacific Rim. As plates began to converge, so (formerly) volcanically active island arcs eventually collided with and became accreted onto the Pacific Rim. Similar processes brought about accretion of continental terranes along the periphery of the Mesozoic Pacific Basin.

A striking demonstration of the importance of terrane dispersal around the Pacific Rim comes from studies of the bivalve genus *Monotis* (Silberling, 1985). The thin shells of this animal, which lived only during the late Triassic, are now found in rocks around the entire Pacific Rim (Plate 5.1). It is likely that these groups of *Monotis* lived only in quite a restricted area of the Pacific so that their present fossil distribution is clear testimony to its subsequent break-up and dispersal.

5.2.2 The Pacifica Conundrum: Geological and Biogeographical Arguments

Controversy attends the reality of Pacifica – a continent-sized chunk of Gondwana alleged to have become detached from the rest around 180 Ma and henceforth disaggregated within the Pacific Basin, with pieces becoming attached to the continental masses around the edge where they appear today as exotic terranes. The collision of these terranes with the Mesozoic Pacific Rim caused uplift of the fold mountain belts which parallel much of its length today (Nur and Ben-Avraham, 1977, 1989).

Most geological objections to the existence and history of Pacifica as proposed by Nur and Ben-Avraham (1977) relate to the origins of exotic or allochthonous terranes along the west coast of the Americas. In the Pacifica model, all these terranes would once have been parts of Gondwana that became gradually disaggregated as they crossed the Pacific during the Mesozoic to be accreted onto the continents forming the contemporary Pacific Rim. Yet Hallam (1986) among others argued that much of the exotic material along the western margins of the Americas is of oceanic origin and even the supposedly allochthonous continental fragments originated close to the cratonic interiors of these continents and/or were emplaced in the Palaeozoic, long before Pacifica was suggested to have existed as an independent entity. Exceptions, which may indeed be fragments of one or more 'Pacifica' micro-continents, include the Amotape–Tahuin terrane of northern Peru (Mégard, 1989), the Cortez terrane of southern California (Howell et al., 1985), and fragments of Gondwana which were dispersed northwards in the western Pacific and accreted onto Asia (Burrett et al., 1991). Despite these, the geological evidence for the existence and subsequent disaggregation of Pacifica is not generally compelling. There were undoubtedly continental remnants in the Mesozoic Pacific Basin but it is unlikely they (had) ever formed a single continental mass.

Yet it was not only geological evidence which suggested that Pacifica may have existed in the

Plate 5.1 The pectinacid bivalve *Monotis* occurs in early Mesozoic (Norian) rocks in many parts of the Pacific Rim. The distribution of the fossils in these places indicates the role of terrane accretion in their formation. (A) *Monotis ochotica* from near Sakawa, Shikoku, Japan. (B) *Monotis ochotica* from the Murihuku Terrane, Nelson, New Zealand
Norm Silberling

Mesozoic Pacific. It has long been recognized that Mesozoic biogeographic tracks crossing the Pacific are not consistent with the classic picture of Pangaea break-up. Consider the dispersal of *Asteraceae* from North America to Chile (*Microseris pygmaea*) and to Australasia (*Microseris scapigera*) (Bachmann, 1987), the existence of Palaeozoic and Mesozoic fossils from the Tethys Sea in the American Cordillera, or the crossing of the late Cretaceous Pacific Ocean from east to west by a variety of Caribbean benthos including bivalves, gastropods, larger foraminifera and calcareous algae (Newton, 1988). The dispersal of these organisms in ocean currents would certainly have been facilitated by the presence of a continent-sized landmass such as Pacifica in the central Pacific. Conversely, an increase in Pacific intraplate volcanism, such as is thought to have occurred during the late Cretaceous, may have provided sufficient insular 'staging posts' to allow organisms to make trans-Pacific journeys of this kind without recourse to Pacifica.

The existence and subsequent break-up of some version of Pacifica would also help resolve other long-standing biogeographical problems, such as the location of a homeland for mammals and flowering plants prior to their conspicuous radiations around the Mesozoic–Cenozoic boundary (66 Ma). Yet a recent review concluded that Pacifica was not a satisfactory solution to the range of biogeographical enigmas associated with the Mesozoic Pacific (Sluys, 1994). The strongest evidence for Pacifica or, more plausibly, a number of smaller 'superterranes' within the Mesozoic Pacific, remains the fact of terrane–craton collision and fold mountain orogeny along much of the contemporary Pacific Rim. A good example of a Mesozoic superterrane is provided by Talkeetna, discussed below – one of the many to have collided with the North American coast over the past 200 million years and to have contributed to the formation of the cordillera.

5.2.3   Cretaceous Terrane–Craton Collision and Mountain Formation: the Story of Talkeetna

Many terranes became accreted during the Mesozoic along the western side of the North American Cordillera (Figure 5.2), some evidently originating at great distances from the cratonic margin. One of the largest and most widespread is the Talkeetna superterrane which consists of three units. The oldest are the Precambrian–Mesozoic Alexander terrane and the Jurassic Peninsular terrane, both in southern Alaska. The youngest is the Wrangellia terrane, fragments of which are found in Alaska, the Queen Charlotte Islands, Vancouver Island and Idaho (Csejtey et al., 1982). The Talkeetna superterrane now extends over 16° of latitude; palaeomagnetic data suggest that its original latitudinal spread was just 5° (Yole and Irving, 1980).

The Talkeetna superterrane comprises largely volcanic rocks (including pillow lavas) overlain by inner platform carbonates, including some fossil coral reefs. Corals in the Wrangellia terrane suggest that it was located as much as 5000 km west of its present position during the Permian (Belasky and Runnegar, 1994).

Talkeetna probably amalgamated on the Kula Plate in the late Jurassic 14–20° south of its present position. It was then carried at least 1500 km towards the North American craton, during which time it became overlain in places by a submarine volcanic arc. When it collided with the craton in the late Cretaceous, it crushed a basin of marine flysch deposits in the process.

5.3   SEA LEVELS

Sea level rose during most of the Mesozoic reaching a high during the mid-Cretaceous (about 90 Ma); global sea level may then have stood 250 m or more above its present level (Haq et al., 1987; Figure 5.3). The rate of Mesozoic sea-level rise in the Pacific evidently varied considerably.

Figure 5.2 (opposite)   Terranes along the western side of the North American craton
After Howell et al. (1985) and Debiche et al. (1987)

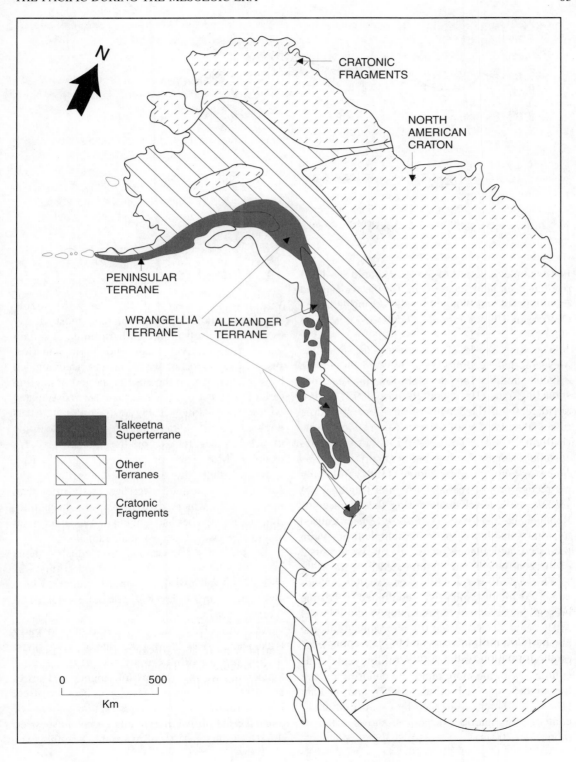

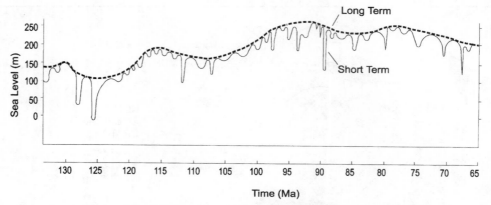

Figure 5.3   Cretaceous sea levels relative to present mean sea level
After Haq et al. (1987)

For example, a period of unusually rapid sea-level rise has been held accountable for the widespread drowning of mid-Pacific reefs during Albian–Cenomanian times (about 100 Ma) (Matthews et al., 1974).

The long sea-level rise during the mid–late Mesozoic has generally been ascribed to tectonic effects, notably an acceleration in sea-floor spreading rate (Hays and Pitman, 1973) and/or an increase in the length of sea-floor spreading ridges associated with the break-up of Gondwana (Hallam, 1984). The possibility of geoidal–eustatic changes having been important at this time was highlighted by Mörner (1981) following his recognition of spatially irregular (regional not global) sea-level changes during the Cretaceous. Relative sea-level rise may also have been caused by the subsidence of continental lithosphere produced, as Pangaea broke up, by the dispersal of the heat which had accumulated beneath it.

The fall of sea level in the late Cretaceous began a trend which has continued until the present. What is of particular interest in the model of Haq et al. (1987) is that the Mesozoic–Cenozoic boundary (~66 Ma) is marked by a fall of sea level of about 80 m lasting just 2–3 million years. This contrasts with the more rapid and enduring falls in the models of Vail et al. (1977) and Hallam (1984): the latter, influenced by the magnitude of contemporary biotic extinctions, shows a fall of more than 200 m.

5.4   CLIMATES

Warming characterized the early Mesozoic (Permian–middle Jurassic), marked most clearly by poleward shifts in the northern limits of certain plants, particularly in Asia and North America. Just as this period of warming was accompanied by sea-level rise, so the ensuing period of cooling (middle Jurassic–early Cretaceous) was associated with sea-level fall (see Figure 5.3) although it is doubted whether a causal connection between the two is as secure as for the Quaternary, for example. Most parts of the modern Pacific Rim were in either low or mid-latitudes during the Jurassic. Low latitudes were generally drier than mid-latitudes which, as demonstrated from tree-ring studies (Creber and Chaloner, 1985), received abundant seasonal (monsoonal) rain.

By the early Cretaceous, monsoonal systems had broken down in many places and aridity was spreading. The zonal system of climate which dominates Earth today was gradually established (Parrish, 1987).

The time from the mid-Cretaceous to early Palaeogene (~106–55 Ma) was one of the warmest in the entire Phanerozoic: average global temperatures were perhaps 6°C higher (Frakes et al., 1992). The poles may have been ice-free. Biotic radiation, most spectacularly that of angiosperms (flowering plants), may have been linked to this period of warmth. Yet this was

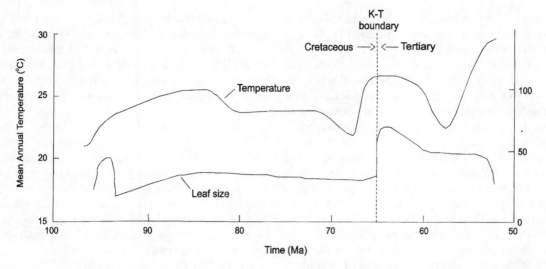

Figure 5.4   Late Mesozoic and early Cenozoic temperatures reconstructed from changes in form of fossil leaf margins (upper curve) together with variations in leaf size
After Upchurch and Wolfe (1987) with permission of Cambridge University Press

neither monotonic nor unbroken. Evidence is available to show that much of the northern hemisphere warmed at a faster rate during the Maastrichtian (latest Cretaceous; Figure 5.4). Yet data have been presented recently to show that this was also a time of cooling climate, and that the cooling marking the Cretaceous–Palaeogene (K–T) boundary was only 'the last in a series of [such] events that affected Cretaceous environments' (Barrera, 1994: 880). The resolution of this apparent contradiction may lie in the interpretation of the proxy temperature data. For example, the interpretation of variations in the character of leaf margins (used to construct the upper curve in Figure 5.4) through time as primarily an effect of palaeotemperature may be too narrow; other climatic factors may have played significant roles. Likewise the ocean-surface temperature proxies used by Barrera may not be applicable widely outside the Southern Ocean from which they derived. This controversy exemplifies the difficulties which continue to dog interpretations of comparable phenomena using different techniques, and flags the likelihood that future interpretations of such phenomena may result in radical re-evaluation of earlier conclusions.

5.5   OCEANOGRAPHIC CHANGES

The changes in the configuration of land and sea, in ocean levels and climates, discussed in the preceding sections, led to oceanographic changes in the Pacific which provided new environments for organic evolution.

In the early Mesozoic, the circulation of the world ocean was simple on account of its dominance of our planet's surface and the concentration of continental lithosphere in Pangaea. At this time, Pangaea prevented the equatorial circulation of water. The pattern of water movement in the contemporary Pacific was thus similar to the modern circulation, dominated by two gyres in the northern and southern parts of the ocean basin (Parrish, 1992).

When the Tethys Sea opened up completely between the two constituent parts of Pangaea – Laurasia and Gondwana – a west-flowing equatorial current was established around the globe. The date of opening of the first so-called Hispanic Corridor linking the Tethys Sea and the Pacific between North and South America has been the subject of some debate (Newton, 1988; Smith and Westermann, 1990; Stanley and

Yancey, 1990). Most authorities regard the earliest ocean crust in the region – the modern Gulf of Mexico – as of middle Jurassic age.

Equatorial ocean circulation had become well established by the early Cretaceous when water flow through the Hispanic Corridor became impeded, though not prevented, by movement of terranes into the Nicaraguan area. Modelling of Cretaceous ocean circulation shows that, except for a blockage of the Hispanic Corridor in the late Cretaceous–Palaeogene, an equatorial flow from the Atlantic to and across the Pacific persisted until only comparatively recently, about 3 Ma (Barron and Peterson, 1989).

Compared to its modern counterpart, the distinctive feature of the Pacific Ocean for most of the Mesozoic was its overall warmth and lower Equator–Pole temperature gradients. Oxygen-isotope data for the northwest Pacific have been interpreted as meaning that mid-Jurassic sea-surface temperatures were ~26°C and reached today's temperatures (~28°C) by the end of the period. Late Cretaceous sea-surface temperatures may have been 4.5°C warmer (Woodruff et al., 1981).

5.6 BIOTIC CHANGES

Although the potential post-Mesozoic diversification of lifeforms was curtailed significantly by the mass extinction marking its end, many modern groups have antecedents dating from this era. The appearance of these groups can be linked to environmental changes brought about by the tectonic, sea-level and climate changes described above.

5.6.1 Evolution of Selected Pacific Basin Plants and Animals

Conditions in most of the Pacific Ocean were particularly conducive to the development of new species at various times during the Mesozoic. Tectonic changes had led to the isolation of previously connected groups of organisms, which subsequently led to them following separate evolutionary paths. A warm stable climate encouraged such developments. Increasing stratification of ocean water would likewise have created new ecological niches which new species would have filled. Frequent transgressions and generally high sea levels promoted the appearance of new species in the shallower parts of the Mesozoic ocean.

Although their ancestors are recognized in the Palaeozoic, all extant conifers originated in the Mesozoic. For example, the Podocarpaceae appeared around the start of the Cretaceous. All living Araucariaceae are today found in the southern hemisphere but during the later Mesozoic they dominated the areas of dry-summer climates in low latitudes, both north and south, and also occupied mid-latitudes (Krassilov, 1978).

Ferns, which appeared first in the Palaeozoic, diversified in the Mesozoic. Large ferns like *Clathropteris* may have dominated Mesozoic forests in Japan and Korea until the appearance and expansion of the angiosperms in Cretaceous times.

One of the most successful plants to have appeared in the Mesozoic was a sphenophyte named *Equisetum*. It is characterized by a long underground rhizome from which vertical shoots grow upwards and an enviable tenacity which allowed it to disperse and adapt to a variety of environments throughout Gondwana. Today *Equisetum* is found in the Americas, eastern Asia, the Philippines, New Guinea, New Caledonia and Fiji: the sole surviving genus of the sphenophytes (Tryon and Tryon, 1982).

Other groups of gymnosperm plants that evolved during the Mesozoic and occurred in the Pacific Basin at that time include cycads and ginkgos. Yet the most important palaeobotanical event to have occurred during the Mesozoic was the origin of the angiosperms which dominate terrestrial plant life today. Angiosperms appeared first, probably in low-latitude Gondwana, during the early Cretaceous and spread polewards, diversifying rapidly as they competed with the existing conifer–fern vegetation (Hickey and Doyle, 1977). By about 93 Ma, angiosperms dominated lowland environments in the northern hemisphere; dominance in the south was attained slightly later. Since many angiosperms developed means of dispersal that relied on insects, birds or mammals, angiosperms exhibit coevolution with animals more than any other group of land plants. And in turn, from the mid-Cretaceous onwards, terrestrial animal life became largely dependent on angiosperms as primary producers.

The main groups of fishes, which appeared first in the Ordovician, generally increased in diversity throughout the Mesozoic, particularly during the period of Cretaceous sea-level rise; at this time, the number of genera of teleost fishes increased from around 20 to more than 400 (Thomson, 1977).

The most successful group of Mesozoic tetrapods – the dinosaurs – inhabited all the main parts of the Pacific Rim (Plate 5.2). Studies of

Plate 5.2   Dinosaur tracks in Tarapaca Province, Chile, date from the late Jurassic and are thought most likely to have been made by a large stegosaur. Since no stegosaur remains are otherwise known from South America, this would be an important discovery if confirmed
R.J. Dingman and the United States Geological Survey

their evolution suggest that this was affected little by climate changes before those associated with the end of the era when dinosaurs became rapidly and comprehensively extinct (see below).

The first mammals in the Pacific Basin were in southern China and although they spread throughout most of the Pacific Rim by the end of the Mesozoic, they did not become ecologically dominant until the Tertiary.

Accompanying and stimulated by the rise of sea level during much of the Mesozoic was a diversification of marine organisms which transformed the Pacific Ocean. Many (now-drowned) atolls and (now-accreted) terranes in the central tropical Pacific served as 'staging posts' for the dispersal of shallow-water fauna across the entire width of the ocean basin at the time. This is different from the situation today owing to the presence of the East Pacific Barrier (the large stretch of landless ocean between the islands of the central and eastern Pacific). Thus only about 4% of the modern coral fauna occurs throughout the ocean basin; in the Mesozoic, although the Pacific was much larger, this figure was much higher (Grigg and Hey, 1992).

The initial opening of the Hispanic Corridor in the middle Jurassic decreased the provincialism of many Pacific marine biota. A good example is the disappearance of the East Pacific Subrealm. This was an area within which marine organisms had evolved in response to very localized environmental conditions but then, as the Hispanic Corridor opened in the middle Jurassic and Tethyan biota swept into the area, its biological distinctiveness soon vanished (Westermann, 1981).

Although most species extinctions have occurred at times of falling or low sea level, the appearance of new species is not closely linked to any particular condition of sea level. Much clearly depends on the local situation and the precise nature of the environmental changes resulting from sea-level change. For example, during the late Cretaceous in the interior of the North American west, new species of ammonites appeared when a new seaway opened up as the result of rising sea level between Alaska and

Texas. A causal connection between ammonite extinction at this location and sea-level fall was suggested by Kennedy and Cobban (1976) although such events cannot now be viewed independently from potential causes having global influence, discussed at the end of this chapter.

Owing to the greater size of the Pacific, which allowed more adaptive opportunities, it is likely that extinction of marine organisms at times of low sea level was less here than in other oceans. Indeed there is evidence, during the Jurassic for instance, that Pacific biota colonized areas elsewhere which had been depopulated as the result of sea-level fall (Hallam, 1981).

When considering the major causes of long-term environmental change during the Mesozoic,

it is tectonics, specifically changes in the configuration of land and ocean, to which we must keep returning. In order to illustrate the environmental changes which resulted from this cause, an example is given below: that of the recently demonstrated connection between Marie Byrd Land in Antarctica and the New Zealand continental sliver.

5.6.2 Biogeographical Consequences of Cretaceous Rifting and Magmatism in the Southernmost Pacific

Much of the Pacific Antarctic coast is that of Marie Byrd Land. Its Mesozoic relationship with New Zealand, which illustrates the nature of

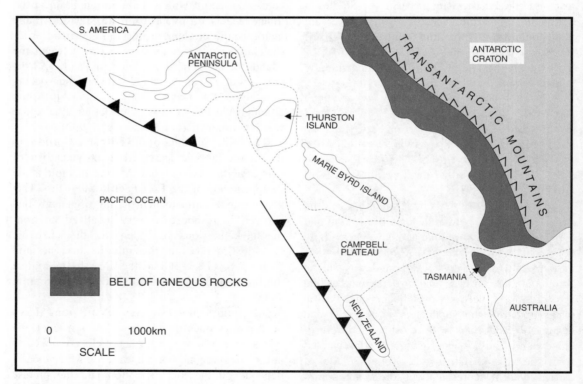

Figure 5.5   The Antarctic Pacific region 175 Ma. Volcanism and uplift associated with subduction occurred along the Pacific margins of this region from southernmost South America to New Zealand. Behind the frontal arc lay a rift zone, the southern border of which was marked by the Transantarctic Mountains

Modified, with permission, from Elliot and Larsen (1993) in Findlay, R.H., Unrug, R., Banks, M.R. and Veevers, J.J. (eds), *Gondwana Eight – assembly, evolution and dispersal*, Proceedings of the eighth Gondwana symposium, Hobart, Tasmania, Australia, 21–24 June 1991. 638 pp. Hfl. 225/US$130.00. A.A. Balkema, PO Box 1675, Rotterdam, Netherlands

contemporary plate movements and associated phenomena, was poorly known until recently (Weaver et al., 1994).

A Palaeozoic connection between New Zealand and Marie Byrd Land is undoubted. Yet the Mesozoic geological history of the two areas is quite distinct. While New Zealand grew largely as the result of the accumulation of forearc sediments and associated volcanism, Marie Byrd Land experienced volcanism, numerous granite intrusions and uplift characteristic of a magmatic arc associated with plate convergence. In this instance, the Phoenix Plate was being subducted beneath the Antarctic Plate (Figure 5.5).

The complication to this simple picture was that the entire Phoenix Plate disappeared as the result of subduction along this convergent margin during the Mesozoic. Specifically, the divergent plate boundary (spreading centre) was itself subducted causing subduction to cease and be replaced by rifting. This was associated initially with mountain building and then, around 84 Ma, by separation between what is now New Zealand and Marie Byrd Land. In effect, New Zealand changed from being on the Antarctic Plate to being on the Pacific Plate (which bordered the northern side of the former Phoenix Plate) and thence was moved further away from Marie Byrd Land.

The biogeographic consequences of the Cretaceous rifting between Marie Byrd Land and New Zealand were profound and have been amply documented for the latter (Stevens, 1980). For most of the time that New Zealand was on the Antarctic Plate, it was under water but, as the spreading ridge between the Phoenix and Pacific plates began to be subducted, it emerged and developed land links with most of the rest of Gondwana. Thus, when New Zealand separated from the rest of Gondwana around 75 Ma, it carried with it representatives of contemporary amphibia, birds and reptiles which subsequently evolved in isolation giving rise to some well known endemic species such as the native frog (*Leiopelma*) and the tuatara (*Sphenodon punctatus*), the sole living representative of the once diverse and widespread order of reptiles, the Rhynchocephalia (Plate 5.3). The ratite birds – notably *moa* and kiwis – also reached New Zealand before it became detached from the rest of Gondwana and seem to have thrived subsequently during its isolation from predators at least until the Holocene (see Chapter 15).

Isolation also led to remnants of Gondwanan forests, notably podocarps, certain ferns and araucarian pines, persisting in New Zealand when elsewhere they evolved in different ways or succumbed to competition from other species. A notable exception is the southern beech (*Nothofagus*), found today in New Guinea, New Caledonia, eastern Australia, Tasmania, New Zealand and western South America; fossils of *Nothofagus* have also been found in West Antarctica. It was clearly an important component of late Mesozoic forests along the high-latitude Pacific coasts of Gondwana (Figure 5.6).

Plate 5.3   The *tuatara* (*Sphenodon punctatus*) is the last of the reptile order Rhynchocephalia, once widespread throughout Gondwana. Today it is found only in New Zealand where its survival has been due to its ability to adapt to the temperate climate and – for most of its occupancy – a lack of predators
David Higham Associates

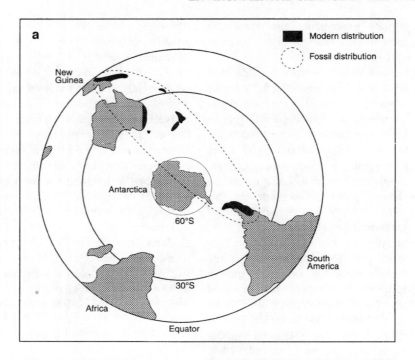

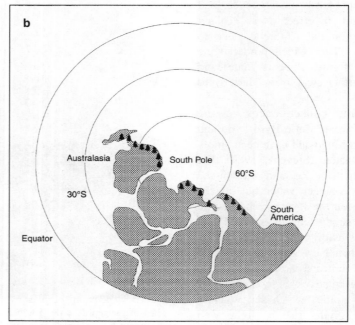

Figure 5.6 (a) Modern and (b) late-Mesozoic distributions of southern beech (*Nothofagus*). It is clear that the modern distribution is inherited from a distribution along the high-latitude Pacific coasts of Gondwana in the southern hemisphere
After Stevens (1980)

5.7 SUMMARY OF MESOZOIC HISTORY BY REGION

The Mesozoic is the earliest time in the history of the Pacific in which its constituent parts, as defined in Chapter 1, can be recognized as reasonably discrete units. The Mesozoic history of each part is therefore summarized below.

5.7.1 Mesozoic Environments of Antarctica

The Transantarctic Mountains, one of the world's major mountain chains, have been uplifted at least 4 km since the late Mesozoic, possibly as the result of moving above anomalously hot asthenosphere in the late Cretaceous (Smith and Drewry, 1984). On the Pacific side of this mountain range, there is evidence for plate convergence accompanied by the docking of terranes (shown in Figure 5.5) throughout the late Mesozoic and early Cenozoic. Collision between exotic terranes and the Antarctic craton compressed basins of marine sediment which form the basement of most of modern West Antarctica.

Post-Pangaea Antarctica did not come to be centred over the South Pole until around 80 Ma. Until then it had experienced a largely temperate climate and had a fauna and flora similar to many other parts of Gondwana (Colbert, 1982; Truswell, 1991). The presence of *Nothofagus* forests in both Australia and South America during the Cretaceous can be explained only by the existence of a forested link through Antarctica at the time (see Figure 5.6).

Various indicators of temperate climate in the Antarctic Peninsula during the late Cretaceous have been provided by pollen and tree-ring analyses (Askin, 1983; Francis, 1986; Plate 5.4). The fact of there still being a (Gondwanan) connection between Antarctica and South America at this time is required to explain the occurrence of

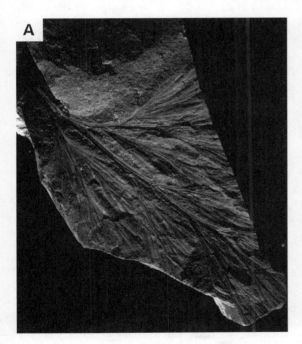

Plate 5.4   Fossil ferns from Alexander Island, Antarctica, are manifest proof that this region was not always as it is now. *Alamatus bifarius* (A) and *Aculea acicularis* (B) are both types of ferns which probably grew in thickets between a patchy overstorey of conifers during the early Cretaceous when this area had a temperate climate
David Cantrill and the British Antarctic Survey

late Cretaceous dinosaur fossils in the same area (Olivero et al., 1991).

5.7.2   Mesozoic Environments of South America

The main parts of the Andes mountains have a simple structure with few allochthonous terranes accreted onto the continental craton compared to the North American cordillera. As for the Cenozoic, the Mesozoic growth of the Andes was due largely to the eastward subduction of oceanic lithosphere. Uplift and folding of backarc sediments began in the mid-Cretaceous since which time both the locus of magmatism (volcanism and intrusion) and the deformation front have moved eastwards (Dalziel and Forsythe, 1985). Unlike the main parts of the Andes, northernmost South America had a late Mesozoic and Cenozoic history of terrane accretion (Mégard, 1989) making this area more similar to the Pacific margin of Central and North America.

Few data referring to pre-Cretaceous palaeotemperatures are available for the Pacific Rim in South America although the occurrence of early Jurassic carbonate sediments as much as 40°S in Chile is suggestive of warmer conditions at the time. The effects of maritime influences in modifying zonal climatic controls along the shores of the post-Pangaea Pacific Rim would have been much as they are today. In South America, these effects would have themselves been progressively modified by the uplift of the Andes producing an increasing degree of altitudinal zonation of climate.

5.7.3   Mesozoic Environments of Central America

Although this part of the Pacific Rim remains poorly known compared to the continental margins north and south, it was affected by plate convergence associated with volcanism and terrane accretion for much of the Mesozoic.

The Nicoya terrane, probably the remnants of a frontal arc originating in the Jurassic, was emplaced across the Panama Isthmus around the end of the Cretaceous and has since become a significant element in maintaining a land connection between North and South America (Howell et al., 1985). The larger Guerrero terrane now outcrops along much of the Mexican coast. It originated as a volcanic island arc, active for most of the Mesozoic, which was probably accreted around its end. This terrane is one of those that make up the southern part of the Mexican Thrust Belt, fringed on its western side by a subduction zone, and comprises only Mesozoic terranes. In contrast, the northern part of the Thrust Belt, where no nearby subduction is occurring (or has been for most of the Mesozoic) comprises only Palaeozoic terranes (Campa U., 1985). In Mexico, the effects of the Laramide Orogeny (see below), which began in the late Cretaceous, culminated in the folding and thrusting now visible in the Sierra Madre Oriental.

Much of this part of the Pacific Rim was covered by ocean during the Cretaceous. Little is known about the Mesozoic climates of the coasts of this Hispanic Corridor although the development of reefs here and in the modern Gulf of Mexico is well represented in the geological record.

5.7.4   Mesozoic Environments of North America

The Mesozoic history of the Pacific Rim in North America is more complex than its counterpart in South America (see above) because the effects of eastward subduction were (and still are) overprinted by both the accretion of a large number of allochthonous terranes onto the craton, and the effects of coastwise transverse (transform) movements. Lithospheric extension in the Basin and Range Province was also of significance, although marginal to the Pacific Basin as defined in this book.

During the early Jurassic, most of the Pacific Rim in North America changed from being a passive continental margin to one marked by plate convergence. This continued throughout the rest of the Mesozoic and saw the disappearance of much of the Kula and Farallon plates. Subduction of these plates beneath the North American continent brought many allochthonous terranes into contact with it. Terrane accretion began along most of the coast of the western United States at least 350 Ma. One of the earliest terranes to be accreted was Sonomia which had docked against the craton in west-central Nevada by the end of the Triassic (Speed, 1979). Another terrane which accreted along this side of the craton at this time was the composite Stikinia terrane, which docked during the Jurassic (see also discussion of Talkeetna above). In southern California, no terranes are thought to have collided with the craton during the Mesozoic, although this was a time which saw many amalgamating offshore in the eastern Pacific. Prominent among these were the Cortez Terrane (a continental fragment), and the Guerrero and Malibu terranes (volcanic arcs) which amalgamated in the Jurassic and Cretaceous prior to collision with the craton (McWilliams and Howell, 1982).

For much of the Mesozoic, dates of terrane accretion are assumed to have coincided with starts of periods of mountain building. For example, the late Jurassic Nevadan Orogeny (150 Ma) has been linked to the accretion of the Western Klamath and other terranes. The mid-Cretaceous Coast Range Orogeny is associated with accretion of seamounts within the Franciscan assemblage (Howell et al., 1985).

From the late Cretaceous until the Palaeogene, the fold mountain belt along the west coast of North America was affected by the Laramide Orogeny: 'a siege of compressive and trans-pressive intraplate deformation' (Coney, 1989: 49). This event probably began with the earlier accretion of exotic terranes along the craton margin and was 'marked especially by crustal buckling and associated fracturing to form giant fault-bounded, basement-cored uplifts separated by intervening basins in which sediment accumulated while deformation was in progress' (Dickinson and Snyder, 1978: 356). There was considerably less magmatism associated with the Laramide Orogeny in the United States than during most other comparable events here and elsewhere. This orogenic episode continued into the Tertiary (Chapter 6).

Late Cretaceous palaeobotanical data for North America have allowed more detailed reconstructions of its vegetation and climate than anywhere else in the Pacific Basin. The warmth of this time is attested to by the poleward displacement of isotherms. The late Cretaceous 20°C isotherm occurred at palaeolatitude 40–50°N; today it occurs about 30°N. The 13°C isotherm occurred 65–75°N whereas today it is found at about 40°N (Upchurch and Wolfe, 1987). The wet and humid nature of the northern part of this region during the later Mesozoic is well known although the southern part was more arid (Ziegler et al., 1983).

5.7.5 Mesozoic Environments of Beringia

This region – a mosaic of craton fragments in a matrix of both continental and oceanic terranes – lies between the North American and Eurasian (Siberian platform) cratons (Figure 5.7). It is likely that an ancient Arctic Alaska–Chukotka superterrane acted as a focus for Mesozoic and Cenozoic terrane accretion in the region (Churkin et al., 1985). Parts of Cordilleria, Talkeetna and other (super)terranes discussed in the preceding sections came to form much of eastern Beringia (Alaska) during the Mesozoic. Other terranes which collided with the existing terrane mosaic during this era include the Angayucham and Innoko oceanic terranes. Western Beringia (northeast Siberia) is similar in terms of terrane organization and accretion history to eastern Beringia. The Cherskiy and Alazeya terranes are two major oceanic terranes which had accreted by the mid-Mesozoic (Fujita and Newberry, 1982).

Much of the region, particularly the ocean floor, is obscured to such an extent by material produced during the process of terrane accretion that its origin is uncertain and correlations across it have proved difficult. One exception is the link demonstrated between the largely oceanic terranes of the Koryak Mountains in northeast Russia and southwest Alaska.

Early Mesozoic climates of Beringia are not well understood, although there are numerous indications – such as the formation of coal during

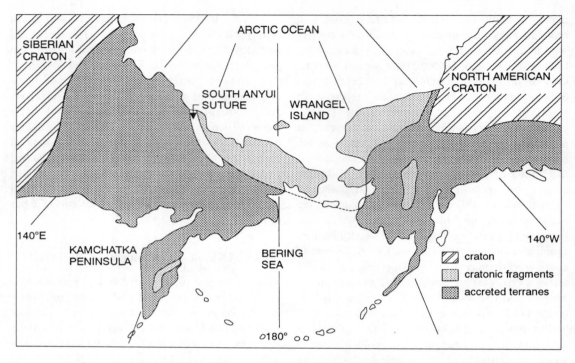

Figure 5.7   Terranes and cratons in Beringia
After Churkin et al. (1985) and Geist et al. (1994)

the late Triassic – that these were warmer and more humid at times than they are today. As evinced by its broad-leaved, evergreen forests (Wolfe, 1994b), the late Cretaceous climate of lowland Beringia was warm. These conditions make it easy to explain how certain dinosaurs, for instance, were able to migrate from East Asia to North America at this time (Cox, 1990). Such questions involving dispersal routes of fauna and flora are still somewhat troubled by a lack of data.

5.7.6   Mesozoic Environments of East Asia

The history of China from the Precambrian to the mid-Cenozoic, when India collided with the craton (45 Ma), is one of accretion of three huge continental fragments (microcontinents). These are separated in places by foldbelts (from less than 100 km to more than 500 km broad) composed of accreted terranes of largely oceanic origin (Howell et al., 1985).

The major Mesozoic event was the collision of the Yangtze (or South China) microcontinent (the youngest of the three) with the Sino-Korea (or North China) microcontinent, which was probably already joined to the continental block (Figure 5.8). This collision occurred along the Qinling–Dabie suture and resulted in 'a spectacular anticlockwise rotation of North China and an overall consolidation of the southeast Asia cratonic "core"' (Metcalfe, 1993: 190).

The Mesozoic history of Japan was dominated by accretion of oceanic terranes, and strike-slip faulting and magmatism related to oblique subduction (Isozaki et al., 1990). Similar statements apply to Taiwan (Ernst et al., 1985). Indo-China – probably a detached piece of Gondwana – collided with the Yangtze microcontinent during the Jurassic and became a focus for terrane accretion in southeast Asia thereafter (Metcalfe, 1993). This cratonic 'core' formed the edge of the Pacific Basin in this region

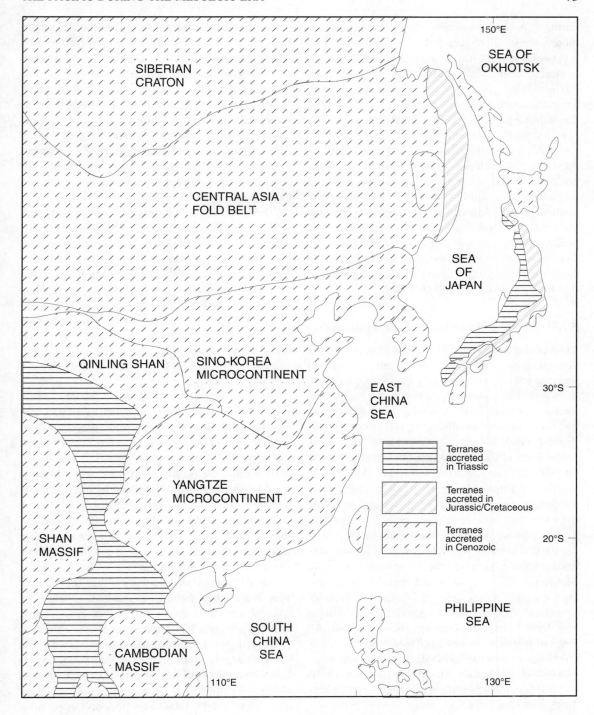

Figure 5.8 Geotectonic units of east Asia
After Zhang (1985)

during the Cretaceous and grew oceanwards as more terranes were accreted.

Owing to its transitional location between continent and ocean basin, Japan provides a particularly good yardstick for measuring Mesozoic environmental change in East Asia and the adjoining Pacific Ocean (Matsumoto, 1978). For example, the presence of Jurassic and Cretaceous limestones along what was then the Pacific coast of Japan suggests that a Mesozoic analogue of the warm Kuroshio Current affected this coast as it does today, and that the gross controls on the climate of the region were also comparable. Notable among these was the occurrence of monsoons. For example, the outer limit of influence of the Asian monsoon during the early Mesozoic, as determined by evaporite distribution, was greater than today.

5.7.7 Mesozoic Environments of Australasia

Most of the large island New Guinea is part of the Australasian continental mass, but its eastern margins are composed largely of oceanic terranes. Subduction was formerly towards the Pacific but then a continent arrived on the downgoing plate and, in an attempt to subduct it, this continent was thrust beneath the oceanic lithosphere which now forms some of the highest mountains in the Pacific half of the island (Allègre, 1988). Since subduction along this line ceased, the partly subducted continent has begun rising, carrying the overlying oceanic terranes with it.

The core of Australia is a Precambrian craton, the eastern boundary of which was an active plate margin during much of the Palaeozoic and early Mesozoic. This brought a number of terranes into contact with the craton although the lack of associated deformation suggests that this margin may have been a transverse (strike-slip) kind. All known terranes were emplaced by the early Cretaceous. Most originated close to each other just off the continental margin so cannot be considered truly exotic (Cawood and Leitch, 1985; Scheibner, 1985).

The subsequent development of a three-pronged rift system – perhaps contemporaneous

with the commencement of uplift of the Australian Eastern Highlands – led to the separation of Australia from Antarctica (see Figure 7.5), and the formation of the Tasman Sea between eastern Australia and a strip of Gondwana which included New Caledonia, the modern Lord Howe Rise and New Zealand. The isolation of these areas continued as a result of sea-floor spreading in the Tasman Sea although subaerial volcanism therein may have created stepping stones for continued biotic interchange during the rest of the Cretaceous (Wilford and Brown, 1994).

The landmass which became New Zealand was wedged between southeast Australia and east Antarctica during the early Cretaceous (see case study above). Its movement away from this location and the subsequent accretion of exotic terranes occupied most of the late Mesozoic. New Zealand probably gained its greatest ever land extent during the early Cretaceous Rangitata Orogeny (Stevens, 1989), the effects of which were also experienced in Marie Byrd Land (Antarctica) and New Caledonia (Spörli and Ballance, 1989).

An important factor in floral and faunal changes during the Mesozoic was the changing climate and sea level. Australia has moved almost 60° of latitude from a position close to the South Pole in the early Cretaceous (see Figure 7.5). Biotas of early Cretaceous southeast Australia would have had to cope with one to two months of continuous darkness and winter temperatures well below freezing. The gradual break-up of Gondwana would have led to oceanographic changes that influenced the climate, as would the largely rising sea levels during this period (see Figure 5.3). Coal found in various places implies that the climate was reasonably humid: a marked absence of tropical or near-tropical fossils (except in New Guinea) suggests that the climate remained cool.

Knowledge of Cretaceous floras in Australia derives largely from studies of sediments in the Gippsland and Otway Basins (Douglas, 1994). Cretaceous forests were dominated by conifers and ferns while other components – pterido-sperms, cycads and ginkgos, for instance – decreased in importance throughout the period.

In Australia, as elsewhere, the rise of angiosperms had a major impact on pre-existing flora.

Information about Mesozoic fauna in Australia is sparse. In one of the most prodigious finds, dinosaur fossils in the Gippsland and Otway Basin sediments are associated with turtles, fish, birds, crustaceans, spiders, bryozoans, possibly earthworms, and representatives of 12 orders of insects (Rich et al., 1988).

At the time of the Rangitata Orogeny, 'long fingers of land may have reached out from New Zealand into the Southwest Pacific' (Stevens and Speden, 1978: 314) along which conifers, ferns and several animal genera (including ratite birds) reached New Zealand. Later arrivals, notably *Nothofagus*, came during the late Cretaceous from South America via the margins of Antarctica. The isolation of New Zealand became complete in the latest Cretaceous, just before marsupials and snakes reached the rest of Australasia.

5.7.8 Mesozoic Environments of the Pacific Ocean and Islands

Owing largely to the paucity of Mesozoic rocks in the Pacific Ocean and Islands, the contemporary history of this region is not well known. From the late Jurassic to middle Cretaceous, the entire central Pacific is thought to have experienced a 'superplume' (volcanic–tectonic) episode which created a large area of anomalously elevated ocean floor named 'Darwin Rise'. Ocean-floor volcanic edifices created at this time underlie many central Pacific islands and island groups today (Menard, 1964).

Also of note is the similarity in process around the Mesozoic Pacific Rim and that in the modern Pacific Ocean and Islands. The comparability in both size and dynamic interaction between the Yangtze microcontinent and the East Asia block during the Mesozoic, and the modern Ontong Java Plateau and the Solomons Arc, is a good example, for 'if one imagines the closing-up of the Coral Sea, this inferred continental fragment wreathed with accreted volcanic arcs would form a major addition to the Australian continent' (Howell and Jones, 1989: 38).

Few modern Pacific islands had land exposed during the Mesozoic so there is no record of contemporary terrestrial biotas. Discussion of oceanographic and ocean-biotic development was given earlier in the chapter.

5.8 THE PACIFIC BASIN AT THE END OF THE MESOZOIC (K–T BOUNDARY)

The (K–T) boundary between the Mesozoic and Cenozoic is one of the most conspicuous time markers in the later history of the Earth. In the Pacific Basin, as elsewhere, it was a time characterized by wholesale extinction of various fauna, notably ammonites and many reptiles, including both orders of dinosaurs. Yet this time was marked not just by the decimation of species but also by the appearance of others. Mammals, particularly placental mammals, diversified extraordinarily just after the K–T boundary and came to dominate vertebrate life.

In the Pacific, as in all the world's oceans, calcareous plankton became extinct at this time, perhaps as a result of abruptly decreased ocean depths resulting from sea-level fall (Hart, 1980). A more recent explanation is that a large influx of nutrients associated with mass mortality of terrestrial biota resulted in a bloom of low-oxygen-tolerant forms which outcompeted the existing marine micro-organisms. Most planktonic foraminifera were depleted around this time and did not recover for several thousand years. Benthic foraminifera recovered more rapidly (Coccioni and Galeotti, 1994).

Significant vegetational disturbance also occurred around the K–T boundary. In many places, it is unclear whether this was sudden or gradual although pollen data, which generally show a rapid depletion of angiosperms followed by a short-lived episode of fern dominance before angiosperms return, suggest the former. Typical is the conclusion of Sweet and Braman (1992: 31) for western Canada that K–T boundary changes 'were not solely owing to a single catastrophic event but, at least in part, resulted from a longer term alteration in the complex series of factors controlling plant community development'. Elsewhere in the Pacific Basin, such as western Beringia, the palaeofloral evidence is best interpreted as 'the result of natural evolutionary processes and climatic change' (Golovneva, 1994: 89) with the idea of extraterrestrial influences specifically dismissed.

The cause of the immense changes in life on Earth around the Cretaceous–Palaeogene boundary has been attributed to a bolide impact (Alvarez et al., 1980). This idea has received support from recent research on osmium isotopes of ocean water which found a pronounced minimum of ^{187}Os/^{186}Os at the K–T boundary consistent with an enormous input of cosmogenic material from a large bolide impact (Peucker-Ehrenbrink et al., 1995). A likely impact site is Chicxulub in Yucatan (Hildebrand et al., 1991) where impact is thought to have sent a large cloud of dust into the atmosphere resulting in several years of darkness: enough perhaps to kill off most photosynthesizing organisms and the animals (except insects) which ultimately depended on them.

The precise links proposed between bolide impact and the K–T boundary extinctions have been questioned. The fossil record on the Antarctic Peninsula, for instance, shows that the disappearance of ammonites here began well before the K–T boundary (when they became globally extinct) possibly in response to global cooling (Zinsmeister and Feldmann, 1996). Elsewhere in the Pacific Basin (as in other parts of the world), the K–T boundary has been recognized as a broad transition zone spanning in China, for example, some 50 000 years (Stets et al., 1996) supporting the view that a bolide impact may have provided only the *coup de grâce* to certain biotic groups already in decline.

More recently, it has been shown that the cooling characterizing the K–T boundary is only one in a series of rapid cooling events to have occurred during the latest Cretaceous (Barrera, 1994). At present this record is significantly at odds with others for the K–T boundary event which see any associated cooling as short-lived and more plausibly an outcome of bolide impact rather than a cause of mass extinctions. In this context, it should yet be noted that the K–T boundary did not signify the end of a 'warm mode' but merely a short-lived interruption as indicated by the continuation of warming during the early Palaeogene (Chapter 6).

PART C

The Pacific During the Tertiary Period

CHAPTER 6

The Pacific Basin During the Palaeogene (66–24.6 Ma)

6.1 INTRODUCTION

The Palaeogene lasted from the K–T (Cretaceous–Tertiary) boundary 66 Ma until the end of the Oligocene 24.6 Ma (Figure 6.1). An understanding of environmental changes during the Palaeogene is particularly important as it marks the climatic transition between the Cretaceous 'hothouse' and the Neogene 'icehouse', a transition which holds important keys to the understanding of long-term climate evolution on Earth (Berggren and Prothero, 1992).

The Palaeogene comprises three epochs: the Palaeocene, marked by the appearance of a remarkable number of new lifeforms; the Eocene, characterized initially by warm conditions which became cooler in a series of steps, associated with extinctions; and the Oligocene, a time of generally cool, dry conditions during which the first continent-scale ice sheet was established in Antarctica.

The configuration of land and ocean in the early Palaeogene Pacific Basin was broadly similar to that of today, which underscores the point that

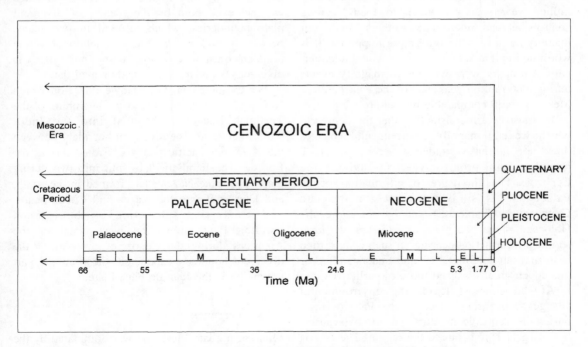

Figure 6.1   Chronology of the Cenozoic Era used in this book

plate-tectonic changes are of less importance in explaining environmental changes during the Cenozoic than during earlier eras. There are some important exceptions to this generalization, notably those referring to the gaps, which widened progressively during the Cenozoic, separating Antarctica from its nearest continental neighbours at the beginning of the Palaeogene (see Figure 3.2).

Plate tectonics also has a subordinate role in the explanation of Cenozoic environmental change in the Pacific Basin compared to the Palaeozoic and Mesozoic, principally because these eras lasted far longer than has the Cenozoic to date. Also the amount of data referring to environmental changes during pre-Cenozoic times is far less in both quantity (per unit time) and precision than that pertaining to the Cenozoic, particularly the Quaternary period. There is no reason to suppose that, once the Cenozoic has lasted 184 million years (as did the Mesozoic) or 340 million years (as did the Palaeozoic), there will be a significant difference in the balance of the gross controls on the environmental changes to have taken place within it. It is just that, because we have only a 66 million year span on which to focus, certain controls become subordinate as others – under our magnifying glass – become more important. It is worth noting that such statements would undoubtedly also apply were we able to magnify events during a 66 million year span of the Palaeozoic or Mesozoic with comparable resolution.

It is also worth considering that the degree to which we can magnify environmental changes does not increase suddenly across the K–T boundary but progressively throughout the Cenozoic. Thus the degree of information for the Palaeogene is similar to that for the late Cretaceous (Chapter 5) yet by the time the Holocene is reached (Chapters 13–15), we have so much information bearing on such a short time span that our interpretation and understanding of this epoch greatly exceed those of earlier ones.

At the coarsest level, the environmental changes which took place in the Pacific Basin during the Cenozoic are attributable to two causes, both largely extraneous to the region. The first is tectonic change, particularly that associated with

the continued break-up of Gondwana and Laurasia, the two components of the Mesozoic supercontinent Pangaea. The second is the long-term change of global climate, which had repercussions for sea-level change, biotic evolution and a host of other factors in the Cenozoic Pacific. The nature and effects of these two primary causes are described for the Palaeogene separately in the following two sections before an account is given of Palaeogene environmental change in each constituent region of the Pacific Basin.

6.2   TECTONIC CHANGE

The smaller plates which had been disappearing during the late Mesozoic continued to do so throughout the Cenozoic as the result of subduction along the periphery of the Pacific Ocean. Owing to continued expansion of the Atlantic, the Pacific Basin itself continued to become smaller, and the dominance of the Pacific Plate was even more pronounced than during the late Mesozoic (see Figure 3.2). This dominance was achieved both by the 'death' of divergent plate boundaries, which resulted in two plates becoming one, and by the subduction of whole plates, commonly along with (part of) the divergent boundary which had created them.

For example, the Lord Howe Plate, on which New Zealand was situated at the beginning of the Tertiary, became subsumed into the Indo-Australian Plate. The divergent boundary between the Kula and Pacific plates also died during the Eocene. The Farallon Plate split into two during the Eocene; the Nazca Plate formed in the south, and later split into the Nazca and Cocos plates. The progressive subduction of the residual Farallon Plate along the west coast of the Americas during the Cenozoic saw part of the divergent boundary subducted; the principal part remaining is the Juan de Fuca Plate.

6.2.1   Terranes

Much of the late Mesozoic was dominated by the movement of exotic terranes, both oceanic and continental, across the Pacific Ocean and their accretion onto the cratons constituting the Rim (Chapter 5). Terrane accretion was a process of less importance during the Cenozoic. Most large continental terranes – fragments of Gondwana in particular – had already been dispersed and accreted. Owing to the increasing Cenozoic dominance of the Pacific Plate, which had originated wholly within the ocean basin, the formation of oceanic terranes became effectively confined to the edges of the Pacific Basin, where most convergent plate boundaries existed.

There are a few exceptions of importance to this picture of reduced terrane accretion during the Cenozoic, most of which occurred during the Palaeogene along the western seaboard of North America as the result of collision between the East Pacific Rise and the North American craton.

Terrane collision and accretion occurred elsewhere in the Pacific Basin during the Palaeogene. The collision of the continental Okhotsk terrane with East Asia caused the Kuril subduction zone to jump eastward trapping parts of the Kula and Pacific plates. The collision between the North Palawan Continental terrane, rifted off south China, and the Philippines island arc caused a reversal of subduction polarity. Within the southwest Pacific Ocean, the collision of the Ontong Java Plateau (a likely continental terrane) with the Solomons Arc along the North Solomon Trench also caused subduction polarity to reverse in part of Papua New Guinea, Solomon Islands and Vanuatu.

The Palaeogene history of New Caledonia was dominated by collision manifested by ophiolite emplacement (Figure 6.2), the environmental consequences of which are worth examining briefly. Since the rocks forming the surface of modern New Caledonia are pieces of ancient ocean floor (ophiolites), they are uncommonly rich in metallic minerals like nickel and chromium (Brothers and Lillie, 1985). The soils which have developed above these rocks contain high levels of substances which are toxic to many plants. As a result, the flora of New Caledonia is likewise unusual, having a high number of endemic plants adapted to living in soils with high levels of these

particular minerals. In this instance, contrary to what might be expected, poor though diverse soils support an unusually high degree of plant diversity which has created an uncommonly great variety of ecological niches which have, in turn, been filled by a highly diverse fauna (Flannery, 1994).

6.2.2 Opening and Closure of Gaps in the Pacific Rim

The modern Bering Strait between east and west Beringia in the North Pacific closed during the late Cretaceous and acted as a land bridge connecting East Asia and North America during most of the Tertiary, as well as the low sea-level stages of the Quaternary. Most Palaeogene migrations of plants and animals across Beringia were from west to east although three mammal groups (edentates, notoungulates and xenungulates) are thought to have migrated from South America through North America and into Asia during this time (Gingerich, 1985).

Although the gap between North and South America closed for a while in the late Cretaceous and the earliest Palaeocene (Briggs, 1987), it opened again shortly afterwards. A water gap remained here for the rest of the Palaeogene; yet ocean-water exchange between the Pacific and Atlantic – as interpreted from the record of biotic exchange (Collins et al., 1996) – became progressively reduced. This exchange did not cease entirely until the Neogene when the gap was completely sealed (Chapter 7). The low-latitude water gap between Australia and southeast Asia, which had existed for some 200 million years, effectively closed during the middle Tertiary as the result of the northward movement of Australia (see Figure 7.5). This significantly reduced ocean-water exchange between the Pacific and Indian oceans compared to earlier times.

The gap between Australia and Antarctica, which began opening in the late Cretaceous, continued opening during the Palaeogene. Its role as a conduit for west–east water movement became greatly enhanced 30–25 Ma as the Tasman Rise subsided. On the other side of the

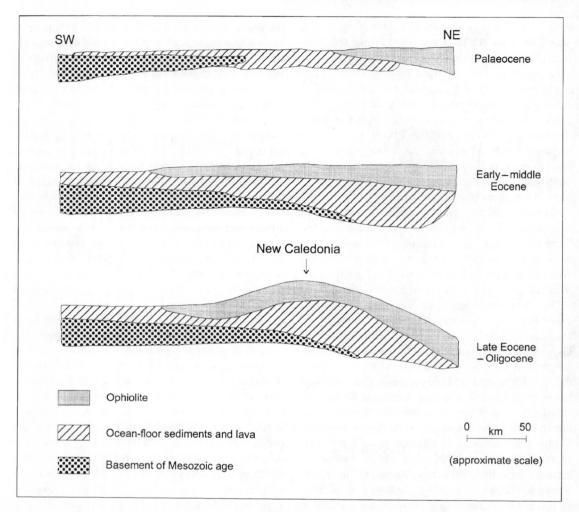

Figure 6.2   The Palaeogene development of New Caledonia showing how the ophiolites, which are widely exposed on the modern island, were emplaced

From *Geology*, Aitchison, J.C., Clarke, G.L., Meffre, S. and Cluzel, D. Modified with permission of the publisher, the Geological Society of America, Boulder, Colorado, USA. Copyright © 1995 Geological Society of America

southernmost Pacific, Drake Passage began opening between South America and Antarctica during the Eocene (Figure 6.3). The opening of these two gaps set the global stage for the most important climate changes of the Cenozoic, discussed at length below.

The combined effect of the closure of two low-latitude gaps and opening of two high-latitude gaps in the Pacific Rim led to the establishment of the modern Pacific Ocean circulation during the Tertiary. Movement of water through the southern gaps led to the development of the Antarctic Circumpolar Current which effectively insulated Antarctica from the influence of warm water from low latitudes. Closure of the low-latitude gaps strengthened gyral circulation in the North and South Pacific resulting in increased north–south and south–north movement of surface water. In the north, this water was returned to the equatorial Pacific by circulation within the closed basin. In

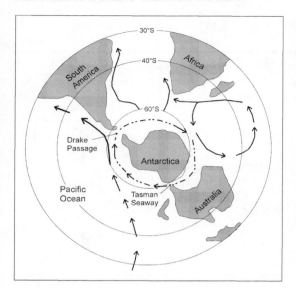

Figure 6.3   Reconstruction of surface (dashed line) and deep-water (solid lines) currents in the Southern Ocean at the end of the Eocene
From various sources

the south, this also happened because of the inability of water circulating in the gyre to penetrate the Antarctic Circumpolar Current and thus exit the Pacific Basin through Drake Passage.

Until these tectonic changes occurred and gyral circulation strengthened, currents within the Pacific Basin had neither been fast enough nor flowed for sufficiently long in a single direction for coral larvae to be transported from the Philippines (where the greatest diversity of early Palaeogene corals existed) north to the Hawaiian islands, which is why no corals older than early Oligocene (34 Ma) are found there (Grigg, 1988). Yet just as new biotic provinces were created by these events, so too did others disintegrate. A good example is that of the highly endemic molluscan faunas of the Weddellian Province in the South Pacific which gradually lost its identity as the gap between Australia and Antarctica widened during the Palaeogene (Zinsmeister, 1982).

The inability of warm surface water in the Pacific to reach high-latitude landmasses contributed to the development of the psychrosphere

(bottom water) in the late Eocene which initiated the modern thermohaline circulation system and led to ice build-up on Antarctica. The establishment and growth of the Antarctic ice sheet 32–28 Ma is the principal feature of Tertiary climate change.

Palaeogene tectonic changes brought about changes of such significance in the Pacific Basin that ocean circulation therein was altered in such a way as to isolate it from the rest of the world's oceans. In a global sense, these changes were part of those which saw climate and ocean circulation change from being dominated by circumequatorial atmosphere–ocean circulation to a situation characterized by circumpolar processes and meridional circulation.

6.3   CLIMATE AND OCEANOGRAPHIC CHANGE

The Mesozoic break-up of Pangaea saw the movement of land from low to high latitudes. Particularly significant was the positioning of Antarctica over the South Pole ~80 Ma. Land, particularly low land, in high latitudes became sites for snow and ice accumulation which increased the Earth's albedo, contributing to global cooling which is the principal feature of Cenozoic climate change outside the tropics. The apparent step-like nature of much Cenozoic cooling may, through the medium of albedo changes, be the result of the attainment of particular thresholds of snow and ice development in high latitudes. During the Cenozoic, land also became progressively concentrated in the northern hemisphere subtropics where subsiding air produced aridity which also increased albedo and contributed to cooling.

Various models of Tertiary sea-level change have been proposed; that favoured by most working in the Pacific Basin today was produced by Haq et al. (1987). This shows that, aside from a rise in the later Palaeocene, sea level has been falling almost continuously since the mid-Cretaceous (~91 Ma) when it was perhaps 250 m or more above its present level (see Figures 5.3 and 6.6).

Table 6.1   North American mammal ages and rates at which new genera have appeared
After Webb, S.D. (1985)

Epoch	Duration (million years)	First appearances	
		Number of genera	Rate (genera/million years)
Pleistocene	1.8	75	41.7
Pliocene	3.5	75	21.4
Miocene	19.3	323	16.7
Oligocene	11.4	185	16.2
Eocene	19.0	249	13.1
Palaeocene	11.0	151	13.7

The earliest Palaeogene saw a warming climate which was recovering from the effects of a rapid cooling at the Cretaceous–Tertiary (K–T) boundary (see Chapter 5). The early Palaeogene was thus hotter and wetter than the preceding few million years had been, conditions which aided the diversification of biota at this time. Several groups of animals radiated spectacularly during the Palaeogene to fill the ecological niches left vacant following the mass extinction at the K–T boundary. Notable among these groups are mammals (Table 6.1), especially placental mammals which today include cows, dogs and humans.

Other groups of animals adapted their behaviour as Palaeogene environments changed. A good example are those marine mammals which developed the ability of filter-feeding during the middle Oligocene perhaps in response to the development of the Antarctic Circumpolar Current and the associated increased biogenic productivity during the early Oligocene (Fordyce, 1992). The success of filter-feeding whales during the remainder of the Cenozoic is testimony to the effectiveness of this adaptation in an ocean where biogenic productivity, as suggested by rates of sedimentation, continued to increase.

The warmth of the early Palaeogene allowed tropical and subtropical plants to live as much as 60° away from the Equator. Broad-leaved deciduous and coniferous forests were confined to higher latitudes. Yet, by the end of the Eocene, these forests began to retreat towards the Equator as cooling, associated with the initiation of Antarctic continental glaciation, began (Wolfe, 1980). Floral diversity rose again during the late Oligocene, probably the result of a short-lived period of warming.

Loss of low-latitude conifer-dominated forests associated with the K–T boundary event allowed ferns and later angiosperms – after they had recovered from the effects of this event – to come to dominate forests in these areas as they do today. Grasses first appeared in the early Eocene.

Cooling is the dominant feature of Cenozoic climate (Figure 6.4). Within the Palaeogene, there were four comparatively rapid periods of change, the nature of which is discussed in the following sections. Most of these events were also marked by mass extinctions yet, despite attempts to demonstrate the contrary, extraterrestrial explanations of these extinctions have proved inadequate (Prothero, 1989).

In addition to the overall cooling, parts of the Pacific Basin also experienced prolonged aridity at various times during the Palaeogene. It seems unlikely that continental glaciation was a feature of the Palaeocene or Eocene in the Pacific Basin

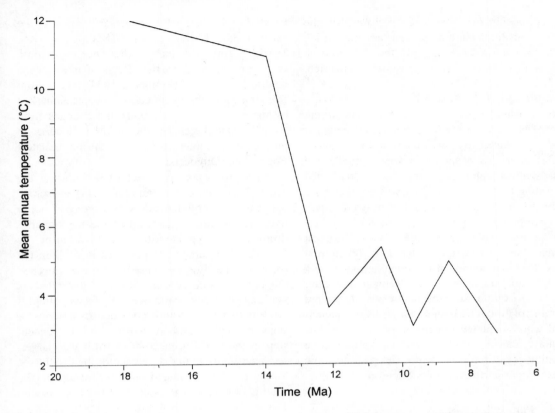

Figure 6.4 The overall trend of Tertiary cooling in the Pacific Basin as shown by CLAMP analyses of fossil-leaf physiognomies at 45°N
After Wolfe (1995). With permission from the *Annual Review of Earth and Planetary Sciences*, Volume 23, © 1995, by Annual Reviews

or elsewhere. The existence of the *Nothofagus* flora – a cool temperate assemblage – on Antarctica during the Eocene implies that it was not glaciated at the time. This is in apparent conflict with the date of 49.4 Ma for a lava overlying glacial deposits on the Antarctic Peninsula (Birkenmajer and Zastawniak, 1989) but these have since been explained convincingly as a product of localized mountain glaciation rather than an indication of continent-wide glaciation (Kennett and Barker, 1990).

It is worth noting at this point that there have been major differences in estimates of Tertiary palaeotemperatures determined by isotopic and palaeontological methods. A good illustration of this refers to ocean-water temperatures during the early Palaeogene. Isotopic data suggested that these differed little from modern temperatures yet examination of the early Palaeogene distributions of mangroves, hermatypic corals and larger foraminifera led to the more cogent conclusion that contemporary temperatures were higher than today (Adams et al., 1990).

6.3.1 The Terminal Palaeocene Event (~55 Ma)

The early Palaeocene rise of temperature which followed its rapid fall at the K–T boundary marked the resumption of the warm phase which had begun earlier in the Cretaceous (Chapter 5). Cooling began again around the mid-Palaeocene and is thought to have been a cause of the conspicuous hiatus in mammalian radiation which began then and did not end until the early Eocene when temperatures rose once more (Gingerich, 1977).

At the end of the Palaeocene, around 55 Ma, a series of major marine faunal extinctions occurred. Molluscan faunas of the south Pacific were almost wholly replaced by warm-water Indo-Pacific species after this event (Zinsmeister, 1982), yet most shallow-water organisms were unaffected suggesting that a bolide impact was not the cause of this event. Nonetheless its effects were rapid; for example, 35–50% of deep-sea foraminiferal taxa became extinct within 3000 years, the most dramatic such extinction of the last 100 million years. The coincidence of these extinctions with a strong reduction of $\delta^{18}O$ led Kennett and Stott (1991) to explain them as the result of a rapid warming of Antarctic bottom waters by as much as 18°C which depleted oxygen in deep water. The ultimate cause of this terminal Palaeocene warming event may have to do with plate-tectonic events in the north Atlantic (Rea et al., 1990).

6.3.2   Late Eocene Events (~50–36 Ma)

Following the terminal Palaeocene warming event, warming continued during the early Eocene reaching a maximum about 52 Ma. Through deep weathering of rocks, this warming has been proposed as the principal cause of the release into the oceans of large amounts of silica which were

fixed abruptly as cooling began in earnest in the early middle Eocene about 50 Ma (McGowran, 1989). Although many other oceanographic indicators fail to show the effects of mid-Eocene cooling so clearly (Prothero, 1994), the fallout from this and subsequent cooling events is marked by marine organic extinctions and changes from warm assemblages in the middle Eocene to progressively cooler assemblages during the latest Eocene and Oligocene.

Since no major ice sheet existed during the early Palaeogene, contemporary oxygen-isotope data can be interpreted solely as a proxy of ocean palaeotemperatures; later, when the ice signal is dominant, oxygen-isotope data cannot be interpreted so simply. Rapid rises in $\delta^{18}O$ in the early middle Eocene and around the Eocene–Oligocene boundary can thus be interpreted as marking periods of rapid ocean-surface cooling. It is clear that this cooling consisted of a series of steps beginning around 50 Ma. The subsequent step in the mid-Eocene (~46–40 Ma) was larger, secondary only to that at the end of the epoch (see Table 6.2). Terminal Eocene cooling in the southwest Pacific is estimated as 3–4°C (Keigwin, 1980).

The effects of cooling during the later Eocene are clear from the terrestrial palaeobiotic record. This time marks the most profound faunal turnover in the history of mammals; the data in Table 6.1 disguise the variation in Eocene first appearances, the rates of which ranged from 20 genera per million years in the middle Eocene to

Table 6.2   Palaeotemperature data characterizing Eocene and Oligocene cooling along the Pacific Rim of North America and Beringia
Derived graphically from Wolfe (1978)

Location	Palaeotemperature (°C)				Total Palaeogene cooling (°C) (minima)
	46 Ma	40 Ma	36 Ma	32.5 Ma	
Northern California	27.5	17	25	12	15.5
Northwest United States	25	14.5	23	10.5	14.5
Southern Alaska	21	12	18	5.5	15.5

just six genera per million years in its latest part (Webb, S.D., 1985).

The earliest direct deep-ocean connection between the southernmost Pacific and South Indian Oceans during the Tertiary occurred with the subsidence of the Tasman Rise 30–25 Ma, associated with the separation of Australia from Antarctica around the end of the Eocene (see Figure 6.3). The subsequent influx of cool Indian Ocean water into the southwest Pacific led to the separation of the modern south Pacific gyre from the Antarctic continent through establishment of the cool Antarctic Circumpolar Current. This allowed Antarctic sea ice to develop for the first time and initiated development of the oceanic psychrosphere (cold bottom water); bottom water temperatures in the Pacific fell rapidly by 4–5°C at this time (Keigwin, 1980).

All these changes were probably linked to late Eocene cooling events which set the stage for initial Antarctic ice-sheet formation by the middle Oligocene (see below). It seems unlikely that large ice sheets existed during the Eocene, so ice-sheet formation cannot justifiably be regarded as a cause (although perhaps as an effect) of Eocene cooling.

Mention needs to be made of the bolide impacts which occurred during the later Eocene and which – not unreasonably given the accord which has emerged among scientists regarding the role of a bolide impact in the K–T boundary extinctions (Chapter 5) – were once regarded as having been the major cause of the stepwise nature of the cooling at this time. Evidence for bolide impacts during the late Eocene is widespread – a crystal-bearing microspherule layer crosses the tropical Pacific from the Philippines to the Caribbean (Keller et al., 1987) – yet comparatively few biotic extinctions were contemporaneous with these impacts and most scientists now regard their occurrence as being coincidental with (rather than explaining) late Eocene stepwise cooling.

6.3.3 The Early Oligocene Deterioration (~34 Ma)

It was once thought that the major cooling phase of the Palaeogene occurred at the end of the Eocene (the 'Terminal Eocene Event') but, as Tertiary stratigraphy has been tidied up, this event is now placed in the early part of the Oligocene epoch. This 'early Oligocene deterioration' was a time of rapid cooling marked by ice build-up on Antarctica. The exact timing of this event remains controversial. From considerations of benthic foraminiferal extinctions, Miller (1992: 173) argued that 'large ice sheets existed in Antarctica during the earliest Oligocene'.

A significant event during the early Oligocene was the effective closure of the last vestige of the Tethys Seaway between Australia and southeast Asia. This was caused by the northward movement of Australia, which also caused the Tasman Seaway to widen. Combined with the opening of the Drake Passage between South America and Antarctica, this event strengthened the Antarctic Circumpolar Current further. This in turn reduced the amount of heat moved from the Equator to the South Pole (that is, it increased the meridional temperature gradient) and contributed to cooling on Antarctica.

During the early Oligocene deterioration, temperatures over the land in high latitudes may have dropped by as much as 10°C producing unusually rapid changes in vegetation; areas that had been occupied by broad-leaved evergreen forests were replaced by temperate broad-leaved deciduous forests of low diversity (Wolfe, 1978).

6.3.4 Antarctic Ice Formation in the Middle Oligocene (~32–28 Ma)

It is plausible to suppose that Antarctic ice sheets grew slowly during the early Oligocene before crossing an accumulation threshold which caused an acceleration of this process during the middle Oligocene and led to the first appearance of a continent-wide ice sheet in Antarctica. Such a threshold may have been associated with isostatic depression of the continent under a growing ice load.

While the early Oligocene climatic deterioration may have brought about countless extinctions and caused massive environmental changes throughout the Pacific Basin (and

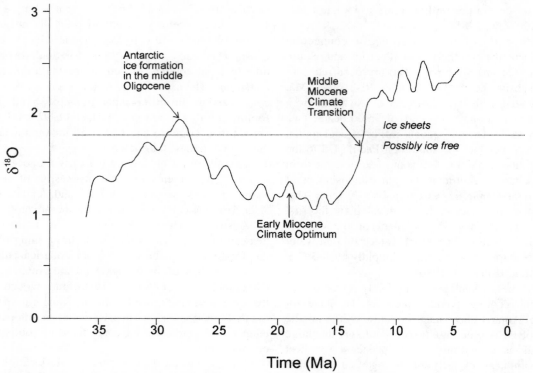

Figure 6.5   Composite benthic foraminiferal oxygen-isotope record for the Pacific during the middle Tertiary. The dotted line is drawn through 1.8‰; values greater than this provide evidence for existence of significant terrestrial ice bodies. Note the evidence for (Antarctic) ice-sheet formation during the mid–late Oligocene, its disappearance in the early Miocene followed by its re-establishment in the mid-Miocene
After Miller et al. (1987) with permission of the American Geophysical Union

elsewhere), it lasted perhaps only a few hundred thousand years and was overshadowed in magnitude by the 'big chill' of the middle Oligocene (Prothero, 1994). The evidence for this is best marked by oxygen-isotope data from benthic foraminifera. $\delta^{18}O$ values exceeded 1.6‰ – the lower-level indicator of continental glaciation – from ~32 to 28 Ma (Figure 6.5).

The conspicuous nature of mid-Oligocene cooling in Antarctica emphasizes the point that it is here that evidence for its immediate effect is clustered. Elsewhere in the world, most terrestrial environments exhibited little response to this event, perhaps because the only biota which survived the early Oligocene deterioration were those adapted to cold conditions and were not therefore unduly affected by the onset of another

cooling event. In fact, ecosystems which had been disrupted by early Oligocene cooling, particularly in the tropics, did not recover until the later Oligocene.

Warming during the late Oligocene and early Miocene may have caused Antarctic ice cover to shrink markedly or even disappear before its major period of growth during the middle Miocene (Chapter 7).

The one terrestrial environment to be profoundly affected by mid-Oligocene ice formation on Antarctica was the world's coastline which experienced the effects of a massive sea-level fall resulting from the huge amount of ocean water which became locked up in Antarctic ice at this time. A rapid sea-level fall ~30 Ma of nearly 150 m (Figure 6.6) is marked by unconformities in

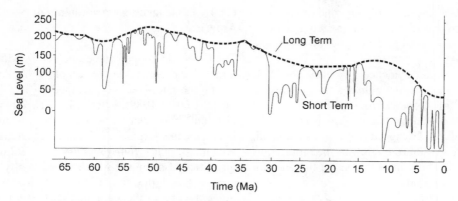

Figure 6.6   Cenozoic sea levels relative to present mean sea level
After Haq et al. (1987)

sedimentary records from many continental shelves and has also been identified in Pacific deep-sea cores from measurements of $\delta^{18}O$.

Most indicators suggest that ice sheets existed only on East Antarctica during the Oligocene and that that part of the Pacific Rim in Antarctica (West Antarctica) was ice-free at this time. Ice began forming in West Antarctica only 10–5 Ma.

6.4   SUMMARY OF PALAEOGENE HISTORY BY REGION

6.4.1   Palaeogene Environments of Antarctica

Although Antarctica was over the South Pole by the start of the Palaeogene, temperatures in the surrounding ocean remained high owing to warm water carried in the south Pacific gyre from low latitudes. For surface and bottom waters respectively, temperatures of 17.5°C and 17°C were calculated for the early Eocene, and 9°C and 11°C for the end of the Eocene (Shackleton and Kennett, 1975). Bottom temperatures fell by at least 5°C after the psychrosphere developed shortly after this time, and still further after the Antarctic Circumpolar Current developed in deep water in the middle Miocene (Chapter 7).

Evidence of late Palaeogene climatic deteriorations in the ocean waters around Antarctica is available from many sources. The record of

calcareous nannoplankton has proved particularly sensitive (Aubry, 1992). The development of cold Antarctic bottom waters in the early Oligocene saw many benthic foraminiferal taxa become extinct around the periphery of the continent (Prothero, 1994). Yet once climate stabilized in the late Oligocene, many new forms of marine micro-organisms appeared in circum-Antarctic waters, giving rise off Pacific Australasian coasts to the earliest groups of modern whales.

Although localized glaciations may have occurred in Antarctica during the Eocene, the middle Oligocene marked the time when a continent-wide ice sheet is thought to have developed in East Antarctica for the first time. Confirmation that ice existed along at least a few parts of the Pacific Rim in (West) Antarctica during the middle Oligocene has come from sedimentary records and records of ice-sheet grounding in the Ross Sea (Anderson and Bartek, 1992). Lavas erupted beneath ice in the mountains of West Antarctica indicate that small ice caps existed in places at this time (see Plate 9.2).

Eocene floras containing *Araucaria* and *Nothofagus* from the Antarctic Peninsula and the Transantarctic Mountains suggest mean annual temperatures of >6°C and cold-month means above 0°C for latitudes 65–75°S (Greenwood and Wing, 1995). The persistence of cool temperate forests across most of Antarctica during the middle and late Eocene cooling is one of the main counters to suggestions that continental glaciation began at this time rather than during the Oligocene (Mohr, 1990).

Although floral evidence provides the best indicator of Palaeogene climates in Antarctica, discoveries of Eocene fossils of marsupial mammals (Woodburne and Zinsmeister, 1982), similar to those living at the time in South America and Australasia, confirm the similarities in contemporary environments before the onset of continental glaciation.

6.4.2   Palaeogene Environments of South America

A marked increase in volcanism in the Peruvian Andes around 40 Ma (Noble et al., 1974) has been linked to changes in the directions of plate movement in the Pacific Basin, which are also clear in its west and southwest parts (see below). This situation was followed by a decline in volcanism which was maintained for much of the Oligocene.

Although the palaeofaunal record for Pacific South America is frustratingly incomplete for much of the Palaeogene, the effects of the early Oligocene climatic deterioration are apparent in the extinction of subtropical woodland browsers and their replacement by grazers adapted to life in the seasonally arid savannas that developed around this time (Prothero, 1994). It is reasonable to suppose that the appearance of new lifeforms and changes in the ecological composition of particular communities in this region during the Palaeogene may have been at least partly in response to vicariance associated with the rise of the Andes (Pascual et al., 1985).

6.4.3   Palaeogene Environments of Central America

Much of this region was submerged during the Palaeogene although the case for the presence of one or more 'filter bridges' which enabled certain animals to reach one of the Americas from the other seems watertight.

6.4.4   Palaeogene Environments of North America

The Laramide Orogeny (see Chapter 5), which had affected the entire west coast of North America during the late Cretaceous, ended by late Eocene times (Coney, 1989). It left a landscape markedly different from that in which it had begun – a landscape now crossed by high, coast-parallel fold mountains, with young volcanoes, yielding vast quantities of sediment which subsequently buried and extended the coastal lowlands to the west. Since the end of the Laramide Orogeny, the fold-mountain belt has been dominated by extensional and transcurrent movements represented today by the San Andreas and Queen Charlotte faults.

Some of the most precise information about Tertiary climates in North America has come from the application of the Climate–Leaf Analysis Multivariate Program (CLAMP) to fossil-leaf assemblages (Wolfe, 1994a). In the mid-latitudes of North America, leaf size increased during the earliest Palaeocene (see Figure 5.4), as did the numbers of species with probable vine habit and drip tips off which abundant rainfall is shed easily. These data have been interpreted as indicating a four-fold increase in precipitation during the earliest Palaeogene and an increase in mean annual temperature of around 10°C compared to the latest Cretaceous across most of the Earth (Wolfe, 1990). Some 0.5–1.0 million years into the Palaeocene, precipitation decreased to around three times its latest Cretaceous values, and mean annual temperatures fell by 5–6°C.

Although experiencing fluctuations of 5–8°C, Eocene temperatures were generally higher than today, varying from 19°C at palaeolatitude 70°N to >27°C at palaeolatitude 45°N. Leaf-margin analysis allows palaeotemperatures to be determined with a high degree of precision; data referring to the two major steps in Eocene cooling were given in Table 6.2. Evidence supporting the idea that arid conditions accompanied periods of cooling comes from coeval alluvial units in Oregon which have been interpreted as the products of increased sediment yields under such conditions (Bestland et al., 1997).

North American land plants were drastically affected by the early Oligocene climatic deterioration. In the northwest United States, for example, tropical rainforests indicating mean annual temperatures of 20–22°C were replaced within a million years by forests indicative of temperatures about 10°C lower (Wolfe, 1978).

Palaeogene faunal migrations into North America – largely from Asia via Beringia – reached a maximum during the early Eocene yet subsequent cooling brought about massive extinctions especially of forest dwellers. Asian faunal immigrants, particularly those adapted to living in more open vegetation types, filled the ecological gaps left in North America by these extinctions.

6.4.5   Palaeogene Environments of Beringia

Beringia formed a dryland connection between North America and East Asia during the entire Palaeogene. Climate thus posed the only barrier to biotic interchange between the two continents at the time. Evidence of heightened interchange, such as occurred during the early Eocene, has thus been taken to corroborate the other evidence for warm conditions while marked reductions have been found to correlate well with cooler conditions, such as those marking the later Eocene stepwise cooling.

During the warmest times of the Palaeogene, Arctic climates were cool-temperate. Despite the prolonged winter darkness, alligators and pond turtles thrived (Prothero, 1994); subtropical flora lived in Alaska during the Palaeocene, and palms grew in Kamchatka during the early Eocene thermal optimum. The early Oligocene deterioration is well marked in the palaeofloral record. For example, Alaskan forest composition changed in ways that suggest mean annual temperature dropped 8°C during the early Oligocene. Temperate climates characterized Beringia during the middle Oligocene; continued cooling saw modern climates established by the Pliocene.

6.4.6   Palaeogene Environments of East Asia

The collision between India and Asia about 50 Ma sent ripples out along the Pacific Rim, particularly in East Asia. These were manifested as an eastward displacement of central China and the opening of the South China Sea. Deformation of sedimentary basins occurred throughout China. Some researchers ascribe the notable plate reorganization in the Pacific Basin 43–42 Ma to this collision (Zhang et al., 1989) although the subduction of the entire North New Guinea Plate and associated divergent boundary (mid-ocean ridge) may also have been an important factor. One indisputable consequence of this reorganization was heightened volcanic activity 41–36 Ma in Japan.

A separate development brought about by the Palaeogene collision between the Okhotsk terrane and the craton (see above) led to the initiation of subduction along the Aleutian and Kuril trenches which in turn caused parts of the Kula and Pacific plates to become trapped behind the modern island arc in the northwest Pacific at this time (Maruyama et al., 1989). By the end of the Oligocene, active subduction and associated processes were occurring along the convergent margin from the Kuriles in the north through Japan, the Ryukyus and east of the Philippine Sea. Several backarc basins began opening during the Miocene. The resulting increases in volcanism and tectonism along the continental margins of the central western Pacific caused changes in the nature of ocean-floor sedimentation offshore. Notable among these were the increased number and thickness of ash layers, and the rise in non-carbonate sedimentation.

East Asia was an important location for Palaeocene biotic diversification, particularly of mammals (Savage and Russell, 1983); the Palaeogene migratory records of these mammals northeast through Beringia and into North America and, during the Neogene, south into Australasia are well known (Briggs, 1987). The appearance of new animal species in East Asia during the early Eocene is also well known. Many – including rabbits and several rhinoceros genera –

also colonized North America via Beringia at this time. Yet, as in North America, the severe cooling during the later Eocene led to numerous extinctions (Prothero, 1994) following which many new forms appeared. Among these was *Paraceratherium*, the largest known land mammal, which was almost 6 m high at the shoulder and weighed perhaps 18 000 kg.

Early Eocene warmth in East Asia is shown dramatically by records of palms growing in north China and northern Japan at the time. In China, cold-month temperature means exceeded 1°C; a broad-leaved evergreen forest grew in much of the region (Greenwood and Wing, 1995); cypress swamps and mangroves were characteristic of much of the mainland coast.

Aridity became widespread in much of the interior of mainland East Asia during the later Palaeogene. Cooling was generally associated with the expansion of these arid lands; pollen data from China, for example, show that an arid belt marked by woody savanna but lacking grasses expanded northwest during the late Eocene. The appearance of temperate deciduous trees and conifers by the mid-Oligocene may have been a response to global climate deterioration and/or more localized climate changes associated with the uplift of the Tibetan Plateau (Leopold et al., 1992).

6.4.7 Palaeogene Environments of Australasia

Early Palaeogene biotas of Australasia indicate a tropical climate for most of this region with floral elements akin to plants living today in Papua New Guinea and New Caledonia. The formation of coal in southeast Australia can be attributed to high rainfalls associated with the development of the Antarctic Circumpolar Current. Some Pacific floras of Australia exhibit signs of increased aridity by the mid-Oligocene (Christophel, 1990).

Despite being farther south than today, no part of Australia experienced freezing temperatures during the Eocene; the presence of crocodiles (Willis and Molnar, 1991) and the thermophilic mangrove palm *Nypa* (Greenwood, 1994) at

palaeolatitudes 50°S have been used in support of this argument. Similar data are available for New Zealand (Pocknall, 1990).

It has been argued that the unusual stability of the Australian biota during the Cenozoic can be attributed – in part at least – to the comparatively stable climate. This argument is founded on the observation that the progressive cooling which characterized most of the Cenozoic throughout the world was offset in Australia by the simultaneous movement of that continent northwards into lower latitudes (Flannery, 1994).

Numerous groups of plants and animals reached Australasia from South America via Antarctica during the early Palaeogene, notably many families of angiosperms and marsupial mammals. A hiatus in immigration during the late Palaeogene can be explained by the development of conditions along the fringe of Antarctica which were less conducive to movements of biota. Indeed, this connection seems never to have been revived; when immigration into Australasia resumed in mid-Miocene times, the new arrivals came exclusively from Asia via New Guinea.

Late Mesozoic heating of the New Zealand microcontinent had led to its elevation; cooling during the Cenozoic led to its subsidence. Fleming (1979) estimated that around 90% of the New Zealand platform was under water by the late Oligocene (30 Ma); even today, after its late Cenozoic history of plate convergence and crustal thickening, it is still more than 70% submerged compared to its late Mesozoic condition (Spörli and Ballance, 1989).

One of the most significant events in the Tertiary history of New Zealand was the marked increase in volcanism associated with the establishment of an active volcanic and tectonic zone through the islands about 42 Ma. This event was also well marked in the south and west Pacific and is discussed at greater length in the following section.

An even more significant event occurred some 25 Ma in New Zealand with the establishment of the modern plate boundary, and the initiation of island-arc volcanism. Late Cenozoic slip along the Alpine Fault divided and dispersed the various

Palaeozoic and Mesozoic terranes constituting New Zealand.

Since much of New Zealand has subsided since the Palaeogene, records of its contemporary environment are patchy, like those of many Pacific islands (see below). Early Eocene warmth is marked by the record of mangroves and coconut palms on land, and by the appearance of larger foraminifera like *Asterocyclina* whose descendants live today only in low latitudes where waters are warmer than 18–20°C (Hornibrook, 1992).

6.4.8 Palaeogene Environments of the Pacific Ocean and Islands

The development of an active convergent plate boundary through New Zealand and much of the southwest Pacific during the latest Eocene was associated with a change in the relative directions of plate movements in the region. It was an event which found expression elsewhere along the Pacific Rim and has been linked to the collision of India with Asia.

Today the most visible evidence for this event is that of the bends in chains of islands whose origins are best explained by plate movement across fixed mantle hotspots. The conspicuous bend in the Hawaiian–Emperor island-seamount chain, close to Kammu and Yuryaku seamounts, dates from some 42 Ma (Dalrymple and Clague, 1976). Bends dating from this time are also visible in island chains in French Polynesia, Kiribati and in the trends of the Eltanin, Mendocino and Udintsev Fracture Zones.

In the southwest Pacific, the aftermath of the 43–42 Ma plate reorganization saw marginal basin development and the formation of new island-arc systems. The Philippine, Bonin, Mariana, Yap, Palau and Tonga trench-arc zones developed after this event along the lines of former north–south trending fracture zones (Hilde et al., 1977). Deformation of existing island arcs to accommodate the change in directions of plate convergence in the region may have spurred development of backarc basins and led to the initial formation of microplates, particularly in the

southwest Pacific. These effects continue to dominate the tectonics of the region.

Palaeogene mountain building along the Pacific seaboard of North America caused increases in sedimentation offshore. A good example is the Aleutian abyssal plain, which was built between the early Eocene and middle Oligocene (Scholl and Creager, 1973).

The distribution of warm-water indicators – mangroves, hermatypic corals and larger foraminifera – in the Palaeogene Pacific confirms the broad patterns of temperature change outlined above (Adams et al., 1990). Coral reefs, for example, which exist today in ocean waters of 18–36°C (optimally 26–28°C) were found 10° of latitude farther north and south in the Palaeocene and early Eocene Pacific Ocean which, like its terrestrial counterparts at this time, was a veritable 'Garden of Eden' in which life flourished by comparison to later Palaeogene times (Prothero, 1994). Estimates of early Eocene ocean temperatures in the south Pacific are 12–15°C warmer than today (Miller et al., 1987).

Later Eocene cooling is well marked in the record of calcareous nannoplankton from the Pacific (Aubry, 1992). Tropical taxa disappeared progressively from north Pacific waters and from high latitudes in the south Pacific. The early Oligocene deterioration is marked by a rash of extinctions, followed in low and mid-latitudes by slow evolutionary turnover (moderate rates of extinction, very low rates of origination of new forms) indicating the gradual biotic adjustment to the new oceanographic regimen. Other marine micro-organisms display similar changes.

Growth of the Antarctic ice sheet during the Tertiary led to changes in the nature of deep-sea sedimentation and oceanographic processes in the Pacific Ocean, particularly its southern part. Biogeographic barriers (Polar Front, Antarctic Divergence, Subtropical Convergence) developed during the late Palaeogene and migrated northwards from Antarctica throughout the rest of the Cenozoic. This led to an expansion of the siliceous ooze belt and an increase in the range of calcareous and ice-rafted sediment facies. An increase in the latitudinal thermal gradient enhanced both ocean and atmospheric circulation leading to greater upwelling and siliceous biogenic productivity in the south Pacific. In the north Pacific, similar changes have been ascribed to Oligocene ice-sheet formation in Antarctica and the development of the psychrosphere. Among these changes was the cessation of carbonate sedimentation on many north Pacific seamounts (Kennett, 1977, 1982).

CHAPTER 7

The Pacific Basin During the Neogene (24.6–1.77 Ma)

7.1 INTRODUCTION

The second division of the Tertiary Period is the Neogene[1], comprising the Miocene and Pliocene epochs (see Figure 6.1). This chapter highlights the main features of Neogene environmental changes in the Pacific Basin, then briefly reviews their contributory causes before going on to describe environmental changes in each of its constituent regions.

7.2 THE MIOCENE

The Miocene epoch was the longest of the Neogene, lasting from 24.6 Ma to 5.3 Ma. The early Miocene was marked by warming in which environments recovered to some extent from the 'big chill' of the middle Oligocene. Yet cooling began once more in the middle Miocene leading to establishment of the 'icehouse', in the grip of which the Earth has been since. These cold conditions were brought on by the formation of a permanent continent-wide ice sheet in Antarctica, an event which was itself the outcome of the development – from ocean surface to ocean bottom – of the Antarctic Circumpolar Current which led to the complete thermal insulation of the continent (see Figure 6.3).

7.2.1 Early Miocene Climatic Optimum[2] (~18–15 Ma)

The maximum ocean temperatures of the entire Neogene were attained in the Pacific Basin during the early Miocene about 18 Ma, in some places apparently 2 or 3 million years later[3]. This warming was accompanied by a sea-level rise which affected all parts of the Pacific Rim (see Figure 6.6). It is uncertain whether or not the Antarctic ice sheet formed in the middle Oligocene (Chapter 6) disappeared at this time; oxygen-isotope data suggest it did (see Figure 6.5).

The biotic consequences of this climatic optimum were particularly clear around the Pacific Rim. For example, broad-leaved tropical forests reached 45°N along the Pacific Rim in North America, about 25° farther north than today (Wolfe, 1994a). In western Beringia, subtropical biota extended throughout the region reaching a peak around 15 Ma (White et al., 1997). Around New Zealand, tropical and subtropical marine fauna migrated south in the early Miocene (Hornibrook, 1992). A climatic optimum ~16 Ma is one of the most widespread and prominent events in the Neogene marine faunal record of Japan and has been correlated with similar events throughout the Pacific Basin (Tsuchi, 1990, 1992).

---

[1] Since it is that most widely followed in the writer's assessment, the definition of 'Neogene' used in this book does not include the Quaternary

[2] The mid-Neogene climatic optimum of Tsuchi (1992)
[3] In reality this optimum is unlikely to have lasted 3 million years; the range of dates could be an artefact of the chronologies used and the times at which they were applied

Yet this period of warming was but a short-lived interruption in the cooling trend begun in the late Palaeogene which culminated with a major expansion of the Antarctic ice sheet during the middle Miocene (see below). The juxtaposition between early Miocene warmth and subsequent cooling may not be coincidental; Mercer and Sutter (1982) argued that a warm period must necessarily precede one of polar ice accumulation.

A cooling interval lasting about 4 million years marks the end of the early Miocene climatic optimum. Termed the 'Monterey Carbon Isotope Excursion', it was characterized by deposition of large amounts of organic, commonly dia-tomaceous, material around the Pacific Rim, notably the Monterey Shales off California (Vincent and Berger, 1985). This deposition is likely to have been an outcome of enhanced biogenic productivity in the oceans, associated with heightened upwelling in places, which reduced levels of atmospheric carbon dioxide and contributed to cooling and the expansion of Antarctic ice.

7.2.2   The Middle Miocene Climate Transition (~16.1–11.1 Ma)

A critical turning point in Cenozoic climate evolution, the middle Miocene climate transition, is one for which evidence has been recognized in every part of the world, particularly in marine sedimentary records (Woodruff et al., 1981; Gasperi and Kennett, 1993).

A notable discovery of work in the 1970s was the rapid worldwide increase in $\delta^{18}O$ values of benthic foraminifera during the early middle Miocene (see Figure 6.5), interpreted as signalling ocean cooling resulting from rapid Antarctic ice-sheet growth. It is now clear that while this period of cooling may have been uncommonly rapid, it was merely part of an overall cooling trend which had been continuing for much of the earlier Cenozoic – an idea compatible with the later discovery that an Antarctic ice sheet formed first during the Palaeogene rather than the Neogene (see Chapter 6).

The formation of a semipermanent ice sheet in East Antarctica during the middle Miocene (~16–14 Ma) is well documented, largely from stratigraphic investigations, particularly those of ice-rafted debris and ice-grounding events, and changes in $\delta^{18}O$ from extratropical ocean sites (Miller et al., 1991; Anderson and Bartek, 1992). An ice sheet formed in East Antarctica several million years before one formed in West Antarctica, that part of the continent bordering the Pacific Ocean. Evidence of largely ice-free conditions yet continued cooling of West Antarctica 16–10 Ma was inferred from changes in clay type (smectite to illite and chlorite) deposited on the adjacent ocean floor. An increase in the presence of ice-rafted debris suggests ice-sheet formation began 10–5 Ma in this region (Leg 113 Shipboard Scientific Party, 1987).

The biotic outcome of the mid-Miocene climate transition can be traced throughout the Pacific Rim. For instance, warm-loving hardwood forests gradually disappeared along northern continental coasts; warm-water foraminifera disappeared along high-to middle-latitude continental margins in the south Pacific. The initial appearance of open grassland (savanna) in various places, such as Australia and temperate North America, indicates increasing aridity associated with this cooling. The development of grasslands was accompanied by the appearance of new animals, notably a variety of herbivores adapted to grassland browsing, and a number of carnivores adapted to devouring these herbivores.

There is evidence that cooling during the mid-Miocene climate transition in middle and high latitudes was accompanied by continued warming in parts of the tropical Pacific; the persistence of coral reefs in Papua New Guinea throughout the Miocene has been hailed as corroboration of this (Feary et al., 1991). The same was not true of the central Pacific Ocean which registered the effects of mid–late Miocene cooling in many ways. A major evolutionary turnover in diatom assemblages, for example, occurred 14.9–12.4 Ma (Barron and Baldauf, 1990).

Increased productivity in the equatorial Pacific during the middle Miocene climate transition is indicated by increased biogenic silica deposition and calcium carbonate supply rates, both effects which are likely to have been the outcome of the strengthened gyral circulation in the Pacific Basin. Towards the end of this period, an increase was recorded in the mean grain size of aeolian material in ocean-floor deposits implying a strengthening of the tradewinds in both hemispheres (Rea et al., 1991, 1993).

Although cooling continued after the middle Miocene transition, its biotic consequences were comparatively subdued, probably because most warm-climate indicator species had already disappeared.

7.2.3 Terminal Miocene Cooling (10–5.3 Ma)

The last ~5 million years of the Miocene was marked by an acceleration in the cooling trend, although this was superimposed on a number of rapid cyclic climate changes tracked by sea-level oscillations. It has been suggested that this cooling phase was associated with the establishment of the West Antarctic ice sheet, an event which would explain contemporaneous changes in the South Pacific. These included northward shifts in the Antarctic Convergence by 300 km, in the extent of ice rafting, and in the area of siliceous biogenic sedimentation (Frakes et al., 1992).

Increased biogenic productivity of upwelling-indicative diatoms like *Thalassionema nitzschioides* 6.4–6.0 Ma led to further 'Monterey-type' deposits being laid down around the margins of the Pacific Basin in such quantities that these are thought to have been the principal cause of the 'carbon shift' at this time.

The terminal Miocene fall of sea level by 40–50 m may have been caused by expansion of Antarctic ice and is thought to have been a major trigger of the Messinian salinity crisis (see below), the effects of which overshadow those of the earlier part of the late Miocene and have compounded the difficulties associated with the unravelling of contemporary climate and environmental history.

7.2.4 The Messinian Salinity Crisis (6.5–4.8 Ma)

Events on the other side of the world affected the Pacific around the Miocene–Pliocene boundary (Hodell et al., 1986). On two occasions in conjunction with local tectonism, sea-level fall caused various seas – the Mediterranean, the Persian Gulf and Red Sea – to be cut off from the other oceans of the world. The water within these seas evaporated. Ocean water which spilled into these basins during intervening interglacial periods also evaporated. When this 'Messinian salinity crisis' ended about 4.8 Ma, the volume of salts laid down in the Mediterranean Sea represented the evaporation of about 40 times its present water volume.

In addition to the evidence from ocean-core analyses, a tangible legacy of the Messinian salinity crisis has been found on the high limestone island Niue in the central south Pacific. Although emerged today, Niue was a subsiding atoll until uplift began about 2.3 Ma. $\delta^{18}O$ and $\delta^{13}C$ values from cores through the Niue limestone can be used as surrogates for sea-level change (Figure 7.1). Pronounced depletions mark the positions of solution unconformities which developed when sea level fell more rapidly than the island was sinking. The ~10 m sea-level fall 6.14–5.75 Ma was coincident with the first phase of evaporite deposition in the Mediterranean. The second conspicuous spike in these data is close to the Miocene–Pliocene boundary and represents a total sea-level fall of at least 30 m beginning 5.36 Ma. This event coincides with the most severe phase of Mediterranean desiccation and has been used to illuminate the role of eustasy in the Messinian salinity crisis (Aharon et al., 1993).

The end of the Messinian salinity crisis, marked by the permanent refilling of the Mediterranean Sea, saw the release of a warm-water mass which may have caused the West Antarctic ice sheet to collapse temporarily (Frakes et al., 1992).

7.3 THE PLIOCENE

The Pliocene lasted from 5.3 Ma to 1.77 Ma and was marked by a change from low-amplitude

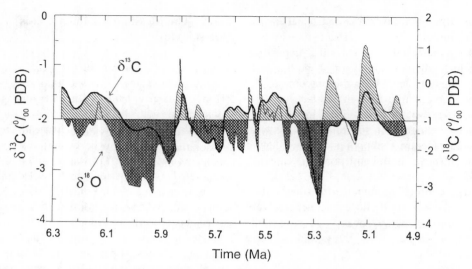

Figure 7.1   Composite $\delta^{18}O$ and $\delta^{13}C$ records from Niue island, south-central Pacific. Two episodes of sea-level fall are shown by pronounced $\delta^{18}O$ and $\delta^{13}C$ depletions beginning around 6.14 Ma and 5.26 Ma. These episodes are considered to be those which triggered two important periods of evaporite deposition in the Mediterranean Sea during the Messinian salinity crisis
From *Geology*, Aharon, P., Goldstein, S.L., Wheeler, C.W. and Jacobson, G. Modified with permission of the publisher, the Geological Society of America, Boulder, Colorado, USA. Copyright © 1993 Geological Society of America

climate variations in its early warmer part to larger-amplitude cyclic oscillations associated with cooler climates in its later part. These latter oscillations are of the kind best explained as the product of orbital changes (Chapter 8) and mark the start of a trend of climate change which has continued ever since.

The end of the Messinian salinity crisis 4.8 Ma indicates the restoration of open-water conditions in the Mediterranean and saw the beginning of a period of warming and sea-level rise which characterizes the early part of the Pliocene (4.8–3.0 Ma) worldwide. If the West Antarctic ice sheet had collapsed, at least partly, after the Messinian salinity crisis (see above), the sedimentary record from the south Pacific indicates that it was in place once more 4.35 Ma (Frakes et al., 1992). About this time, sea level reached a maximum of ~25 m higher than today, largely because the Antarctic ice sheet(s) was less extensive than at present. Sea-surface temperatures in the tropics were about the same, but considerably higher in higher latitudes (Dowsett et al., 1994).

At this point, it is worth noting the most recent occasion on which plate tectonics had a Pacific-wide effect. This was the time of change in the direction of motion of the Pacific Plate 3.9–3.4 Ma recognized by Harbert and Cox (1989) from a study of the Pacific–Antarctic mid-ocean ridge and surrounding ocean floor. Effects have been identified in North America, Japan and New Zealand (see below).

Cooling characterized the later Pliocene (3.3–1.77 Ma), probably in response to an increase in the extent of Antarctic ice and, later, development of terrestrial ice sheets in the northern hemisphere. It is during this time that orbitally driven glacial–interglacial oscillations of climate, which dominate the record of Quaternary climate change (Chapters 8–15), become conspicuous for the first time. Later Pliocene sea level appeared to have oscillated in a similar way. Several transgressions on the north Alaskan coast were of such magnitude that they caused significant reductions of Arctic sea ice (Carter et al., 1986), indicating a connection between climate and sea level which had hitherto been securely established only for the Quaternary.

An increased amount of information bearing on the Pliocene compared to earlier epochs allows greater resolution of the palaeoclimate record, although individual observations are not always easy to reconcile with the general trends observed. For example, short-lived episodes of cooling occurred about 4.7 and 3.5 Ma in the south Pacific, when a significant northward increase in the deposition of sediments originating in the Antarctic has been measured (Ciesielski and Weaver, 1974). The latter event is believed to signal the episode of Antarctic ice expansion which triggered the late Pliocene cooling, but the significance of the former is uncertain. The question of whether Antarctica was completely deglaciated during the mid-Pliocene is reviewed below. The initial formation of northern-hemisphere ice sheets in the late Pliocene is discussed thereafter.

7.3.1 Mid-Pliocene Antarctic Deglaciation?

While it is clear that the Earth was warmer than at present during the early part of the Pliocene (4.8–3.0 Ma), controversy surrounds the question of whether or not it became sufficiently warm to allow significant deglaciation of Antarctica in the later part of this period. Until a decade or so ago, it was generally assumed that the East Antarctic ice sheet had persisted since its growth spurt in the middle Miocene (see above). Yet there is evidence which could be interpreted as meaning that temperate climate conditions prevailed in Antarctica in the middle Pliocene.

The critical discovery, which weighted much opinion in favour of mid-Pliocene Antarctic deglaciation, was of marine diatoms, notably *Thalassiosira insigna* and *Thalassiosira vulnifica*, found in glacial till of the Sirius Group at several locations along the Transantarctic Mountains (Webb et al., 1986) and later elsewhere (LeMasurier et al., 1994; Webb et al., 1996). These diatoms, which lived together only within the age range 3.1–2.5 Ma, were interpreted as having been eroded, transported and deposited by glacier ice

from a marine basin somewhere in the interior of East Antarctica. This means that the centre of East Antarctica, today covered by the largest and thickest ice sheet in the world, must have been ice-free and below sea level around 3 Ma, a date confirmed by K-Ar dates of associated volcanic ash (Barrett et al., 1992).

Other compelling evidence of mid-Pliocene deglaciation involves the growth of southern beech (*Nothofagus*) forests in many parts of Antarctica around 3 Ma, implying that a considerable land area had been uncovered by deglaciation (Webb and Harwood, 1987; Fleming and Barron, 1996).

The implications of mid-Pliocene Antarctic deglaciation were spelled out by Barrett et al. (1992: 818):

> It now seems likely that for most of the past 40 million years, the Antarctic region was characterized by waxing and waning temperate ice sheets of continental proportions. The present polar ice sheet may not have formed until the latest Pliocene, about the same time as the first northern ice sheets developed. This different climate history will require us to revise our ideas of the evolution of the flora and fauna of the region, and of the role of the ice sheet in past ocean circulation.

Yet, since the original suggestion of mid-Pliocene deglaciation of Antarctica was made, there have been many contradictory opinions expressed. Many oppose the idea of Pliocene deglaciation of Antarctica because of the conspicuous lack of evidence in places where it would be expected to show up, principally in the ocean-floor sediment record (Kennett and Hodell, 1993) and in the glacial history of Wright Valley (Hall et al., 1997). Other compelling evidence against massive Antarctic deglaciation during the Pliocene is provided by a study of volcanic ashes and associated landforms in southern Victoria Land (Marchant et al., 1996). Ashes ranging in age from 15.15 Ma to 4.33 Ma exhibit no signs of authigenic clay formation, which would be expected had deglaciation occurred subsequently. On the contrary, all indications are that they have been preserved in a cold desert environment since their formation. In nearby Arena Valley, ash-avalanche deposits have not been modified much since being laid down at least 7.42 Ma, an observation also consistent with the persistence of the cold desert conditions under which they formed.

Much of the recent argument has focused on the dating of the Sirius Group. The original biostratigraphic (diatom) dating was confirmed by K-Ar dating of a volcanic ash within the Sirius Group found in the CIROS-2 drillhole in the Ross Sea (Barrett et al., 1992). Ages of other Sirius Group deposits have generally been reckoned of similar age; those at Mount Murphy (see Plate 9.2) were dated to 2.4–2.0 Ma (LeMasurier et al., 1994). Yet the most recent age determinations of the Sirius Group tills, made by measuring cosmogenic ^{3}He and ^{21}Ne, are minima of ~6 Ma and ~6.5 Ma which contradict the idea of Antarctic deglaciation any time since the middle Miocene (Bruno et al., 1997). This conclusion would be compatible with the recently expressed view that the marine diatoms at one location at least are windblown additions overlying the Sirius Group tills (Stroeven et al., 1996).

With such evidence to hand, and the possibility that the Sirius Group has been assigned too recent a date by some investigators, the view taken in this book is that the East Antarctic ice sheet has indeed persisted since the mid-Miocene in much its present form. The implications of this position merit comment. If in fact the dating of the Sirius Group (by many means) is wrong, and it was only recognized as such because of conflicting evidence, it is likely that many other dates or chronologies are also in error but have not yet been uncovered because they do not appear to contradict others. Yet, if one day they are found to be in error then regional correlations, such as are attempted throughout this book, will need re-evaluation. Clearly a degree of humility is helpful in studies of past environments.

7.3.2 Onset of Northern-Hemisphere Glaciation

A conspicuous increase in δ^{18}O in benthic foraminifera and a pronounced shift in $CaCO_3$

preservation occurred about 2.4 Ma and have been widely linked to the beginning of northern-hemisphere glaciation (Raymo, 1994; Wang, 1994). Supportive evidence is available along the margins of the Pacific Basin. For example, shallow-marine sedimentary records have been interpreted to show that the Arctic Ocean lacked complete ice cover until at least 2.4 Ma. On the Loess Plateau of north central China, the earliest loess – signalling severe cold and dry climate conditions – formed ~2.5 Ma (Figure 7.2).

Why ice began forming so much later in the northern hemisphere compared to the southern hemisphere is uncertain. One explanation relates to the closure of the gaps between Australia and southeast Asia, and North and South America in the late Tertiary which would probably have resulted in the intensification of north-flowing currents – the Kuroshio in the Pacific and the Gulf Stream in the Atlantic – which in turn would probably have increased precipitation in high-latitude areas of the northern hemisphere. Much of this precipitation is likely to have fallen as snow causing ice to accumulate. Alternatively, the onset of northern hemisphere glaciation may have been triggered by the impact of the large meteorite, some 0.5 km in diameter, which hit the southeast Pacific around 2.3 Ma and may have brought about a temporary cooling of the Earth's climate and disrupted biological productivity (Kyte, 1988).

7.4 CAUSES OF NEOGENE CLIMATE CHANGE

While many likely causes of climate changes within the Tertiary can be identified, the dominant cause of the major cooling of Cenozoic climate remains unresolved. Most authorities would regard it as linked in some way to tectonic changes but the importance of these depends on a correct understanding of the role of extraplanetary changes, particularly those in the Earth's orbit around the Sun.

At present the most plausible Earth-based cause of Tertiary climate change involves the positioning of the Antarctic continent over the

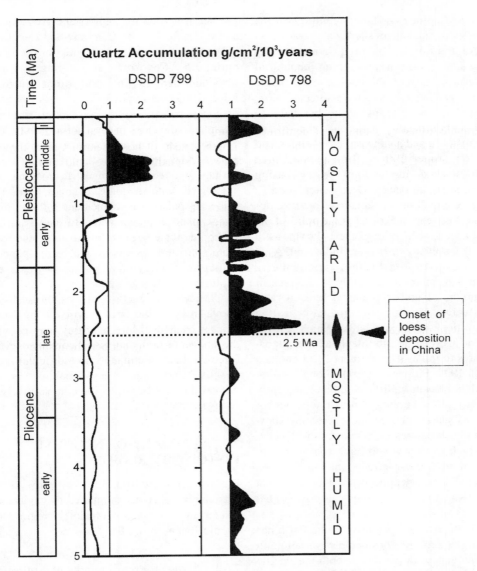

Figure 7.2   Rates of quartz accumulation on the deep-ocean floor at DSDP sites 798 and 799 in the Sea of Japan can be linked to late Cenozoic changes in the climate of Asia. See also Table 9.6

Reprinted from *Palaeogeography, Palaeoclimatology, Palaeoecology*, 108, Dersch, M. and Stein, R. 'Late Cenozoic records of eolian quartz flux in the Sea of Japan (ODP Leg 128, Sites 798 and 799) and paleoclimate in Asia', 523–535. Copyright 1994, with permission from Elsevier Science

South Pole and its subsequent separation from Australia and South America, which led to the establishment of the Antarctic Circumpolar Current and the effective insulation of Antarctica from warm waters thereby allowing ice build-up

(Chapter 6). Yet, by itself, this explanation has proved inconsistent with some models of global climate evolution (Oglesby, 1989).

Closure of other gaps around the Pacific Rim may also have been significant. North and South

America were joined at the beginning of the Tertiary yet separated again shortly afterwards. As the result of uplift of the Panama Isthmus, this gap closed finally 3.7–3.1 Ma stopping equatorial water exchange between the Pacific and Atlantic Oceans, an event which may have directed more moisture polewards allowing ice to build up on high-latitude continents. A second gap – that between Australia and southeast Asia – which had existed for some 200 million years, also effectively closed during the mid-Tertiary causing intensification of the Pacific Ocean circulation.

Another compelling explanation for development of the Neogene 'icehouse' is the uplift of the Tibetan Plateau, which began 52–44 Ma as a result of the collision between India and Asia. This event disrupted global atmospheric circulation and, more significantly in the present context, produced higher chemical erosion rates (as shown by the Neogene increase in the $^{87}Sr/^{86}Sr$ ratio; see Figure 4.3), which caused atmospheric CO_2 to become depleted and global temperatures to fall as a consequence (Raymo and Ruddiman, 1992; Filippelli, 1997). Global cooling initiated by Tibetan uplift would subsequently have led to more cooling through a series of positive feedbacks. For example, increasing snow cover would have increased albedo on the Tibetan Plateau which in turn would have contributed to global cooling. Similar effects may have been occasioned by Tertiary uplift of the Sierra Nevada and Colorado Plateau in the western United States.

The past history of volcanism on the Earth has been characterized by considerable variation in eruption frequency. In particular, both the Pacific islands and the Pacific Rim have experienced pulses of volcanic activity separated by periods of relative quiescence. These pulses of activity appear to have been approximately synchronous. Results of deep-sea drilling in the north Pacific show a massive increase in regional volcanism coincident with the onset of northern-hemisphere glaciation (Rea et al., 1993), prompting calls for renewed investigations into the possible link between volcanism and Cenozoic climate change (Raymo, 1994).

Data compiled for eight island arcs around the entire Pacific Rim by Cambray and Cadet (1994) show that two major volcanic pulses occurred during the Cenozoic: one during the middle Miocene (18–13 Ma), the other during the Pliocene–Quaternary (5–0 Ma). The occurrence of these pulses around the entire rim and at intraplate volcanoes in the Pacific islands (Figure 7.3) suggests that the cause of their increased activity was global rather than regional in nature, perhaps related to changes in magma supply associated with tectonic changes. The middle Miocene pulse was coincident with a time of major plate reorganization in the Pacific: the Philippine Plate began to move north; the Farallon Plate split into the Cocos and Nazca plates. The Pliocene–Quaternary pulse is less easily explicable in such terms.

Clearly other factors besides those discussed above may have been involved in Neogene climate changes. More plausibly, two or more of the above factors may have combined either to force the Earth's climate into a different mode or to push particular climate parameters across critical thresholds, producing comparatively rapid changes.

7.5 SUMMARY OF NEOGENE HISTORY BY REGION

Enough information is available to allow reasonably coherent summaries of environmental change in each of the constituent regions of the Pacific Basin during the Neogene.

7.5.1 Neogene Environments of Antarctica

The Antarctic Plate remained stationary throughout most of the Neogene and Quaternary, so contemporary volcanic activity is difficult to explain by plate motions. For this reason and because of the observed systematic migrations of eruptive activity in much of West Antarctica, fracture propagation is considered a more likely cause (Le Masurier and Rex, 1991).

A major ice sheet began forming in East Antarctica only during the middle Oligocene

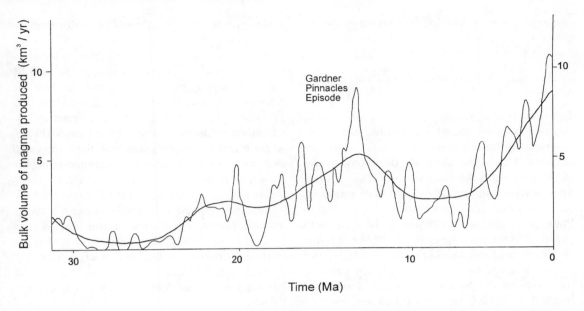

Figure 7.3   Magmatic production in the Hawaii islands. Note the pulses of activity and periods of relative quiescence
From *Geology*, Vogt, P.R. Modified with permission of the publisher, the Geological Society of America, Boulder, Colorado, USA. Copyright © 1979 Geological Society of America

around 32–28 Ma. On balance, it seems more likely that this ice sheet (largely) disappeared during the warmth of the early Miocene yet became re-established during the middle Miocene (16–14 Ma). An ice sheet developed first in West Antarctica during the terminal Miocene cooling 10.0–5.3 Ma. Despite evidence to the contrary, reviewed above, the view taken here is that the East Antarctic ice sheet has persisted in much its same form to the present day.

7.5.2   Neogene Environments of South America

Throughout the Neogene, the subduction of the Farallon–Nazca Plate beneath the South American Plate was the main control of tectonic and magmatic activity in Pacific South America. Subduction of the aseismic Nazca Ridge has been held accountable for Plio-Quaternary variations in uplift along this coast, although the ridge axis does not coincide with that of maximum uplift suggesting that other factors

may also be important (Macharé and Ortlieb, 1992).

In Pacific South America, Patagonian glaciers extended beyond the Andes for the first time 7–5.2 Ma (Mercer and Sutter, 1982). The 3.5 Ma cooling minimum (see above) is also represented by increased glaciation in southernmost South America (Mercer, 1976).

The arid zone along the Chile–Peru coast has been largely arid since the Eocene. Yet the modern hyper-arid conditions developed only during the middle Miocene climate transition when there was an intensification in the Peru Current and increased upwelling offshore as a result of contemporary climate changes (see above). It is also likely that the Andes at this time had been raised just high enough to create an effective rain shadow (Alpers and Brimhall, 1988).

In northern South America, such indications of global events were more subdued. For example, mid-Pliocene vegetation patterns and altitudinal ranges were similar to today, excepting the presence of North American immigrant species particularly *Alnus* and *Quercus* (Hooghiemstra, 1989).

7.5.3 Neogene Environments of Central America

The mid-Tertiary history of much of Central America was dominated by deposition of great volumes of ignimbrites, which bury most of western Mexico and comprise the Sierra Madre Occidental (see following section). The northern limit of volcanism migrated southwards during the Neogene as the divergent boundary of the Cocos Plate was gradually subducted in the north resulting in the opening of the Gulf of California (Barrash and Venkatakrishnan, 1982). An important consequence of this activity was the formation of the Valley of Mexico. Lavas had blocked the valleys of rivers which had formerly flowed south to the ocean so these now formed lakes in the valley centre; it was around these lakes that the Aztecs were to establish themselves during the late Holocene and, by constructing artificial fields (*chinampas*) within the lakes, were able to produce enough food to sustain massive urban populations by the time of Spanish arrival in the area (see Chapter 15).

One important outcome of the associated tectonic activity in this region during the middle Miocene was a rapid bathymetric change in Panama 12.9–11.8 Ma caused by uplift of a sill, an event which abruptly ended interchange of intermediate (depth) equatorial water between the Atlantic and Pacific. Closure was not completed until 3.7–3.1 Ma. The gradual inhibition of water exchange is documented by $\delta^{18}O$ values from DSDP cores on either side of the Panama Isthmus (Figure 7.4). As the two oceans became isolated from each other, so the two continents became progressively linked, an event which led to increasing interaction between biotas which had evolved largely separately, except for a few million years in the late Cretaceous and earliest Palaeocene.

The outcome of the closure of the ocean gap between North and South America has become known as the Great American Biotic Interchange (Stehli and Webb, 1985). This involved the mixing of the two biotas and the extirpation of many 'naive' elements in both continents. Little is

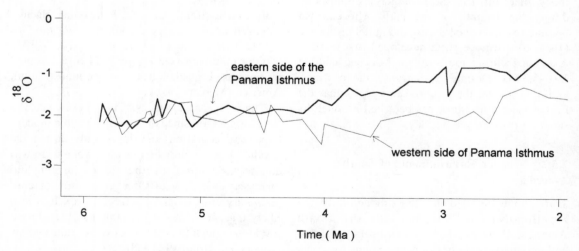

Figure 7.4   Planktonic foraminiferal $\delta^{18}O$ values for latest Miocene and Pliocene sediments from DSDP sites 502 and 503 on either side of the Isthmus of Panama. Note how the records are similar in the late Miocene when there was little restriction to exchange of water between the two sites. Then, during the earliest Pliocene, shoaling begins to inhibit water exchange significantly and the two records begin to diverge. When the emergence of the Panama Isthmus was complete by 3.1 Ma, the divergence of the records is clear
After Savin and Douglas (1985) with permission of Plenum Publishing Corporation

known of the environmental impacts of this interchange, which would be difficult to separate from those occurring for other reasons.

Although other parts of the Pacific Basin register the effects of mid-Pliocene warmth clearly, pollen data from parts of Pacific Mexico suggest that the area covered by tropical rainforest at this time was less than today with a greater-than-modern coverage of deciduous forest (Graham, 1989). More data would help our understanding of the significance of such studies.

7.5.4   Neogene Environments of North America

Most parts of Pacific North America continued to be affected by tectonic processes associated with plate convergence and terrane accretion during the Neogene. The upwarping of the entire Rocky Mountains and adjacent areas was marked by faulting and volcanism which transformed environments and led to high rates of floral extinction in particular. Uplift led to the accentuation of differences in climate between east and west sides of mountain ranges. It also caused rates of erosion to rise along the western seaboard of North America, a change reflected in increased late Neogene ocean-floor sedimentation in the northeast Pacific. Perhaps the greatest amount of uplift was in the far north of this region where pre-Pleistocene glaciomarine strata in the St Elias Mountains, which extend from Yukon into Alaska, have been raised 2 km (Plafker, 1981).

The middle Miocene volcanic pulse (see above) registered clearly along the Pacific margin of North America as the result of the uprising of an elongated mantle diapir to the east of the subducting Farallon Plate (Barrash and Venkatakrishnan, 1982). In terms of environmental change, it was the formation of the plateau basalts in the northwest United States which were most significant; the Columbia River Basalts, for example, erupted from dykes 17–13.5 Ma. Similar activity characterized this period in the Great Basin and elsewhere in North America.

Later Neogene volcanic history in this region is explicable by the continuing subduction of the last vestiges of the Farallon Plate in the northwest United States and southwest Canada, and the development of largely strike-slip movements elsewhere.

The change in motion of the Pacific Plate 3.9–3.4 Ma produced a change in the character of plate movement in several parts of the Pacific Basin including coastal California (Harbert and Cox, 1989). Here the change is registered by compression across the transform (transpression) boundary, manifested in the landscape by the emergence of the San Joaquin valley, the uplift and folding of several upland ranges having trends parallel to this boundary, and by formation of thrust faults in the San Francisco Bay area along which the epicentres of many of this area's recent, most destructive earthquakes have been located (see Plate 15.17).

The early Miocene climatic optimum is well marked in palaeoclimate data for North America. Much of the warmth was accompanied by aridity. In response, grasslands spread throughout most of the region that they now cover by the end of the Miocene.

Tectonic effects notwithstanding, the general trend of Pliocene climate from the vegetation record of the western United States is one of warmer, wetter conditions 4.8–2.4 Ma and more arid conditions thereafter (Thompson, 1991). Early Pliocene warmth is indicated in many pollen records although the levels of summer rainfall are uncertain: analyses of cores off the California coast suggest that modern summer drought conditions were already established while onland records imply that summer rainfall was greater.

Glacial deposits in the Sierra Nevada dating from 3.1–2.7 Ma foreshadow the start of northern-hemisphere glaciation in this region (Curray, 1966) a few hundred thousand years later. The drier, possibly cooler, interval 2.4 Ma in the Tule Lake core in northern California is conspicuous within the warmer Pliocene (Bradbury, 1992) and may signal a threshold event associated with the start of northern-hemisphere glaciation. In the United States this was associated with increased aridity until at least 2 Ma. Thereafter at Tule Lake and elsewhere, moister conditions returned.

7.5.5 Neogene Environments of Beringia

A tectonic change of comparatively minor oceanographic but great biogeographic significance was the opening of the Bering Strait between East Asia and North America around 3.7 Ma, a date constrained by the time of an invasion of North America by Asian microtine rodents (Repenning et al., 1990). Before that time, movements of plants and animals between these landmasses had been comparatively unrestricted; after the Bering Strait opened, exchange became less frequent until around 2.5 Ma when an upsurge in biotic interchange has been interpreted – like similar events thereafter – as marking the start of a temporary emergence of the area in response to sea-level fall.

Analysis of fossil-leaf physiognomy using CLAMP (see Chapter 6) shows the early Miocene climate optimum well in Beringia (Wolfe, 1994b; see Figure 6.4). The subsequent cooling associated with build-up of the Antarctic ice sheet is marked by a temperature decline between about 13 Ma and 6 Ma, when summer temperatures reached their present level; winter temperatures continued to decline during the Pliocene. A similar picture emerges from the synthesis of pollen records for Alaska and northwest Canada published recently by White et al. (1997) except that their Miocene climatic optimum occurs ~15 Ma, notably later than that elsewhere in the Pacific Basin. These data indicate temperature decline in the mid-Miocene, and register the effects of terminal Miocene cooling associated with the Messinian salinity crisis by an increase in herbs and shrubs suggestive of a colder, drier, more continental climate. This change is attributed not to global climate events but to the effects of mountain building in the area, an interpretation which emphasizes the point that such records are an expression of global changes overprinted, and in some cases outweighed, by local changes.

The presence of pack ice around Beringian coasts and traces of ice-rafted debris in sediments from the floor of the north Pacific have suggested to several investigators that land ice (more extensive than just small ice caps feeding mountain glaciers) was present in Beringia well before the full development of a continental ice sheet in this hemisphere. Data are from Alaska, the Kamchatka Peninsula and Sakhalin Island (Mercer and Sutter, 1982; Zubakov and Borzhenkova, 1983; Serova, 1985; Plate 7.1).

Yet early Neogene ice bodies in Beringia must have been localized for indicators of ice-free conditions are abundant throughout the region until the time of initial northern hemisphere ice-sheet formation (Matthews and Ovenden, 1990). For example, loess deposits formed 3 Ma in central Alaska (Westgate et al., 1990) were probably associated with mountain glaciers rather than an extensive ice sheet because the climate at the time was generally warmer than at present, characterized by open boreal forest rather than the forest–shrub tundra which developed only later in the Pliocene (Brigham-Grette and Carter, 1992). This view is supported by studies from other Beringian sites which suggest that the Arctic Ocean was either seasonally or entirely ice-free before 2.6 Ma (Carter et al., 1986). Subsequent cooling was associated with the intensification of northern hemisphere glaciation. A marked upsurge in ice-rafted debris in the north Pacific dates from about 2.4 Ma and signals the onset of continental glaciation in Beringia.

7.5.6 Neogene Environments of East Asia

Although backarc (marginal) basin opening began in this region during the Palaeogene (see above), the Sea of Japan opened only in the early Miocene. This event led to the bending of the Japanese islands around the location of central Honshu in a comparatively short time 16–15 Ma (Hirooka, 1992). Later the change in motion of the Pacific Plate 3.9–3.4 Ma caused an increase in subduction rate along the Pacific coast of Japan of about 0.5 cm/year which resulted in deformation visible in both ocean-floor sections on the Sea of Japan and onland (Harbert and Cox, 1989).

The most complete record of non-marine sedimentation for the late Neogene and Quaternary is found on the Loess Plateau of north China. Although geographically marginal to the area

Plate 7.1   Ice-rafted dropstone and associated diamictite deforming a laminated turbidite bed in the upper part of
the Yakataga Reef section, eastern Gulf of Alaska
Reprinted from *Marine Micropaleontology*, 27, Martin B. Lagoe and Sarah D. Zellers, 'Depositional and
microfaunal response to Pliocene climate change and tectonics in the eastern Gulf of Alaska', 121–140, Copyright
(1996), with permission from Elsevier Science

under study, reconstructions of late Neogene climate around the Pacific Rim and offshore areas to the east cannot ignore information from this important area. The loess is the product of dust deposition during cold arid conditions. The cessation of loess deposition during warm humid periods is marked by palaeosols in the stratigraphy (Kukla and An, 1989). The earliest extensive loess dates from ~2.35 Ma and marks the onset of extremely cold, dry conditions, an event signalled by the replacement of forest with steppe in many parts of East Asia (Ning et al., 1993).

The nature of the sediments accumulating in the Sea of Japan has long been regarded as a key to changing climate in the continental hinterland. In particular, variable accumulation rates of wind-transported quartz have been linked to the vege-

tation of the hinterland: low quartz accumulation is equated with low supply resulting from increased vegetation cover inhibiting the removal of quartz from surface sediments by wind (see Figure 7.3). Thus middle Pliocene (4.3–2.5 Ma) sediments exhibit low quartz accumulation rates which are equated with a warm humid environment inland. The situation changes around 2.5 Ma, also the time at which loess deposition began in China, with an increase in quartz accumulation rates signalling the onset of colder, more arid conditions in East Asia during the latest Pliocene (Dersch and Stein, 1994).

Climate deteriorations around 3 Ma and around 2.5 Ma have been inferred from the plant macrofossil record in Japan. These are times of major extinctions of warm-indicator species and their replacement by those diagnostic of cooler temperate climates (Momohara, 1994). Yet the Pliocene palaeoclimate record in Japan is also sufficiently detailed to show clear evidence of oscillating temperature; Iwauchi (1994) recorded three warm and two cold periods in the late Pliocene. Climate oscillations also left their mark on areas offshore (Dersch and Stein, 1994).

The extraordinarily informative record from Lake Biwa in Japan begins in the middle Pliocene. Thick forests of evergreen broad-leaved trees dominated the vegetation of the vast alluvial plain surrounding the lake. Elephants (*Stegodon*) roamed the area; a discovery of bones in AD 1804 led to an imaginative reconstruction (Plate 7.2). The nature of the mammals suggests that land connections existed with the Asian continent to both the north and the south. By the later Pliocene, movement of the lake northwest on

account of a tectonic shift was accompanied by the development of mixed forest in a large part of the basin. Such forest favoured smaller varieties of elephant; the contemporaneous increase in numbers of cool temperate mammals, particularly Nihowan deer, suggests that the southern land connection with Asia had been broken while the northern route was still open (Kamei, 1984). Pollen in Lake Biwa cores indicate a warmer-than-modern climate about 3 Ma (Fuji, 1988).

7.5.7 Neogene Environments of Australasia

Volcanism occurred in most parts of Australasia during the Tertiary. The conspicuous intensification of volcanism around New Zealand in the early Miocene, accompanied by the first development of terrestrial volcanoes, has been attributed to a shift of the line of convergence

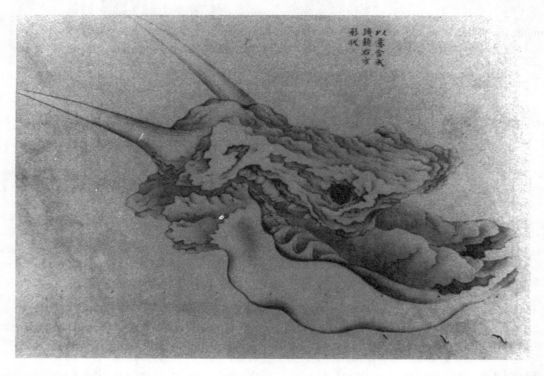

Plate 7.2   The 'dragon's bone', an imaginative reconstruction from what are now known to be Pliocene elephant bones found near Lake Biwa, Japan (from Kamei, 1984). The association of dragon stories with bone caches in long-settled East Asia, along with the recognized place of dragon bones in its traditional pharmacopoeia, led to the discovery of many important early human sites including that at 'dragon-bone hill' near Zhoukoutien

between the Pacific and Indo-Australian plates close to its present position off the east coast of North Island (Cole, 1985).

The effects of the 3.9–3.4 Ma change in motion of the Pacific Plate are apparent more in New Zealand than elsewhere in this region (Harbert and Cox, 1989). They may include sediment deformation and faulting associated with compression arising from a shift of about 10° in subduction azimuth along the Hikurangi Trench. Associated uplift in parts of North Island and along the Alpine Fault in South Island may also be an outcome of this event.

The Tertiary landscape history of most of Australasia was dominated by its move into low latitudes. Changes in the type of clay sediments washed off the continent and deposited in the deep ocean calibrate this process (Figure 7.5). Desertification in northern Australia increased and open grassland developed on the continent for the first time during the middle Miocene climate transition (see above) as a result of Antarctic ice build-up. It is possible that this process was partly a response to mountain building in New Guinea and the rain shadow cast to the south by this event (Flannery, 1994).

An increase in aridity affected the Australian continent during the remainder of the Cenozoic but most notably during terminal Miocene cooling. Increasing aridity has been well tracked by vegetation changes, most notably the change from temperate rainforest, dominated by southern beech (*Nothofagus*), to sclerophyllous forest, dominated by *Eucalyptus* and associated with other taxa indicative of drier conditions such as Asteraceae and Poaceae (grasses). In southeast Australia, a short-lived recovery of *Nothofagus* during the Pliocene probably signals a marine trangression and consequent higher precipitation (Bishop, 1994). Increasing aridity in Australia may have led to the extermination of much *Nothofagus* from eastern Australia by the mid-Pleistocene but this process was matched by its expansion into New Guinea where, owing to the effects of mountain building from the late Miocene onwards, habitats became available for it to colonize (Hope, 1996).

While the last appearances of a number of warm-water marine species around New Zealand occur ~3.1 Ma and may signal the onset of cooling in the later Pliocene, the first appearance of a subantarctic scallop (*Chlamys delicatula*) and crab (*Jacquinotia edwardsii*) in deposits of North Island some 2.4 Ma are the first unequivocal signs of this event (Pillans, 1991). Cooling was accompanied by increasing aridity which increased ocean-floor sedimentation around New Zealand at this time (Shane et al., 1995). Onland cooling in New Zealand is also indicated by vegetation changes, particularly as *Nothofagus brassi*, an indicator of warm conditions, was

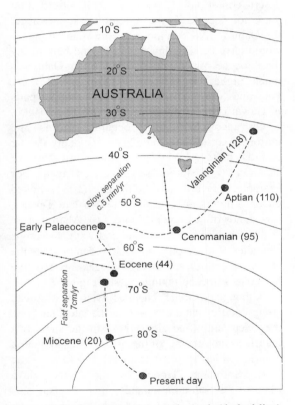

Figure 7.5 Apparent polar wander path (dashed line) of the South Pole relative to Australia during the late Cretaceous and Cenozoic. The earliest interval is when Australia was part of Gondwana, the later ones when Australia was separating (slowly or rapidly) from Antarctica

After Quilty (1994) with permission of Cambridge University Press replaced by the temperate indicator *Nothofagus fusca* during the late Pliocene (Nelson et al., 1988).

7.5.8 Neogene Environments of the Pacific Ocean and Islands

Largely because of the youth of Pacific islands, few studies have been particularly illuminating about Neogene environments. One exception is a study made of plant moulds in Hawaiian volcanoes which interpreted a paucity of moulds in lavas from the Waianae and Koolau volcanoes on Oahu as indicating that they erupted into near-desert conditions 3.9–1.8 Ma. This observation does not necessarily have regional significance, for Oahu – like the modern Hawaiian islands – would have been within the tradewind belt during the Pliocene. What it does signify is that Oahu was much higher, by perhaps 2000 m, when these volcanoes were active, rather like the upper parts of Hawai'i (the Big Island) today (Walker, 1995).

On account of their oceanic origins and uniform long-term tectonic histories, many Pacific islands have proved excellent indicators of Tertiary (and later) sea-level changes. For example, Lincoln and Schlanger (1987) linked unconformities on Midway Atoll to late Tertiary low stands of sea level (Figure 7.6). Aharon et al. (1993) analysed the carbon- and oxygen-isotope character of limestones on Niue to demonstrate the effect of rapid sea-level changes around the Miocene–Pliocene boundary (see Figure 7.1).

Although tectonic changes were widespread along Pacific island arcs during the late Cenozoic, they were amplified along those in the southwest Pacific as an outcome of the reversal in subduction polarity which resulted from the blocking of the North Solomons Trench by the Ontong Java Plateau. In Solomon Islands, this reversal saw the beginning of northward subduction along a trench south(west) of the Solomons Arc about 11 Ma. This caused compression of the arc which led to its rapid uplift and also caused a line of volcanoes to become established from the Shortland Islands in the northwest to Guadalcanal (Coulson, 1985). The blockage of the North

Solomons Trench by the Ontong Java Plateau had repercussions across a large region of the southwest Pacific, with a similar reversal of subduction polarity occurring in Vanuatu, and an anticlockwise rotation of the Fiji Platform farther east.

Most of the general information about Neogene climate changes in the Pacific Basin, discussed above, was gleaned from a study of ocean-floor sediments and will not be repeated here. Worth noting, however, is the record of terminal Miocene cooling in south Pacific ocean-sediment records marked by increased biogenic deposition associated with enhanced upwelling along ocean-water convergences. This is especially noticeable on the Lord Howe Rise and the Ontong Java Plateau (Kennett and von der Borch, 1985).

The movement of Australia northwards during the Palaeogene resulted in the opening of the Tasmanian Passage and the constriction of the Indonesian Seaway. This led to reduced exchange of warm water between the Pacific and Indian oceans, and the accumulation of warm water in the equatorial Pacific around 9.9–7.5 Ma (Gasperi and Kennett, 1993). It is likely that El Niño events intensified on account of the reduced exchange when these gaps closed.

Even more comprehensive was the closure of the gap between North and South America 3.7–3.1 Ma. Most studies have focused on the ensuing terrestrial interchange of biotas (Stehli and Webb, 1985). For this reason, an example is given below of the effects on marine animals in the Pacific; these effects resulted not only from the cessation of ocean exchange but also from the resulting change of environments.

7.5.9 Marine Animals of the Neogene Pacific Ocean

The open seaway between North and South America during the middle Tertiary introduced a number of new marine animals to Pacific waters. Among these were an order of herbivorous mammals called *Sirenia* or sea cows which had entered the Pacific by the early middle Miocene.

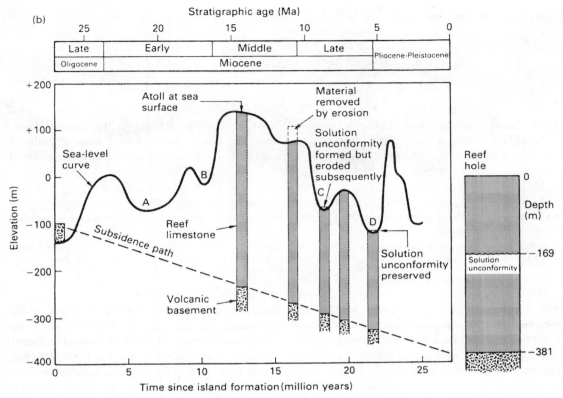

Figure 7.6   Tertiary sea-level change (after Vail and Hardenbol, 1979) compared with the reef stratigraphy of Midway Island (after Lincoln and Schlanger, 1987). Solution unconformities formed during low sea-level stages A, B and C were obliterated during lowering of the atoll surface by subaerial processes during low sea-level stage D. The solution unconformity formed at this stage has remained preserved since subsidence carried it below the reach of subsequent periods of atoll surface lowering

From Nunn (1994a)

These *Sirenia* were adapted to warm-water living and fed on seagrasses in sheltered bays along the then deeply embayed California–Mexico coast. Subsequent uplift led to a straightening of this coast and required sirenians to migrate northwards in search of more suitable feeding grounds, and to adapt their feeding habits. Only one species of sirenian adapted successfully to the cold conditions of the northernmost Pacific. This species adapted to feeding on marine algae and left descendants (*Hydrodamalis*) in the later Cenozoic. By the late Pliocene, hydrodamalines lived along north Pacific coasts from Japan to Mexico yet were gradually rendered extinct by humans; the last population was extirpated in the

Bering Sea between AD 1741 and 1768 (Domning, 1987; Plate 7.3).

Another example is provided by monk seals (*Monachus* spp.) which live today in the central Pacific only in the northwestern islands of the Hawaii group (Repenning and Ray, 1977). In evolutionary terms, the Hawaiian monk seal is more primitive than any of its relatives, living or fossil. This suggests that this seal, which originated in the central Atlantic, passed through the seaway between North and South America some time before 14 Ma (the age of the oldest known fossil monk seal). Once in the Pacific, some monk seals went south, following the nutrient-rich Peru Current to the Antarctic, where

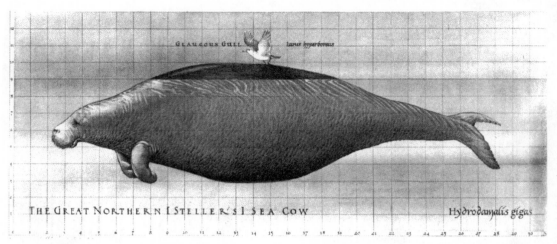

Plate 7.3   Steller's sea cow (*Hydrodamalis gigas*), the largest of the Sirenia in the north Pacific, reached lengths of 8 m (~25 feet). A sought-after source of food for fur hunters, the animals became extinct in about 1768
Daryl Domning

their descendants live still. Other monk seals followed the then course of the North Equatorial Current to Hawaii. Of the linear group of Hawaiian islands, only those north and west of Laysan, most of which are now submerged, would have existed 14 Ma. At the time the monk seals reached these islands, they lay within the North Equatorial Current but, since that time, they have been moved northwest out of it on the mobile Pacific Plate. The result is that the monk seals, which have occasionally tried unsuccessfully to migrate south back into the warm current, have become progressively fewer in number owing in part to their inability to cope with the cooler conditions and the comparative barrenness of the ocean waters they inhabit.

PART D

The Pacific During the Pleistocene Epoch

CHAPTER 8

Understanding the Chronology of the Pleistocene (1.77–0.012 Ma)

... existentialists declare
That they are in complete despair,
Yet go on writing.
W.H. Auden, *Under Which Lyre* (1946)

8.1 ORGANIZATION OF THE PLEISTOCENE

The Pleistocene Epoch comprises the main part of the Quaternary Period. The other part, the Holocene (0.012 Ma–present), is discussed in Part E (Chapters 12–15) of this book. The relationship of these divisions to the entire Cenozoic Era is shown in Figure 6.1.

A multiplicity of chronological schemes for the Pleistocene no doubt puts many budding researchers off beginning to study this most worthwhile of fields. Even seasoned investigators may become confused when they move to study the subject in a new part of the world where stage names may be different and/or differently defined. This is especially true when terrestrial workers become involved in the world of the marine geologist or *vice versa*. To avoid such confusion, stage names used for the late Quaternary in this book are listed along with some common (regional) equivalents in Table 8.1. A standard Pleistocene chronology for the entire Pacific Basin is premised on the idea that, notwithstanding minor local variations, the Pleistocene climate of the region has varied synchronously. The stage names adopted are thus deliberately free of regional connotations.

The degree of detail given about the Quaternary (Pleistocene and Holocene) in the rest of this book is justified in three ways.

Firstly, the Quaternary was a time of climate change which was expressed in the environment in ways which may have been unprecedented in the history of the Earth. Of course, given our inability to focus as clearly on earlier times as we can on the Quaternary, it would be foolhardy to insist on this. Indeed, recent discoveries about the nature of cooling during the latest Miocene and late Pliocene (Chapter 7) suggest that the nature of climate change at these times may have been analogous to that of the Quaternary. Whatever the reality, the fact remains that the unusual nature of Quaternary climate change is in itself sufficient justification for a detailed discussion.

Secondly, the Quaternary is that period of time about which we have most information: an overwhelming amount compared to that for earlier times. The potential of using that information to shed light on the ways in which Pacific Basin environments have changed in the past is vast, and cannot therefore be ignored or underrated.

Finally, an understanding of the Quaternary history of the Pacific Basin has direct implications for an understanding of its modern environments and their sustainable management. In the last few years it has become clear that an understanding of past environmental changes and their causes is perhaps the most important preliminary to understanding properly the nature of the environmental changes which may occur in the

Table 8.1   Names used for divisions of the late Quaternary in this book, their absolute ages (from oxygen isotopes), selected local stage names regarded as approximately synonymous, and association with the oxygen-isotope chronology (Table 8.2)
Major sources of data are Arkhipov et al. (1986), Bowen et al. (1986), Korotkii (1985), Salinger (1983), Sibrava (1986) and Suggate (1990)

Age (cal. yr BP)	Stage name	Synonyms	Oxygen-isotope stage
	Holocene	Aranui, Flandrian, postglacial	Stage 1
12 050			
	Last Glacial	Devensian, Otira, Tali, Valdai, Weichsel, Wisconsin, Würm, Zyrianka	Stage 1–5d
111 790			
	Last Interglacial	Eem (Eemian), Ipswichian, Kaihinu, Kazantsevo, Mikulino, Oturi, Sangamon	Stage 5e
128 000			
	Penultimate Glacial	Late Illinoian, Lushan, Moscow, Ossor, Waimea, Warthe	Stage 6
194 000			
	Penultimate Interglacial	Illinoian, Karoro, Odintsovo, Saale, Terangi	Stage 7
258 000			
	Antepenultimate Glacial	Dneiper, Drenthe, Early Illinoian, Taku, Waimaunga	Stage 8
313 000			
	Antepenultimate Interglacial	Domnitz, Romny, Waiwhera	Stage 9
359 000			
	(glacial)	Fuhne, Nemona, Orchik	Stage 10
386 000			
	(interglacial)	Holstein, Lichvin, Ob (Tobol?)	Stage 11
430 000			
	(glacial)	Oka	Stage 12
486 000			
	(interglacial)		Stage 13
521 000			
	(glacial)		Stage 14
544 000			
	(interglacial)	Cromerian	Stage 15
589 000			
	(glacial)	Beestonian	Stage 16
622 000			
	(interglacial)	Pastonian	Stage 17
658 000			
	(glacial)		Stage 18
695 000			
	(interglacial)		Stage 19
780 000			
	(glacial)		Stage 20
797 000			
	(interglacial)	Arten	Stage 21
849 000			

future as the consequence of causes with which we are much more familiar.

8.2 QUATERNARY CYCLES

Many environmental variables followed cyclic patterns of change during the Quaternary. The most fundamental control of this cyclicity is the variation in the amount of solar radiation at the Earth's surface arising from variations in the Earth's orbit around the Sun. The cyclicity is expressed most clearly by temperature changes – from glacial to interglacial to glacial – and by the sea-level changes driven by them – from low to high to low.

There are three basic orbital cycles. The orbital-eccentricity cycle involves the stretching of the Earth's orbit every 100 000 years or so. Then there is the 41 000-year axial-tilt (obliquity) cycle, the product of the Earth rolling like a ship. Finally, there is the 23 000-year precessional cycle which causes the time when the Earth is nearest the Sun (perihelion) to vary as the result of a wobble, as on a spinning top. The combination of the three cycles – 'stretch, roll and wobble' in Nigel Calder's (1974) memorable phrase – allowed Milankovitch (1930) to 'predict' variations in solar radiation over the past 650 000 years. Milankovitch's predictions waited several decades before being demonstrated as correct by ocean-core data and firmly established orbital variations as the 'pacemaker of the ice ages' (Hays et al., 1976).

During the Quaternary, the relative dominance of particular orbital cycles changed. From the establishment of the first northern hemisphere ice sheets about 2.4 Ma until about 0.85 Ma[1], the dominant control was the 41 000-year axial-tilt cycle (Ruddiman et al., 1986). Since 0.85 Ma, the orbital-eccentricity cycle has dominated, glacials occurring every 100 000 years or so. Yet some environmental variables have also been found to change according to the 23 000-year precessional

cycle (Frakes et al., 1992). Still other cycles become evident when the spotlight moves to the most recent Quaternary interglacial, the Holocene (see Part E of this book).

8.3 DIVISION OF THE PLEISTOCENE

The Pleistocene began 1.77 Ma at the top (end) of the Olduvai (C2n) subchron and was marked by a reversal in the normal polarity of the Earth's magnetic field. This time is also well marked in the biochronology of planktonic foraminifera being the first appearance datum (FAD) of *Globigerina cariacoensis* and the last appearance datum (LAD) of *Globigerina extremus*.

The Pleistocene is subdivided in this book largely for ease of discussion. Minor adjustments to chronologies derived from terrestrial data have been made to make these coincident with orbital cycles. Such 'orbital tuning' is now well established and underpins the recent, most authoritative chronologies for the late Cenozoic (Berggren et al., 1995; Shackleton et al., 1995) which are used as the basis for subdividing the Pleistocene here.

The basic division of the Pleistocene used is threefold. The early Pleistocene began 1.77 Ma and ended 0.78 Ma. The latter time marks the boundary between the Matuyama and Brunhes chrons when magnetic polarity changed from reversed to normal (see Table 2.3). The middle Pleistocene began 0.78 Ma and ended 0.128 Ma at the beginning of the Last Interglacial. The late Pleistocene began 0.128 Ma and ended 0.012 Ma at the start of the Holocene, the interglacial in which we are now living.

For the sake of easy reference, dates of Quaternary oxygen-isotope stages, referred to throughout this part of the book, are listed in Table 8.2. Not everyone agrees on the ages of these stages so it is deemed important to state those ages used here. Some oxygen-isotope stages are delimited by biochronological markers listed in Table 8.3.

---

[1] In the view of Frakes et al. (1992), this transition was more gradual, lasting from ~0.9 to 0.6 Ma. It is not used as a chronological marker in this book for that reason

Table 8.2   Ages of Quaternary oxygen-isotope stages used in this book. Geomagnetic controls on age are given as footnotes and tabulated in Table 2.3, biochronological controls used are listed in Table 8.3
Stage boundaries 1/2 to 5d/5e are in calendar years BP marking events as noted from Martinson et al. (1987); all other stage boundaries in Ma recalculated after Williams et al. (1988), Shackleton et al. (1990) and Berggren et al. (1995)

Stage boundary	Age (Ma except where stated otherwise)	Stage boundary	Age (Ma except where stated otherwise)
1/2	12 050 cal. yr BP (event 2.0)	30/31[3]	1.053
2/3	24 110 cal. yr BP (event 3.0)	31/32	1.091
3/4	58 960 cal. yr BP (event 4.0)	32/33	1.132
4/5a	73 910 cal. yr BP (event 5.0)	33/34	1.175
5a/5b	79 250 cal. yr BP (event 5.1)	34/35	1.204
5b/5c	90 950 cal. yr BP (event 5.2)	35/36[4]	1.240
5c/5d	99 380 cal. yr BP (event 5.3)	36/37	1.253
5d/5e	110 790 cal. yr BP (event 5.4)	37/38	1.272
		38/39	1.285
5e/6	0.128	39/40	1.302
6/7	0.194	40/41	1.309
7/8	0.258	41/42	1.340
8/9	0.313	42/43	1.352
9/10	0.359	43/44	1.370
10/11	0.386	44/45	1.384
11/12	0.430	45/46	1.396
12/13	0.486	46/47	1.406
13/14	0.521	47/48	1.428
14/15	0.544	48/49	1.438
15/16	0.589	49/50	1.449
16/17	0.622	50/51	1.459
17/18	0.658	51/52	1.494
18/19	0.695	52/53	1.524
19/20[1]	0.780	53/54	1.546
20/21	0.797	54/55	1.557
21/22	0.849	55/56	1.571
22/23	0.881	56/57	1.598
23/24	0.903	57/58	1.619
24/25	0.922	58/59	1.626
25/26	0.958	59/60	1.671
26/27[2]	0.974	60/61	1.691
27/28	1.003	61/62	1.716
28/29	1.015	62/63	1.740
29/30	1.034	63/64[5]	1.770

[1] Base of the Brunhes chron (0.78 Ma)
[2] Middle stage 27 is the top of the Jaramillo subchron (0.99 Ma)
[3] Middle stage 31 is the base of the Jaramillo subchron (1.07 Ma)
[4] Base of the Cobb Mountain subchron (1.24 Ma)
[5] Top of the Olduvai subchron (1.77 Ma)

Table 8.3 Selected highlights of Quaternary biochronologies used in establishing the oxygen-isotope chronology used in this book (see Table 8.2) (FAD, first appearance datum; LAD, last appearance datum)
After Berggren et al. (1995)

Datum event	Correlation with oxygen-isotope stratigraphy	Age (Ma)
1. Calcareous nannofossils		
Emiliania huxleyi FAD	Stage 8	0.26
Pseudoemiliania lacunosa LAD	Stage 12	0.46
2. Planktonic foraminifera		
Globorotalia flexuosa LAD	Stage 4	0.068
Bolliella calida FAD	Stage 7	0.22
Globoquadrina pseudofoliata LAD	Stage 7	0.22
Globorotalia flexuosa FAD	Stage 11	0.401
Globorotalia hirsuta FAD	Stage 12	0.45

CHAPTER 9

The Early and Middle Pleistocene (1.77–0.128 Ma) in the Pacific Basin

9.1 INTRODUCTION

The amount of information available for understanding what happened during the early and (to a lesser extent) the middle Pleistocene is much less than for the latest Pleistocene and Holocene. It is therefore difficult to identify with comparable precision those early and middle Pleistocene events to have affected the entire Pacific Basin. This difficulty is compounded by the comparatively low amplitude and shorter duration of the climate cycles during the early and middle Pleistocene (see Chapter 8). Despite these problems, a coherent picture of environmental change in the Pacific Basin emerges for the early and middle Pleistocene.

Following a short review of the character of the Pacific Basin at the start of the Pleistocene, this chapter discusses early and middle Pleistocene environmental changes in the Pacific Basin by each of its constituent regions. There follows a summary of the principal environmental changes to have occurred during the early and middle Pleistocene in the Pacific Basin, and a preview of the discussion of the late Pleistocene in the following chapters.

9.2 THE PACIFIC BASIN AT THE BEGINNING OF THE PLEISTOCENE

The configuration of land and ocean in the Pacific Basin at the beginning of the Pleistocene was similar to the present configuration, especially if the comparison is made with the entire Phanerozoic history of the region in mind. The outlines of the continents were similar, the topography of the ocean floor was almost the same.

Ocean gaps which had existed during the Neogene along the Bering Strait, through Panama, and between Australia and southeast Asia had effectively closed by its end, conferring on the north Pacific Ocean in particular a degree of isolation unsurpassed in its earlier history. Conversely, ocean gaps in the southernmost Pacific – the Tasman Seaway in the west and Drake Passage in the east – which had opened during the Tertiary, were now well established conduits for deep-water movement and exerted important controls on the climate and ocean circulation of the Pacific Basin.

Most mountain ranges in the Pacific Rim were established by the beginning of the Pleistocene; many continued to grow subsequently, producing or enhancing associated environmental changes. Lowland areas of the Pacific Rim, notably that in southern China, were likewise established by the beginning of the Pleistocene but fluctuated markedly in extent thereafter as sea level alternately rose and fell.

The vast majority of Pacific islands which exist today had some expression at the beginning of the Pleistocene. Many islands either grew or became reduced in size during this time as a result of tectonism or volcanism.

The similarity in gross form between the modern Pacific Basin and that which existed at the beginning of the Quaternary demonstrates the point that there is little need – as was the case for earlier times – to consider long-term processes such as plate movements as agents of environ-mental change over this time period.

9.3 OVERVIEW OF EARLY AND MIDDLE PLEISTOCENE ENVIRONMENTAL CHANGE IN THE PACIFIC BASIN

The terrestrial legacy of glacial–interglacial alternations experienced during the early and middle Pleistocene has generally been obliterated by later processes so it is the oxygen-isotope record from deep-ocean cores to which we must turn in order to recognize them (see Table 8.2). Nowhere in the terrestrial records of environmental change is the erasure of earlier change by later processes as apparent as with glaciation. Former ice limits in a particular area can be mapped relatively easily if no ice covered this area subsequently but, where there has been a multiplicity of ice advances – as is the case in most such parts of the Pacific Rim – the identification of the earliest poses a challenge which may of necessity prove insurmountable.

Sea level oscillated with the same frequency as did temperature throughout the Pleistocene. Yet the record of early and middle Pleistocene sea-level change is patchy for the same reasons as is that of climate change. The most recent swings of sea level have tended to remove or disguise – almost beyond recognition – the record of earlier fluctuations. The main exceptions are places where the land has been rising or sinking at rates which have prevented the removal of such records during later transgressions or regressions respectively (see Figure 7.6).

The effects of tectonism and volcanism are easier to detect for many parts of the early and middle Pleistocene in the Pacific Basin. Although in most cases the environmental effects of these phenomena were of only local extent, they often allow worthwhile insights into the nature of early and middle Pleistocene environments.

9.3.1 Climate and Ice

The cryosphere – that part of the Earth covered with ice – expanded during Quaternary glacials

and shrunk during Quaternary interglacials. The global cryosphere typically expanded during glacials from around $15 \times 10^6 \, \text{km}^2$ to around $40 \times 10^6 \, \text{km}^2$ including $14 \times 10^6 \, \text{km}^2$ for Antarctica; the respective figures for water stored as ice are $30 \times 10^6 \, \text{km}^3$ and $90 \times 10^6 \, \text{km}^3$ (Williams et al., 1993).

During Pleistocene glacials, the main ice sheets in the Pacific Basin were in West Antarctica, which exists today in much the same form, and in North America, where ice sheets are largely absent today. The westernmost of the North American ice sheets was the Cordilleran, centred around the Rocky Mountains. This ran into the Keewatin ice sheet, covering most of western Canada, and several others. Together, these ice sheets are referred to as the Laurentide ice sheet (see Figure 11.2).

Despite its high latitude, most of northeast Asia (including western Beringia) experienced only periglacial conditions during Pleistocene glacials. Isolated ice bodies developed in and around mountain ranges here – along the spine of the Kamchatka Peninsula for example – as they did in other mountainous parts of the Pacific Rim. The largest southern hemisphere ice sheet was centred on the southernmost Andes. New Zealand was extensively glaciated during Pleistocene ice ages; smaller ice bodies existed on the islands New Guinea, Tasmania and Hawai'i.

Changes in climate and terrestrial ice extent during the Pleistocene had major impacts on biota[1]. During glacials, many thermophilic species became extinct as a result of cooling. Migrations of other species into the tropics at these times would also have had an impact on existing tropical biota. During interglacials, biota migrated back out of the tropics and into the warming and newly ice-free higher latitudes. Refugia – where particular species had weathered the glacial without becoming extinct or having to migrate – existed both within the tropics and in ice-free

---

[1] These impacts of Pleistocene glacial–interglacial alternations were not as great in the Pacific Basin as in Europe and much of Asia where north–south migration was hampered by east–west trending mountain ranges and seas

higher latitudes. In the Pacific Basin, Beringia and those parts of adjoining northeast Asia which were not glaciated provided refugia for particular species of plants and animals.

9.3.2 Sea Level

Early Pleistocene sea-level oscillations had smaller amplitudes and frequencies than those of the later Pleistocene (see Chapter 8). The largest body of information about early and middle Pleistocene sea levels has come from oxygen-isotope data. While there are dangers in uncritically regarding the latter as proxies of the former, this procedure has yielded the best information about sea-level change during the early and middle Pleistocene for which few other records are available.

The clearest terrestrial records of early to middle Pleistocene sea levels come from studies of rapidly rising coasts in the south and west

Pacific. Table 9.1 shows the most dependable heights of late middle Pleistocene sea levels (relative to the present) in the Pacific Basin. There is considerable variation in both the magnitude and timing of these, which seems unable at present to be reduced. This variation may be real, indicating that the oceanic geoid at these times had configurations which differed from its present one (Nunn, 1986), or it may be an outcome of insufficiently accurate dating.

9.3.3 Tectonics and Volcanism

Tectonic changes were important causes of environmental change in many parts of the Pacific Basin during the early and middle Pleistocene. In a general sense, rates of uplift were greatest near convergent plate boundaries while rates of subsidence – typically an order of magnitude or more lower – were sustained longest in intraplate situations. Intraregional variations in uplift rates

Table 9.1   Age and heights (relative to the present) of principal late middle Pleistocene sea-level maxima and minimum. For each location/stand combination, the upper figure is age (ka), the lower is elevation (m)

Location	Sea-level stand				
	High	High	High	High	Low
Huon Peninsula I, Papua New Guinea[1]	325 0	240 −30	215 −6	175 −24	150 −145
Huon Peninsula II, Papua New Guinea[2]					135 −130
Ryukyu Islands[3]			200 +10		
Tongatapu, Tonga[4]			225 +4		
Vanuatu[5]				175 −55	

[1] Chappell (1983)
[2] Orbitally tuned by Chappell and Shackleton (1986)
[3] Konishi et al. (1974)
[4] Taylor (1978)
[5] Lecolle and Bernat (1985)

Table 9.2   Representative rates of long-term uplift and subsidence operating during the Quaternary in the Pacific Basin. For details of Quaternary uplift along the Pacific coast of North America, see Table 9.3. For details of uplift along the Huon Peninsula, see Table 9.7. For details of uplift along the west coast of South Island, New Zealand, see Table 9.8. For Holocene magnitudes of Holocene coseismic uplift and subsidence in the Pacific Basin, see Table 15.11

Location	Uplift/subsidence rate (mm/year)	Source of information
UPLIFT		
Antarctica		
Marie Byrd Land	0.122	LeMasurier and Rex (1983)
Transantarctic Mountains	0.1	Fitzgerald et al. (1986)
South America		
Bolivian Andes	0.7	Goudie (1995)
Chala Bay, Peru	0.46	Goy et al. (1992)
Central America		
Baja California	0	Muhs et al. (1992a)
North America		
Coast, British Columbia	0.2–0.43	Parrish (1983)
Saint Elias Mountains	0.3	Parrish (1981)
East Asia		
Hateruma, Ryukyu Islands	0.1–0.3	Pirazzoli and Kawana (1986)
Australasia		
Huon Peninsula, New Guinea	1.0–3.0	Goudie (1995)
New Zealand (general)	0.2–0.5	Pillans et al. (1992)
Southern Alps, New Zealand	10.0	Goudie (1995)
Pacific Islands		
Anaa Atoll, Tuamotus	0.1	Pirazzoli et al. (1987)
Maré, Loyalty Islands	0.7	Dubois et al. (1974)
Torres Islands, Vanuatu	0.7–0.9	Taylor et al. (1985)
SUBSIDENCE		
Antarctica		
Marie Byrd Land	0.138	LeMasurier and Rex (1983)
North America		
South California (basins)	1.2–2.0	Kukal (1990)
Beringia		
Kamchatka	4.0	Kukal (1990)
East Asia		
Honshu, Japan (general)	1.0–1.2	Tjia (1970)
Australasia		
West North Island, New Zealand	0.1	Pillans (1986)
Pacific Islands		
Enewetak, Marshall Islands	0.1–0.2	Yonekura (1983)
Hawai'i, Hawaii Islands	4.4	Scott and Rotondo (1983)
Mururoa, Gambier Islands	0.12	Labeyrie et al. (1969)

have been noted in some regions and, like variations in volcanism, have been linked to changes in plate convergence rates. Selected rates of Quaternary uplift and subsidence in the Pacific Basin are given in Table 9.2.

Late Cenozoic variations in the frequency of volcanic eruptions were discussed in Chapter 7. The entire Quaternary falls within the last major volcanic pulse 5–0 Ma in the Pacific Basin (Cambray and Cadet, 1994). Within this pulse, variations in the frequency of volcanic eruptions in the Pacific Basin are thought to be linked to tectonic changes along convergent plate boundaries. This appears to be a Pacific-wide phenomenon and emphasizes the point that lithospheric deformation assists magma reach the surface. This process is explicit in several parts of the Pacific Rim. In Peru, for example, volcanism associated with extension occurs when the rate of convergence between the Nazca and South American plates is reduced. In the western United States and Japan, increases in eruptive frequency are linked to accelerations in subduction rates.

9.4   EARLY AND MIDDLE PLEISTOCENE ENVIRONMENTS OF ANTARCTICA

9.4.1   Climate and Ice

The modern East Antarctic ice sheet was established around 16 Ma. The smaller ice sheet in West Antarctica, which fronts the Pacific Ocean, was established 10–5 Ma. The East Antarctic ice sheet has almost certainly persisted since then although there is some question about whether this is also true of the West Antarctic ice sheet (see Chapters 10 and 11), most of which is grounded below sea level and is thus more vulnerable to fundamental change when sea level changes rapidly.

Although in East Antarctica and strictly applicable as a palaeoclimate indicator only to that region (which does not border the Pacific Basin), the results of ice coring at Vostok have been widely hailed as demonstrating a clear link

between terrestrial and oceanic climate changes (Lorius et al., 1985). A recent core has extended the Vostok record back 200 000 years (Jouzel et al., 1993). The record for the middle Pleistocene is shown in Figure 9.1 and confirms a prolonged cold period in East Antarctica 200–140 ka during the Penultimate Glacial.

For several reasons, including the suspicion that it has been a more transitory entity, the West Antarctic ice sheet has been the target of fewer palaeoclimate investigations than the East Antarctic ice sheet. Most work in West Antarctica has concentrated on its more accessible margins, mainly along the Antarctic Peninsula in the east and the border of the Ross Sea (especially around McMurdo Sound) in the west. Valley glaciers – some now gone (Plate 9.1) – feeding the Ross Ice Shelf from the Transantarctic Mountains have

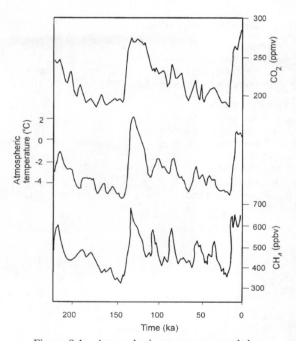

Figure 9.1   Atmospheric temperature and the concentrations of carbon dioxide (CO_2) and methane (CH_4) over the past 220 000 years. All data from analyses of ice cores from Vostok, East Antarctica. Reprinted with permission from *Nature* (Jouzel et al., Extending the Vostok ice-core record of palaeoclimate to the penultimate glacial period, volume 364, 407–212). Copyright (1993) Macmillan Magazines Limited

Plate 9.1   The Dais in the Wright Valley in Antarctica, one of the dry valleys entering the Ross Sea. Dates from moraines in these valleys have been extremely informative about the chronology of ice-sheet changes in both East Antarctica and the Ross Sea
Official United States Navy Photo by David Thompson

been especially sensitive to fluctuations of the East Antarctic ice sheet associated with Pleistocene climate changes. Some older moraines associated with these glaciers date from the early and middle Pleistocene. Using cosmogenic ^{3}He, the age of the Taylor IVb drift has been estimated as 1.2 ± 0.2 Ma. The Taylor IVa and III drifts date from the middle Pleistocene while the younger Meyer and Danum drifts, fringing the Beardmore and Hatherton glaciers respectively (see Figure

11.1), both perhaps formed during the Penultimate Glacial (Denton et al., 1989a; Brook et al., 1993).

9.4.2   Sea Level

An understanding of early and middle Pleistocene sea-level changes around Antarctica depends on records preserved in presently ice-free areas of the continent's periphery and in ocean-floor sediments. Yet there are distinct problems interpreting this record, in particular linking sea-level changes to ice-volume and climate changes. For example, a fall of sea level will produce an increase in the lateral extent of the Antarctic ice sheets, an increase in their convexity and thickening of ice cover in coastal areas which could readily be mistaken for a first-order response to climate change.

An exception is provided by the Ross Ice Shelf, the mass of currently floating ice filling the Ross Embayment. During Pleistocene glacials, low sea levels caused this ice shelf to become grounded on the floor of the embayment. Consequent thickening of the valley glaciers feeding it produced lateral moraines, which are now found along the sides of these glaciers. Dates for some of those which formed during the early and middle Pleistocene were given in the previous section.

Rapid uplift of parts of Pacific Antarctica during the Pleistocene has further confused the record of contemporary sea-level changes. For example, some of the early work on the dry valleys around McMurdo Sound interpreted the record of marine invasions in terms of sea-level change. Such interpretations have been revised following recognition that the area is being uplifted at around 0.15 mm/year.

Most observed shoreline displacements along the coast of Pacific Antarctica probably include the effects of both sea-level and tectonic changes. Few attempts have been made to disentangle the two yet there is some potential for isolating early and middle Pleistocene sea levels from such an exercise. One worthwhile site might be on Adelaide Island, where subpiedmont surfaces, cut at sea level, are now emerged several hundred metres, and talus slopes, now as high as 91 m above sea level, were trimmed by the ocean to rest at unusually high angles (Dewar, 1967a,b).

9.4.3   Tectonics and Volcanism

Despite isostatic depression caused by the weight of overlying ice, several parts of Antarctica were affected by uplift during the Pleistocene (Tingey, 1985). Unusually, little of that uplift appears to have been associated with either earthquake or volcanic activity, an observation which has been used to argue for elastic flexure of the continental plate(s) comprising Antarctica and, specifically, as an explanation for the uplift along the Transantarctic Mountains (Stern and ten Brink, 1989). A good example of the effects of this uplift along the Pacific Antarctic coast is from Wright Valley (see Plate 9.1), one of the much-studied dry valleys in McMurdo Sound, where gravels containing Pliocene marine diatoms deposited in 100 m of water are now 165 m above sea level. Even if these diatoms are not *in situ* (as appears likely, see Chapter 7), the rate of uplift of 0.15 mm/year which has been calculated to explain their emergence (Denton et al., 1991) is similar to rates derived for other parts of Pacific Antarctica by other means (see Table 9.2).

In West Antarctica, lithospheric extension continued to dominate during the Quaternary as it had earlier. The propagation of linear volcanic chains in Marie Byrd Land linked to extensional block faulting (LeMasurier and Rex, 1991; Plate 9.2) is one manifestation of this. Both uplift and subsidence characterized parts of the coast of Marie Byrd Land, which constitutes the central part of the Pacific front of Antarctica, during the Quaternary (see Table 9.2).

9.5   EARLY AND MIDDLE PLEISTOCENE ENVIRONMENTS OF SOUTH AMERICA

9.5.1   Climate and Ice

The main ice-covered areas of the Andes today include the Cordilleras Occidental, Central and Oriental in the north, the Cordillera Blanca in

Plate 9.2   Basaltic hyaloclastites at Mount Murphy, Marie Byrd Land, West Antarctica. Mount Murphy is a Tertiary volcano which experienced volcanic activity beneath ice about 27 Ma at the time of the middle Oligocene climatic deterioration (which indicates that mountain ice caps existed in West Antarctica at the time) and then again about 0.9 Ma. The most recent eruptive activity occurred along the top of the ridge in this photo
Wesley LeMasurier

Peru, and in the south in Patagonia. During Pleistocene glacials, Andean snowlines were lowered several hundred metres and large ice caps, elongated north–south, developed.

In contrast to the northern hemisphere, where late Pleistocene glaciations were generally greater in extent than those of the early and middle Pleistocene, the glaciation of greatest extent in Patagonia occurred 1.2–1.0 Ma (Mercer, 1976). Although evidence of former glacial extent using moraines is notoriously difficult to interpret correctly, several groups of early Pleistocene moraines associated with this glaciation were mapped in southern Argentina by Sylwan (1989). Many of the difficulties which previous researchers encountered trying to distinguish moraines of particular ages were overcome by

Porter (1981) in his study of glacial deposits in the Chilean Lakes region. Using a variety of weathering and erosion criteria, Porter mapped four 'drifts' representing early to middle Pleistocene first-order glaciations (Figure 9.2). Contemporary glaciations probably occurred elsewhere in the Andes (Clapperton, 1993a).

The most widespread middle Pleistocene glacial event in the central Andes probably occurred during the Penultimate Glacial although, as Clapperton (1993a: 366) cautioned, 'there is little direct proof that any sedimentary formations assigned to the Penultimate Glacial were actually deposited at that time'; it is simply that they are older than those of the Last Glacial and unlikely to be of early Pleistocene age. Among the best candidates for a middle Pleistocene glacial deposit

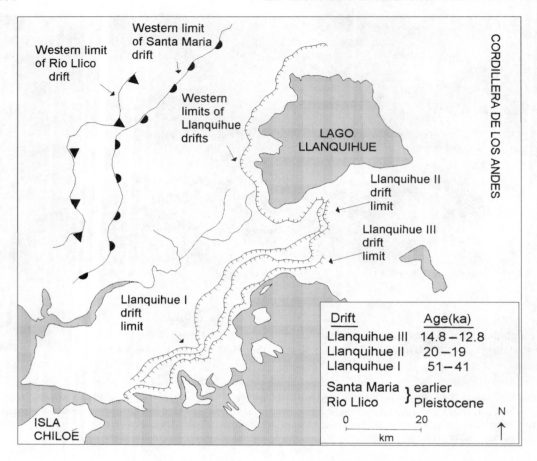

Figure 9.2   Limits of Pleistocene glacial drifts around Lago Llanquihue in the Chilean Lakes
After Porter (1981) and Clapperton (1993a)

farther south is the till sheet, perhaps 170–130 000 years old, which outcrops widely in Patagonia (Clapperton, 1983).

The warmest time of the entire Quaternary in northern Chile (and probably the entire southern hemisphere) occurred ~0.4 Ma (oxygen-isotope stage 11). This conclusion was reached following discovery of an anomalously warm-water molluscan assemblage in emerged nearshore deposits on the Mejillones Peninsula (23°S). The coexistence of cool-water fauna suggests that coastal upwelling was still prevalent at this time and that the warmest waters were to be found in lagoons and protected bays (Ortlieb et al., 1996a).

Glacial sediments were not the only kind to

have formed during the Pleistocene in the Andes. A loess named *cangagua* mantles several areas of the northern Andes. The contained fossils include those of megafauna such as *Mastodon* and *Equus*, and a humid-loving beetle *Deltochinum deltipes*, the presence of which suggests that parts of the *cangagua*, which may be early to middle Pleistocene in age, formed in an interglacial environment where temperatures were slightly higher than today (Iriondo, 1994).

9.5.2   Sea Level

The widespread tectonic instability of most of the Pacific coast of South America during the

Pleistocene means that the record of contemporary sea-level changes is more difficult to interpret than along stable coasts. Both the great magnitude of uplift and its variations over comparatively short distances have frustrated attempts to reconstruct the early and middle Pleistocene history of sea-level change in this region.

Specifically, the problem is that there have been so many uplift events in the region that often many more surfaces have been uplifted than there have been interglacials. This makes correlation with high sea-level stands difficult, especially given the frequent absence of diagnostic deposits on the emerged surfaces. A recent attempt to overcome this problem was based at a part of the Peru coast where Pleistocene uplift rates were much lower than elsewhere (Goy et al., 1992). Early Pleistocene marine terraces exist 200–274 m above sea level. Terraces formed during the last four interglacials are found at 20–200 m. Investigations of a similar site in northern Chile led Ortlieb et al. (1996b) to conclude that there had probably been two sea-level maxima during the Penultimate Interglacial.

It is clear in both South America and Antarctica where, for different reasons, there is a lack of intelligible data relating to early and middle Pleistocene sea-level history, that the forthcoming understanding of this will depend on data acquired elsewhere in the world. Although presently uncontroversial, this approach may eventually prove problematic because it does not allow for independent demonstration. A few decades ago, a similar approach involving the 'correlation' of shorelines worldwide on the basis of their elevation led to the collapse of an orthodoxy which had ruled geomorphology for 40 years (Chorley, 1963).

9.5.3 Tectonics and Volcanism

Much of Pacific South America was characterized by comparatively rapid uplift during the Pleistocene, a process which has been continuing for several million years as a consequence of the subduction of the Nazca Plate beneath the South American Plate at a rate of around 8 cm/year (Pardo and Molnar, 1987). The highest Pleistocene shoreline is now 780 m above sea level in the San Juan-Marcona region of Peru; its unusually rapid uplift has been attributed to the subduction of the aseismic Nazca Ridge (Macharé and Ortlieb, 1992).

The 27 marine terraces bordering Chala Bay in southern Peru provide one of the best chronologies of Pleistocene uplift on account of the preservation of alluvial and colluvial units with each terrace (Goy et al., 1992). The terraces were grouped and assigned to various Pleistocene interglacial and interstadial stages. Shorelines emerged 184 m, 168 m and 121 m are thought to date from the three youngest middle Pleistocene interglacials. This assumption leads to the derivation of a Pleistocene uplift rate of 0.46 mm/year which is greater than in surrounding coastal areas (0.27–0.35 mm/year) but less than in places where the aseismic Nazca Ridge is being subducted (maximum 0.74 mm/year).

Another important site occurs farther south in the Hornitos area where Ortlieb et al. (1996b) found evidence that local tectonic processes (uplift and warping) have been superimposed on larger-scale regional uplift producing uplift averaging 0.24 mm/year for the last 330 ka.

Considerable warping is evident elsewhere along this coast. A middle Pleistocene terrace along the Altos de Talinay in northern Chile varies in elevation from 220 m to 675 m above sea level within just 50 km (Ota et al., 1995). The absence of warping of lower terraces has been interpreted to mean that middle Pleistocene coupling of the Nazca and South America plates was more intense than that during the later Pleistocene – a situation which is the reverse of that along most circum-Pacific subduction zones. The same explanation can be used for the 'canyon-cutting event' in the Patagonian Andes during the Penultimate Interglacial. An increased rate of uplift is believed to have caused incision of a new valley network, which produced new outlets for the discharge of meltwater from the icefields and thus changed the region's landscape profoundly (Rabassa and Clapperton, 1990).

How representative coastal uplift is of that inland along the main axis of the Andes is

debatable but the prevailing view at present is that
– at least in their south-central part – the Andes
have not been uplifted much since the last major
episode in the late Miocene (Kött et al., 1995).

Volcanism occurred along most parts of the
Pacific Rim in South America during the early and
middle Pleistocene. Indeed, the oldest known
volcanic structures in this region formed only
during the later Pleistocene. The periods of
greatest activity in the Peruvian Andes occurred
in response to increased lithospheric compression
(Petford and Atherton, 1992). Volcanism was also
associated with extension, particularly at times
when the rate of convergence between the Nazca
and South American plates fell (Mégard et al.,
1984).

9.6   EARLY AND MIDDLE PLEISTOCENE ENVIRONMENTS OF CENTRAL AMERICA

The major tectonic and volcanic changes to occur
during the Quaternary in Central America were the
outcome of the subduction of the Cocos Plate just
offshore its Pacific margins.

Two volcanic chains developed in Central
America during the Quaternary. The southern
one stretches from central Costa Rica north to
Tacaná Volcano on the Mexico–Guatemala
border. The trend of the northern 'trans-Mexico'
chain is not parallel to the trench axis leading to
suggestions that it may follow one or more
fractures produced as a result of the complex
nature of plate convergence across this region
(Dengo, 1985).

9.7   EARLY AND MIDDLE PLEISTOCENE ENVIRONMENTS OF NORTH AMERICA

9.7.1   Climate and Ice

The landscape of North America owes most of its
present outline to the effects of intermittent
glaciation and periglaciation during the
Pleistocene. Although local glaciations are known
from before the start of the Pleistocene, the oldest

continental glaciation in North America dates
from its early part (Mathews and Rouse, 1986).
During most Pleistocene glacial maxima, the
Cordilleran ice sheet covered the entire Pacific
Rim in Canada and part of the northwest United
States (see Figure 11.2). As elsewhere, much of
the early and middle Pleistocene record has been
obliterated by the effects of more extensive glacial
fluctuations during the late Pleistocene. In most of
Pacific Canada, for example, reasonable evidence
of ice advance is available only for the last three
glaciations.

Interglacial records in most of North America
are obscured for much the same reasons as is the
record of early and middle Pleistocene glacial
advances. One exception comes from the record of
localized glaciation in the Sierra Nevada where
several early and middle Pleistocene glacial drifts
are overlain by lavas; K-Ar dates from these
provide minimum ages for the associated glaciations, which appear to have been synchronous with
expansions of the Laurentide ice sheet to the north
(Izett and Naeser, 1976). Another striking
exception is provided by the three periods of
speleothem formation in caves in the Rocky and
Mackenzie Mountains >350 ka, 320–275 ka and
235–185 ka, thought to correspond to middle
Pleistocene interglacials when conditions were
wet enough for speleothems to form (Harmon et
al., 1977).

In areas south of those covered by ice sheets
during contemporary glaciations, the record of
early Pleistocene climate changes is fragmentary
and frequently overprinted by large intraregional
variations attributable to tectonics, an example of
which is discussed below.

9.7.2   Climate change influenced by Tectonics in the Great Basin

The southern part of the Pacific Rim in the United
States lay south of the limit of continental ice
sheets during the Pleistocene. Yet the area
contains two large mountain ranges – the Sierra
Nevada and the Transverse Ranges – which
experienced considerable uplift during this epoch.
The south-central Sierra Nevada has risen around

600 m in the past 2 million years; the Transverse Ranges have risen as much as 3000 m in half that time. Many climate changes recorded during the early and middle Pleistocene in this part of the Pacific Rim are attributable to the effects of this uplift rather than the effects of regional climate changes.

The first hint that uplift may have been the dominant cause of Pleistocene climate change in this area was found by Winograd et al. (1985) who showed that the amount of deuterium in groundwaters of the Great Basin – sandwiched between the Sierra Nevada in the west and the Rocky Mountains in the east – had decreased throughout the Pleistocene. This they attributed to the effects of progressive uplift of the mountains which today cause deuterium in west–east air streams to become significantly depleted.

The rise of these mountain ranges would also have drastically reduced the amount of precipitation reaching the Great Basin both by increasing orographic precipitation on west-facing slopes and by blocking the eastward progress of Pacific storms. Increasing Pleistocene aridity of the Great Basin is marked most clearly in records from old lake deposits, notably Lake Tecopa, which was dry 1.6–0.86 Ma.

Recent work from nearby Owens Lake has illuminated the Quaternary history of the Great Basin still further, particularly in regard to its history of effective precipitation (Smith et al., 1997). The first signs of glaciation in the adjacent Sierra Nevada date from >800 ka, a time which saw Owens Lake choked with sediment. Later, 810–645 ka, the lake became several metres deep and supported a biota characteristic of fresh, often cool, water. The lake became shallower 645–450 ka but then deeper for most of the time 450–5 ka. Superimposed on these trends are changes in water volume which follow the 100 000-year orbital-eccentricity cycle.

The palaeoclimate record obtained from diatom analysis of a core from Tule Lake farther north in the Great Basin has also proved illuminating (Bradbury, 1992). Variations in the proportions of the fresh-water diatom species *Fragilaria*, indicating fresh-water swamp conditions, and

Stephanodiscus show a good correlation with oceanic oxygen-isotope records. Particularly during the middle Pleistocene, *Fragilaria* showed pulse-like increases during arid glacial climates, while *Stephanodiscus* was associated with warmer, generally moister conditions.

9.7.3 Sea Level

Most work on pre-Holocene sea-level change in this region has focused on the Last Interglacial, discussed in Chapter 10. Comparatively little is known about the earlier Pleistocene largely because of an absence of material suitable for uranium-series dating in particular. One exception is the work of Ludwig et al. (1992) who dated emerged shorelines along the California coast by correlating $^{87}Sr/^{86}Sr$ variations in cover deposits with better dated submerged reefs off the islands of Hawaii. By combining these data with probable uplift rates, four shorelines were dated to the early and middle Pleistocene.

9.7.4 Tectonics and Volcanism

The effects of Pleistocene uplift of mountain ranges on the climate history of the Great Basin were discussed above. The effects of Pleistocene uplift are also well marked along the coast (Table 9.3). The effects of lateral movements are less conspicuous in the area, and have had less of an effect on its gross Pleistocene landscape development.

Successive periods of growth and decay of the Cordilleran ice sheet were responsible for the isostatic depression and rebound of the lithosphere in Canada which produced emerged shorelines up to 200 m above their modern counterparts along the coast of British Columbia, for example (Clague, 1991).

Tectonic changes associated with isostatic recovery following deglaciation are quite different from those in most other parts of the Pacific Basin where deformation or rupture occurred because of stress between or within lithospheric plates. Uplift following deglaciation affects large areas – some 14×10^6 of North America during the Holocene –

Table 9.3   Selected rates of Quaternary uplift along the Pacific Rim of North America

Location	Latitude	Uplift rate (mm/year)	Source of data
San Diego	32°50'N	0.13–0.14	Kern and Rockwell (1992)
California, typical	~33°N	0.15–0.35	Muhs et al. (1992a)
Transverse Ranges[1]	34°N	3.0	Smith et al. (1993)
San Luis	35°15'N	0.14–0.23	Hanson et al. (1992)
San Simeon	35°45'N	0.12–0.27	Hanson et al. (1992)
Sierra Nevada[1]	37°N	0.3	Smith et al. (1993)
Cascadia frontal arc	43°N	0.45–1.08	Muhs et al. (1992a)
Vancouver Island	50°N	0.2	Parrish (1983)
Coastal mountains, British Columbia	52°N	0.2–0.43	Parrish (1983)
Queen Charlotte Islands	53°N	0.1	Parrish (1983)

[1] Inland mountains, not coastal

and is not as variable in time and space as uplift caused in other ways. Yet rebound following recession of the Last Glacial Cordilleran ice sheet destroyed the tectonic legacy of earlier glacial isostatic changes. It is therefore only for the Holocene that such tectonic effects can be clearly identified and interpreted.

The complex plate interactions which occurred (and continue to occur) along the Pacific coast of Canada in the early and middle Pleistocene left a strong imprint on the landscape (Adams and Clague, 1993). Uplift and warping combined to produce incision of river valleys and to cause low-level surfaces and deposits to be elevated. Many such deposits lie on peneplains cut by rivers draining the Cordillera in pre-Pleistocene times.

As is clear from Figure 9.3, Pleistocene and Holocene volcanism in the Pacific Rim part of the United States followed a pattern established by about 5 Ma (see Chapter 7). As in West Antarctica and Central America, this pattern is based on a straight-line grid believed to be a response to structural deformation (Luedke and Smith, 1991). For much of the Quaternary, volcanism along the Pacific Rim in Canada was concentrated in four major belts. In each, volcanic activity has been caused by a different mechanism (Adams and Clague, 1993). The Anahim belt in southern British Columbia marks a hotspot which has been crossed by the North American Plate. In the same area the Garibaldi belt is a volcanic arc related to

the nearby subduction of the Juan de Fuca Plate. The Stikine belt of northern British Columbia is associated with lithospheric extension. The Wrangell belt in the north is an extension of the Aleutian island arc.

9.8   EARLY AND MIDDLE PLEISTOCENE ENVIRONMENTS OF BERINGIA

9.8.1   Climate and Ice

Although northern hemisphere glaciation began around 2.4 Ma, the time at which an ice sheet developed first in the Arctic Ocean and Bering Strait is less certain. Evidence from animal remains in northern Alaska suggests that sea ice covered much of the Arctic Ocean for less than one month each year 2.5–1.6 Ma (Repenning et al., 1987). These comparatively mild conditions were followed by cooler conditions associated with the development of a perennial sea-ice cover around 0.4 Ma (Scott et al., 1989).

The free movement of terrestrial biota across the Bering land bridge during the Pleistocene was impeded both by interglacial high sea levels and by glacial ice cover. Knowledge of the timing of Quaternary biotic interchange through Beringia thus has considerable potential to throw light on the chronology of changes in these various climate controls. It is possible that the Bering land bridge

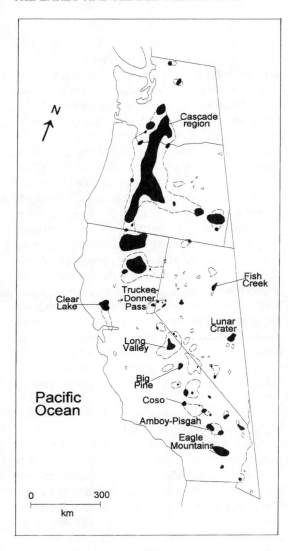

Figure 9.3   Pleistocene and Holocene volcanism (black) in the Pacific Rim sector of the United States compared to that which occurred during the earlier Cenozoic (bounded by broken lines)
From *Quaternary Nonglacial Geology; Conterminous U.S.* (Morrison, R.B., editor), Luedke, R.G. and Smith, R.L. Modified with permission of the publisher, the Geological Society of America, Boulder, Colorado, USA. Copyright © 1991 Geological Society of America

Table 9.4   Times of known movements of microtine rodents across the Bering land bridge during the early and middle Pleistocene
Repenning (1996)

Genus	Time of movement (Ma)	Glacial stage
Lasiopodomys	0.85	Isotope stage 22
Allophaiomys	2.1–1.9	Latest Pliocene

suggestion is that the major mammalian migrations across the land bridge occurred at times of maximum glaciation on the North American continent (see Table 9.4).

The Kulukbuk Formation of southwest Alaska has been dated to the Penultimate Glacial by its relationship with the Old Crow tephra (135 ± 5 ka). Palaeoenvironmental analyses suggest that the contemporary environment was cold and dry, with sparse herb-dominated tundra (Waythomas, 1996).

9.8.2   Sea Level and Tectonics

In considering the movement of plants and animals from East Asia to North America across the Bering land bridge, the timing of sea-level maxima (which may have prevented this movement) has been a major concern. Table 9.4 lists two times of known movements of biota across the Bering land bridge during the early and middle Pleistocene. The sea level must have been low at these times, probably as the result of the prevalence of glacial conditions, as the timing suggests, yet these could not have been so severe that Arctic ice moved through the Bering Strait to block the land bridge.

As for northern North America, the destructive effect of Last Glacial ice on earlier records of both glaciation and associated tectonic changes makes an understanding of these extremely challenging in Beringia. Evidence for uplift associated with plate convergence is best exposed along the coasts but, compared to most other parts of the Pacific Rim, these have been little investigated.

remained free of continental ice during some early and middle Pleistocene glacials (Repenning, 1993). The principal evidence in support of this

9.9   EARLY AND MIDDLE PLEISTOCENE ENVIRONMENTS OF EAST ASIA

9.9.1   Climate and Ice

Much information about climate changes in East Asia during the early and middle Pleistocene comes from the history of loess deposition, which began in north China around 2.4 Ma (Kukla and An, 1989; see Figure 7.2). Loess accumulated when conditions in the inland source areas were dry, conditions commonly attained during Pleistocene glacials. Early Pleistocene cooling has been inferred in this way from a major period of loess deposition on the Loess Plateau in north China 1.22–1.13 Ma.

Independent dating of Chinese loess layers using magnetic susceptibility has allowed correlations to be made with the oxygen-isotope chronology and the Pleistocene glacial–interglacial record (Table 9.5). Yet since loess–soil alternations record changing vegetation patterns, which would be expected to lag significantly behind global temperature changes and incorporate the effects of other variables (notably precipitation), there is a less-than-perfect agreement. Some palaeosols formed during the early or late parts of glacial periods when the climate was presumably moist enough for the development of vegetation sufficient to protect the ground surface from erosion despite low temperatures. Conversely, some loess layers formed during the early parts of interglacials when conditions presumably became so dry that vegetation was not established irrespective of the effects of temperature rise.

An influx of cold-water ostracode species in the Sea of Japan 1.2 Ma (Cronin, 1992) suggests that this early Pleistocene cooling recorded in the Loess Plateau was experienced throughout the Pacific Rim in East Asia. Analysis of the accumulation rates of quartz and opal in the Sea of Japan shows that they vary according to a 41 000-year cycle, interpreted as signalling alternating glacial–interglacial (arid–wet) conditions in the continental hinterland corresponding with the axial-tilt cycle (Table 9.6).

Early Pleistocene (1.6–1.1 Ma) vegetation changes in Japan involve elements of both warming and cooling linked to glacial–interglacial alternations. Later in the early Pleistocene (1.1–0.8 Ma), these alternations become better defined and the influence of the 100 000-year orbital-eccentricity cycle becomes clear. Warm temperate forest developed in western Japan about 0.4 Ma represents the warmest climate during the entire Pleistocene (Momohara, 1994).

Although linked intermittently to the Asian mainland during times of low sea level, the islands of Japan were effectively isolated from the continent for much of the Quaternary. Owing to an absence of glacial refugia, this situation led to the extirpation of particular biota which survived (longer) on the mainland. For example, trees of the genera *Gingko* and *Metasequoia* became extinct in Japan during the early Pleistocene both as a result of cooling and because of the increasing rate of land uplift associated with the islands' location along a convergent plate boundary (Kaizuka, 1980; Momohara, 1994). Pygmy elephants (*Stegodon*) migrated first from the Asian mainland into Japan during the Pliocene but became extinct there in the middle Pleistocene while surviving on the mainland into the Holocene (see Chapter 15).

Palynological studies in Taiwan, which like Japan was connected intermittently to the Asian mainland during the Pleistocene, have shown what were perhaps the most pronounced responses to early and middle Pleistocene climate change in this region. Since Taiwan, unlike most of the rest of this region, lies within the zone of influence of warm-water currents, its vegetation responded comparatively promptly to changes in their temperature during the Pleistocene; these changes likewise occurred comparatively promptly in response to global temperature changes. Periods of Pleistocene cooling in Taiwan were represented by increases in the extent of subalpine mountain forest and lowland deciduous broad-leaved forests. Warming saw the reduction, even the disappearance, of these elements and the introduction of evergreen forests dominated by *Castanopsis*. The most prominent cooling event during the early Pleistocene occurred 1.7–1.1 Ma (maximum cooling 1.3–1.1 Ma); during the middle Pleistocene, cooling occurred 0.9–0.7 Ma.

Table 9.5   Ages of Xifeng (China) loess deposits and their links to Quaternary glacials. S0–S4 are soil units, L1–L5 are loess units
After Hovan et al. (1989), Kukla and An (1989) and data in Table 8.2

Unit	Aeolian age (ka)	Oxygen-isotope stage	Time period
S0	12	2	Late Last Glacial
L1 (LL1)	33	3	Last Glacial
L1 (SS1)	50	3	Last Glacial
L1 (LL2)	85	5a	Last Glacial
S1	120	5e	Last Interglacial
L2 (LL1)	151	6	Penultimate Glacial
L2 (SS1)	172	6	Penultimate Glacial
L2 (LL2)	187	6	Penultimate Glacial
L2 (SS2)	197	7	Late Penultimate Interglacial
L2 (LL3)	209	7	Late Penultimate Interglacial
S2 (SS1)	229	7	Penultimate Interglacial
S2 (LL1)	244	7	Penultimate Interglacial
S2 (SS2)	259	7	Penultimate Interglacial
L3	291	8	Antepenultimate Glacial
S3 (SS1)	300	8	Early Antepenultimate Glacial
S3 (LL1)	310	8	Early Antepenultimate Glacial
S3 (SS2)	336	9	Antepenultimate Interglacial
L4	363	10	(Glacial)
S4	427	11	(Early interglacial)
L5 (LL1)	458	12	(Glacial)
L5 (SS1)	468	12	(Glacial)
L5 (LL2)	476	12	(Glacial)

Several warm episodes have been identified, notably that 0.98–0.91 Ma within the Jaramillo palaeomagnetic event during the middle and later parts of which maximum warmth – indicated by the dominance of *Castanopsis* pollen – was attained (Liew, 1991).

9.9.2   Sea Level

One of the few records of early and middle Pleistocene sea-level changes on the continental Pacific coast of East Asia is from Cape Medvezhii

in eastern Sikhote-Aline. The 10 m terrace here was dated using thermoluminescence to 0.22 Ma (Korotkii, 1985). Another high-latitude marine terrace sequence studied in this region occurs on Sakhalin island. Deposits on a series of surfaces believed to have been formed during successive interglacial sea-level maxima illuminate the nature of early and middle Pleistocene climate changes in the region (Korotky et al., 1994). Deposits on the 60–80 m Boshnykovskaya Terrace show a transition from cool early Pleistocene times to a warm climate at the beginning of the middle

Table 9.6   Quartz and opal accumulation rates (g/100 000 years) in the Sea of Japan during the early Pleistocene (1.33–1.1 Ma) assuming constant sedimentation rates within a single 41 000-year cycle. Note the variability between glacial and interglacial rates which reflect changes in the climate and vegetation of the continental hinterland. Likely ages are derived by interpolation, not measured directly. See also Figure 7.2
After Dersch and Stein (1994)

Possible age (Ma)	Climatic condition	Quartz accumulation rate	Opal accumulation rate
1.12	Interglacial	1.00	1.46
1.15	Glacial	1.22	0.66
1.17	Interglacial	1.35	1.26
1.19	Glacial	1.73	0.80
1.22	Interglacial	1.14	1.55
1.24	Glacial	1.34	0.90
1.26	Interglacial	1.23	1.54
1.28	Glacial	1.54	0.78
1.31	Interglacial	1.45	1.65
1.33	Glacial	1.73	1.06

Pleistocene. The 20–30 m Sergeevskaya Terrace holds a record from the second half of the middle Pleistocene showing a transition from a cold to a warm climate.

Micropalaeontological analyses of cores from the ocean floor off east China have demonstrated that a series of transgressions occurred during the Pleistocene; the earliest of these dates from 0.3 Ma (Pinxian et al., 1981). Work on the marine terraces of the Korean Peninsula has also demonstrated an association between emerged marine terraces and high interglacial sea levels during the early part of the Pleistocene; the 50–60 m Eupcheon Terrace and 35–42 m Wolseong Terrace are examples (Young, 1994).

A lack of stable shorelines in Japan means that their principal value lies not in helping understand Pleistocene sea-level changes but in determining uplift rates and deformation patterns. A good summary was given by Yoko Ota and Akio Omura (1991).

Although it has long been clear that lowered sea level during Pleistocene glacials would have restricted the movement of Pacific Ocean water, particularly the warm Kuroshio Current, into the

East China Sea and Japan Sea (Kobayashi, 1985), significant quantitative support has been lacking until recently. For the South China Sea, it has been suggested that increased turbidity during times of low glacial sea level in the middle (and late) Pleistocene arose from both a reduction in the influence of the Kuroshio and increased sediment input from the drier continental areas to the west. Such changes dramatically affected the abundances of deep-dwelling *Florisphaera profunda*, a nannoplankton species which has been used successfully as an indicator of palaeoturbidity throughout this region's oceans (Ahagon et al., 1993).

9.9.3   Tectonics

A contrast between rates of uplift in Japan during the Pleistocene and Neogene has been recognized for some time. Matsuda (1976) typified the Pleistocene as 'the age of mm/year tectonics' and the Neogene as 'the age of mm/100 year tectonics'. The high uplift rate during the Pleistocene outstripped that of denudation and produced the deep valleys that characterize many parts of the Japanese uplands today. A similar situation pertained in Taiwan during the Quaternary for comparable geotectonic reasons.

Being located on a young, active island arc, the Ryukyu islands of Japan are rising rapidly. Uplift timing and magnitude are known more precisely on these islands owing to the abundance of emerged coral reefs which did not develop around the other islands in Japan because ocean waters were too cool. Many coral-reef terraces in the Ryukyus date from the early Pleistocene and provide a unique record of subsequent tectonic and sea-level changes. The reason why the record does not extend back any farther in time is of great interest, and is discussed in the following section.

9.9.4   Initiation of Pleistocene Coral-Reef Growth in the Ryukyu Islands, Japan

A fine example of the influence of tectonics on environmental change during the early Pleistocene is provided by the initiation of coral-reef growth in the Ryukyu islands (Koba, 1992). Pleistocene reef growth began 0.7–0.6 Ma in the Ryukyus as a result of a critical mass of the warm Kuroshio Current reaching the Okinawa Trough (Figure 9.4). The present Kuroshio crosses the Ryukyu arc through the Yonaguni Depression, runs along the northwest side of the Okinawa Trough, and then back into the Pacific through the Tokara Structural Channel. The modern 20°C isotherm – the lowest temperature which corals here can tolerate – lies on the northwest side of the Okinawa Trough in the coldest months allowing corals to grow today throughout the Ryukyu group.

The barrier to early Pleistocene reef growth in the Ryukyus was the Okinawa–China 'land bridge' which had been in place since the Eocene when an earlier phase of reef-building in the Ryukyus ended. Although this land bridge did not exclude water from the Okinawa Trough, the Yonaguni Depression was rendered so shallow that the integrity of the Kuroshio Current was destroyed. Then around 0.7 Ma, a collision between the Luzon island arc and Taiwan caused the Yonaguni Depression to widen at the same time as sea level rose rapidly. The combination of

rifting and sea-level rise established the modern Kuroshio flow through the Yonaguni Depression. The flow of tropical water into the Okinawa Trough allowed coral reefs to develop there for the first time in the Pleistocene. Since most of those reefs are now emerged, they have been comparatively easy to identify; electron-spin resonance was used to date them (Koba, 1992).

9.10   EARLY AND MIDDLE PLEISTOCENE ENVIRONMENTS OF AUSTRALASIA

9.10.1   Climate and Ice

Ice accumulation during Pleistocene glacials occurred outside high and middle latitudes only in those places where sufficiently low temperatures (and high precipitation) were achieved as a result of high elevation. In low-latitude Australasia, only the highest peaks of the island New Guinea are sufficiently high and wet today for ice to exist there. The situation was probably the same during much of the Pleistocene.

The key to determining the age of the oldest ice in New Guinea lies in appreciating that those active volcanoes – abundant on the island – which erupt beneath ice produce palagonite. This has been dated to around 0.7 Ma on Mount Giluwe and 0.3 Ma in the Gogon valley (Löffler et al., 1980), both places which are ice-free today. It is unlikely that the associated glaciations were particularly extensive. Most of Australia was affected only by frost action during Pleistocene glacials. Ice developed solely in Tasmania on two to four occasions during the early and middle Pleistocene (Colhoun, 1991).

The increasing aridity during the Quaternary in Australia (including its Pacific fringe) began in the Neogene. Large mammals adapted to browsing and grazing in arid conditions had evolved by the start of the Pleistocene. In New South Wales, this time was marked by a change from subtropical forest dominated by *Nothofagus* and associated with moderate rainfall to a semiarid vegetation assemblage, associated with *Eucalyptus* (Martin, 1989; McEwen Mason, 1991). A change from

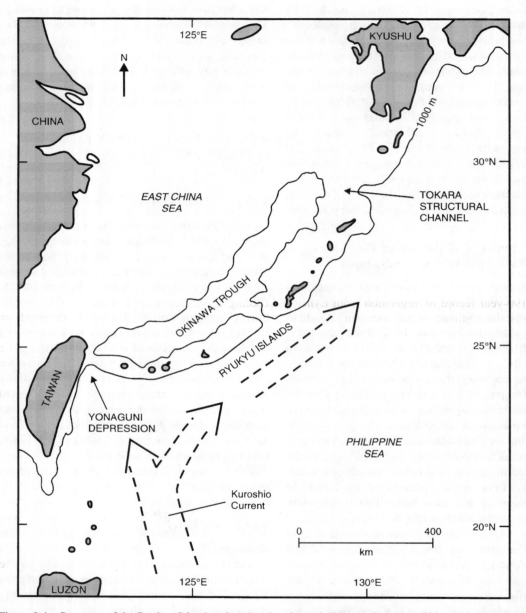

Figure 9.4   Structure of the Ryukyu Island region showing the main islands. Only the 1000 m isobath is shown. See
Figure 11.6 for related information
From various sources, especially Koba (1992) with permission of the Japanese Association for Quaternary Research

moist to drier conditions in the early to middle Pleistocene was also registered in Tasmania (Hill and MacPhail, 1985) and in Tasman Sea sediments (Hesse, 1993).

As in many low-latitude areas, parts of Pacific Australia experienced an alternation between arid and pluvial (wetter) phases during the Pleistocene which correlate tolerably well with the succession of glacial–interglacial alternations in high latitudes (Nanson et al., 1992). Particularly well documented is the middle Pleistocene pluvial 270–220 ka, a conspicuous incident in the history of the Lake Eyre Basin (Gardner et al., 1987). The effects of rapidly alternating climates on biota were occasionally severe although those, such as *Eucalyptus*, able to adapt, migrate and/or disperse readily, achieved dominance by default in many ecosystems (Hope, 1994).

Much information about the Penultimate Glacial in Pacific Australia has been gained from the ~190-year record of vegetation from Lynch's Crater in the northeast of the continent (Kershaw, 1976). The overall rise in sclerophyllic taxa, particularly *Casuarina* ~190–140 ka, indicates increasing aridity during the early and middle parts of the Penultimate Glacial yet this tendency was broken by an abrupt rise about 165 ka, interpreted as signalling a comparatively rapid fall in precipitation (see Figure 10.3).

Some of the keenest insights into the Pleistocene climate history of New Zealand have come from analyses of cores from adjacent ocean floors. For example, periods of increased silt accumulation in cores from DSDP site 594, 250 km east of South Island, were linked to glacial periods when the amounts of erosion onshore were much greater than during interglacial periods, both because of ice scour and because of a reduction in vegetation cover resulting from the cold dry conditions at such times. As many as 12 major expansions of mountain ice in South Island during the past 730 000 years have been identified in this manner. Through links to the oxygen-isotope record, the climate changes represented by these glacial fluctuations are considered near-synchronous with those of the northern hemisphere (Nelson et al., 1985).

Pollen from the last four Pleistocene glacial–interglacial cycles in the same DSDP core was analysed by Heusser and van de Geer (1994). This pollen had been carried offshore by the prevailing westerly winds and by rivers draining South Island. Results show that forest dominated during interglacials and was far less extensive during glacials, owing to increased ice cover and aridity. Variations in several tree and grass species are shown in Figure 9.5.

Rapid uplift of parts of New Zealand during much of the late Cenozoic has lifted deep-sea sediments out of the water allowing the details of earlier climate change to be discerned more fully here than in many other parts of the Pacific Basin. A good example is provided by the study of middle Pleistocene climate and sea-level changes at Landguard Bluff near Wanganui (Pillans et al., 1988). Pollen from alluvium, interbedded with shallow-water marine sediments, indicates a 3°C cooling associated with a sea-level fall of 32 m during oxygen-isotope stage 7b (centred on 230 ka). This event, which represents a short-lived cooling episode within the Penultimate Interglacial, has global correlates.

Loess deposits in New Zealand were also produced during glacial periods, when increased wind speeds moved material exposed on the surfaces of the devegetated periglacial landscape and emerged continental shelves. An early Pleistocene loess in Wairarapa has recently been shown to have accumulated 1.00–0.87 Ma (Shane et al., 1995).

9.10.2 Sea Level

One of the best-known sites for sea-level studies in the world is that on the Huon Peninsula on the island New Guinea. Described originally by Bloom et al. (1974) and Chappell (1974), discussions about the Huon and the chronology of its emerged marine terraces have continued, culminating in the reports of 1992/1993 fieldwork (Ota, 1994). The importance of the Huon sequence is that, by assuming that it has been rising continuously and in only one direction during the (late) Quaternary, the tectonic component of

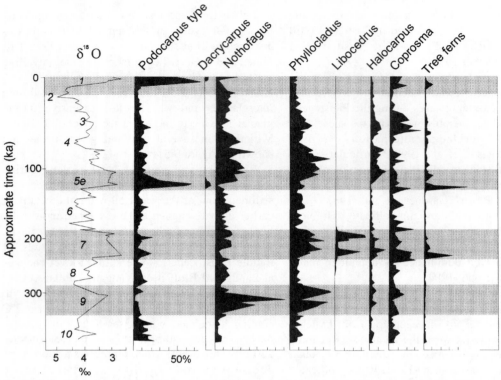

Figure 9.5   Variations in quantities of selected herbaceous pollen types from New Zealand during the middle and late Pleistocene

Reprinted from *Quaternary Science Reviews*, 13, Heusser, L.E. and van de Geer, G. 'Direct correlation of terrestrial and marine palaeoclimatic records from four glacial–interglacial cycles – DSDP site 594, southwest Pacific', 273–282. Copyright 1994, with permission from Elsevier Science

shoreline displacement can be calculated, thereby allowing the isolation of the eustatic (sea-level) component. This led to the production of a sea-level curve for much of the late Quaternary which has been revised several times and orbitally tuned so that sea-level variations coincide with solar-radiation cycles (see Chapter 8). The result has been a sea-level reference curve (see Figure 10.1) which has been employed in the interpretation of displaced shorelines worldwide.

Studies on the Huon Peninsula revolutionized the study of Pleistocene sea-level change by demonstrating that the terrestrial record is occasionally sufficiently complete to be reconcilable with the marine (ocean-floor) record. The significance of the site is greatest in the immediate region where the likelihood of factors interfering with correlation between various displaced shorelines is least. Selected correlations between Huon Peninsula sea-level stages and those in adjoining parts of the Pacific Basin were shown in Table 9.1.

Some of the most stable coasts in this region are found along the edge of the Australian craton. For example, the coastline south of Sydney provides 'a compelling array of evidence attesting to long-term tectonic stability' (Young and Bryant, 1993: 267). Emerged shore platforms here include one 9.5–10.0 m above present mean sea level believed to have been cut during an early Pleistocene high sea-level stand, and another 7–8 m above mean sea level which may have formed 350–250 ka (Plate 9.3).

Plate 9.3   Platform at 7–8 m cutting across the vertically dipping Wagonga Beds at Mystery Bay, coastal New
South Wales, Australia, may have been formed during the Penultimate Glaciation (247–340 ka)
Ted Bryant

Tephra and loess have been used effectively to unravel the Pleistocene climate and sea-level history of New Zealand. For example, a staircase of erosional terraces dating from 0.68 Ma in South Taranaki was interpreted in terms of sea-level change using dated tephra and knowledge of uplift history by Brad Pillans (1983). Dates from loess and tephra allowed Pillans et al. (1988) to demonstrate that there had been two Penultimate Interglacial sea-level maxima in central New Zealand, approximately 240 ka and 210 ka.

Lowering of sea level during Pleistocene glacials exposed the Torres Straits allowing biotic exchange between New Guinea and Australia. The oceanographic aspects of this change are less well known, although it is plausible to suppose that it caused the warm East Australian Current to strengthen which may explain why the Great Barrier Reef developed only during the later part of the early Pleistocene (Andres et al., 1997).

9.10.3   Tectonics and Volcanism

The island New Guinea exhibited some of the most rapid uplift rates in the Pacific Basin during the Pleistocene. Uplift of the Saruwaged Range, for example, at 3 mm/year is capable of having formed these >4 km mountains in just 1.2 million years. Active rifting has further contributed to the increasing complexity of climate and ecosystems on this island during the Pleistocene (Hope, 1996).

An increase in uplift rate between Pleistocene and Holocene on parts of the Huon Peninsula has been noted (Table 9.7); this is analogous to the situation in New Zealand (see below) and elsewhere in the Pacific Basin, and may be linked to Pacific-wide changes in rates of plate convergence.

The islands of New Zealand experienced considerable deformation, both vertical and horizontal, throughout the Quaternary. On account of a comparative plethora of available dating options, the early and middle Pleistocene is

Table 9.7   Uplift rates at selected places along the
Huon Peninsula, Papua New Guinea
Ota et al. (1993)

Location	Pleistocene uplift rate (mm/year)	Holocene uplift rate (mm/year)
Blucher	2.2–2.4	2.2–2.3
Hubegong	3.2–3.4	3.2–3.4
Kanzarua	2.7–2.9	2.8–3.0
Kwambu	1.8–1.9	1.9–2.0

unusually well illuminated (Pillans, 1983, 1986; Bull and Cooper, 1986). In North Island, rates are highest in the east facing the Hikurangi convergence zone offshore and reached 2–3 mm/year in several places during the middle and late Pleistocene and Holocene. Along the west coast of New Zealand's South Island, uplift is associated with compression between the Indo-Australian and Pacific plates. At three places, a distinct change in uplift rate occurred in the late middle Pleistocene (Table 9.8). These changes were approximately synchronous with changes in uplift rate on the Huon Peninsula in Papua New Guinea (see above).

Active volcanism in the island arcs of the western Pacific during the early and middle Pleistocene produced major changes to the environments of these areas which have not yet been fully understood. Some of the best known examples come from New Guinea where the shield volcano of Mount Giluwe (4340 m), for instance, formed only around 0.5 Ma (Hope, 1996).

Despite its intraplate location, there is a belt of volcanoes – many of which were active during the

Quaternary – running parallel to the east coast of Australia.

In contrast, most volcanism in New Zealand during this time was associated with plate-boundary processes, the main area of Quaternary activity in North Island being the Taupo Volcanic Zone, the axis of which parallels that of the Hikurangi Trench offshore to the east. As in many such volcanic zones, late Quaternary volcanoes and their products obscure the record of earlier volcanism although it is apparent that middle Pleistocene eruptions here were similarly voluminous and associated – as today – with an opening rift valley (Cole, 1985). As in a few other parts of the Pacific Basin, late Quaternary volcanism in New Zealand created new land surfaces on which the records of subsequent climate changes can be observed without concern that these records may be confused with older ones. A good example is the Mamaku ignimbrite plateau in North Island, an 1800 km^2 area erupted 140 ka (Kennedy, 1994). In the following 10 000 years, this plateau was rapidly dissected and lowered – possibly an indication (at least in part) of a more arid climate than that of today.

9.11   EARLY AND MIDDLE PLEISTOCENE ENVIRONMENTS OF THE PACIFIC OCEAN AND ISLANDS

9.11.1   Climate and Ice

Many ocean-core analyses from the margins of the Pacific Ocean have shed light on the early and middle Pleistocene history of the Pacific Rim and were discussed above. Evidence of the early

Table 9.8   Uplift rates at selected places along the west coast of South Island, New Zealand
Bull and Cooper (1986)

Location	Early and middle Pleistocene uplift rate (mm/year)	Age of change in uplift rate (ka)	Late Pleistocene uplift rate (mm/year)
Fox-Franz Josef	3.2	134	7.8
Haast River	3.6	136	5.8
Kaniere	2.6	138	5.5

Pleistocene cooling event, which affected other parts of the Pacific Basin 1.3–1.2 Ma, was obtained from radiolarian data in the north Pacific Ocean implying that it was a Pacific-wide, perhaps even global event. Away from the continental borders of the Pacific, in the central part of the ocean, evidence derived from analyses of ocean cores remains the most powerful tool for understanding Pleistocene climate changes in the region. For example, the correlation between high levels of $CaCO_3$ production and glacial periods is common to all central Pacific cores. Other techniques are less diagnostic of palaeoclimate changes in the central Pacific as opposed to its periphery.

Some of the most influential studies in the central equatorial Pacific have been on the 950 000-year palaeoclimate record from core RC11-210 at 1°49′N, 140°03′W (Chuey et al., 1987; Pisias and Rea, 1988; Rea et al., 1991). Oxygen-isotope and $CaCO_3$ records from RC11-210 are shown in Figure 9.6A and B respectively. It seems clear that glacial conditions around the Pacific Rim, particularly the northern part, during the Pleistocene led to stronger wind-driven ocean circulation which enhanced biological productivity, the maxima of which are marked in core RC11-210 by the organic carbon maxima and $CaCO_3$ minima. The ratio between the two has been plotted and interpreted in terms of climate change (Figure 9.6C).

Windblown material (loess[2]) is also present on many parts of the Pacific ocean floor. In the equatorial Pacific south of the ITCZ, the accumulation of loess over the past 300 000 years was found – unexpectedly – to have been greatest during interglacials, possibly because of increased volcanism around adjacent parts of the Pacific Rim. Yet in the North Pacific, as along most parts of the Pacific Rim, loess accumulation was greatest during glacials when conditions on adjoining continents were drier and wind intensity was greater (Olivarez et al., 1991). Loess in ocean

cores is not always so easy to interpret. Variations in the amount of aeolian material in RC11-210 were analysed but proved difficult to reconcile with the history of aridity in presumed continental source areas, either northwest South America or Australia (Chuey et al., 1987).

Not surprisingly, the sequence of Pleistocene climate change on many Pacific islands remote from the Pacific Rim has been controlled less by events taking place there than by those occurring in the central part of the Pacific Basin. A good illustration is provided by islands in the Pacific tradewind belts, such as those of the Hawaii group, much of the rainfall on which is (and has been) produced by orographic effects. Falls of sea level during Pleistocene glacials would have greatly increased the heights of such islands and, as a consequence, parts may have been much wetter during Pleistocene glacials than during interglacials (Gavenda, 1992). Tectonic change amplified these differences: the island Molokai, which presently receives some 3000 mm of rain annually, may have received around 12 000 mm before the island subsided (Stearns and Macdonald, 1947).

Yet just as sea-level fall may have led to increased rainfall in parts of the windward sides of such islands, so it may also have amplified the dryness of their leeward flanks. This condition may have been further exacerbated by the weakening of the tradewind system during Pleistocene glacials.

Hawai'i is the only island in the central Pacific known to have been glaciated during the Pleistocene. The oldest known glacial is marked by the Pohakuloa Formation deposited 278.5 ka (Porter, 1979). The Waihu Formation dates a later glacial to 175–170 ka. More details are given in Chapter 11.

Many islands in the equatorial Pacific have thick deposits of phosphate (phosphorite) that accumulated when great numbers of seabirds lived there and the climate was so dry that the guano they produced was not leached from the system as it is today. Phosphate has not been forming on those islands recently for this reason. In addition, seabird numbers are low today, even where human

---

[2] Broadly defined, as in the 'desert' loess of Smalley and Vita-Finzi (1968), to include aeolian sediments other than those produced by glacial processes

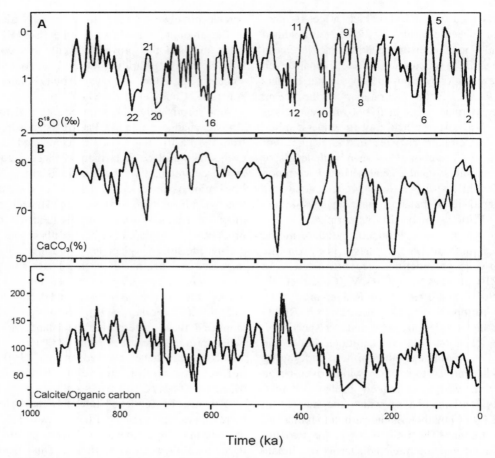

Figure 9.6 Palaeoclimate information for the Pleistocene recorded from central equatorial Pacific ocean core RC11-210 with likely climatic correlations added. (A) $\delta^{18}O$ record (Chuey et al., 1987). (B) $CaCO_3$ record (Chuey et al., 1987). (C) The ratio of calcite to organic carbon (Rea et al., 1991)

predation has been slight. Yet during Pleistocene glacials, the climate was drier and tropical ocean circulation was more intense. This is thought to have been responsible for greater upwelling resulting in increased ocean-water productivity around these islands which could therefore have sustained far larger seabird colonies. The ages of guano-derived Pacific island phosphates would thus be expected to coincide with Pleistocene glacials (Table 9.9). The suggestion by Bourrouilh-Le Jan (1989) that guano may not have been the sole source of phosphate for phosphorite development on certain high Pacific islands – some may have been produced from the weathering of underlying

volcanic rocks – may be one cause of disagreements between times of phosphate formation and glacial conditions in Table 9.9.

9.11.2 Sea Level

Most investigations of sea-level history in the Pacific islands have focused on Last Interglacial and later times. Displaced early and middle Pleistocene shorelines undoubtedly exist, particularly in the islands along the convergent plate boundaries of the southwest Pacific. Examples include the island 'Eua in Tonga where the Last Interglacial shoreline is 6–10 m and at least six

Table 9.9 Times of formation of Pacific island phosphates during the Quaternary. All ages were determined using U/Th; no error terms are given
After Roe and Burnett (1985) and Roe et al. (1983)

Island	Age (ka BP)	Oxygen-isotope stage	Time period
Kiritimati (Christmas)	55	4 (cool)	Middle Last Glacial
Malden	117	5e (warm)	Early Last Glacial
Mataiva	>300	?	?
Mataiva	210	7 (warm?)	Penultimate Interglacial
Mataiva	130	5e (warm)	Early Last Interglacial
Mataiva	127	5e (warm)	Early Last Interglacial
Nauru	>300	?	?
Nauru	145	6 (cool)	Penultimate Glacial
Tabuaeran (Fanning)	22	2 (cool)	Last Glacial Maximum
Tuvuca (Tuvutha)	>300	?	?
Vanuavatu	110	5d (cool)	Early Last Glacial

higher, older surfaces exist up to 230 m (Hoffmeister, 1932; Taylor, 1978).

In the high limestone Lau islands of eastern Fiji, there are many coral-reef terraces which developed during sea-level maxima associated with Pleistocene interglacials (Plate 9.4). Table 9.10 shows the suggested chronology developed using regional correlation and dates from phosphates and lavas associated with the terraces (Nunn, 1996, 1998b). An emerged Penultimate Interglacial reef has been described from nearby Kadavu island (Nunn and Omura, 1999). Reefs drowned when sea level rose at the end of Pleistocene glacials are found on the rapidly subsiding flanks of the Hawaiian ridge. The −950 m terrace off Hawai'i has been dated to 250 ka; that at −1650 m off Maui dates from 340 ka (Jones, 1995).

An innovative approach to ascertaining the timing of sea-level maxima during the early and middle Pleistocene has been to look at the ages of corals below unconformities in reef cores (see also ·Figure 7.6). An example is given in the following section.

9.11.3 Pleistocene Coral-Reef Growth in the Northern Cook Islands

Coral reefs grew in the tropical Pacific throughout the late Cenozoic. During the Pleistocene, their distribution was constrained principally by sea-surface temperature; most corals will flourish only in waters of 25–29°C. The other important constraint on their modern distribution has been sea-level changes associated with glacial–interglacial alternations; the upward growth of some reefs during postglacial times failed to keep pace with rising sea levels and the reefs drowned.

Analyses of cores drilled through the coral caps of Pukapuka and Rakahanga atolls in the northern Cook Islands focused on determining times of interglacial high sea-level stands from dates of the oldest *in situ* lagoonal corals below solution unconformities (Gray et al., 1992). This connection is premised on the idea that sea-level fall at the start of a glacial period cut the connection between the ocean and the atoll lagoon, killing lagoonal corals. The dates of these will therefore approximate the time of the end of the interglacial high sea level at which they were growing. In this way, five periods of high sea level were identified here, the best defined being in the early and middle Pleistocene 650–460 ka, 460–300 ka and 230–180 ka.

The advantage of determining Pleistocene sea-level history from a submerged reef rather than an emerged reef is that assumptions about the former's tectonic history are more plausible than the latter's. There can be little doubt that most

Plate 9.4   One of the most unusual expressions of Quaternary uplift of Mago island in the northern Lau group is the series of at least three lagoons in different stages of emergence along its northeast coast. The modern lagoon is shown on the left, ringed by reef. The lower emerged lagoon is shown in the centre, its reef rings emerged as solution ridges 6–10 m high. The higher lagoon is off to the right; its floor is emerged ~30 m, and it is fringed by solution ridges towering as much as 25 m above this (see also book cover)

Table 9.10   Chronology of emerged terraces in the Lau islands
After Nunn (1996)

Terrace	Mean height above sea level (m)	Age (ka BP)
Vunirewa Terrace	10	106
Bureta Terrace	14	118
Maruna Terrace	20	124
Nalami Terrace	25	175
Koroqara Terrace	33	215
Nativativa Terrace	55	240
Vagadra Terrace	70	325

ancient mid-ocean atolls have been subsiding slowly and continuously for most of the late Cenozoic whereas it is often unclear whether uplift of an emerged reef series has been either monotonic or unidirectional, as is commonly assumed (Nunn, 1994a).

9.11.4   Tectonics and Volcanism

Both uplift and subsidence brought about great changes in certain parts of the Pacific Ocean during the Quaternary. The greatest amounts of uplift occurred along convergent plate boundaries. For example, ocean-floor sediments in Solomon Islands have been uplifted more than 2.5 km since the early Pleistocene. Elsewhere, emerged coral reefs preserve sites of Pleistocene uplift which can be mapped and dated with reasonable precision. Much work of this kind has been carried out in the Ryukyu islands of Japan, and in Papua New Guinea (see above). A selection of dates for early and middle

Table 9.11   Selected early and middle Quaternary reef terraces in the Pacific Islands compared to those in the Ryukyus, Japan, and Huon Peninsula, Papua New Guinea

Island	Terrace height (m a.m.s.l.)	Terrace age (ka BP)	Uplift since formation (m)	Source of information
Pacific Islands				
Efate, Vanuatu	250	300?	300?	Lecolle et al. (1990)
Efate, Vanuatu	120	175.2	175?	Lecolle and Bernat (1985)
Lifou, Loyalty Islands	3.5	180	25	Marshall and Launay (1978)
Makatea, Tuamotu Islands	20	200	8	Montaggioni et al. (1985)
Maré, Loyalty Islands	12	205	12	Gaven and Bourrouilh-Le Jan (1981)
Nagigia, Kadavu, Fiji	2.8	209.2	3–9	This book
Tongatapu, Tonga	4.0	225	4–10	Taylor (1978)
West Pacific Rim				
Hateruma, Ryukyus, Japan	60	200+	48+	Omura (1984)
Hubegong, Huon Peninsula, Papua New Guinea	406	118–126	400?	Ota et al. (1993)

Pleistocene reef terraces in the Pacific islands is given in Table 9.11.

It was during the middle Pleistocene that lithospheric flexure produced by volcanic loading first became apparent in the Hawaiian islands (Grigg and Jones, 1997). Uplift associated with this flexure reached 91 m on Lanai island and is still continuing.

Most active volcanism in the Pacific Ocean and Islands during the late Pleistocene was confined to plate boundaries and intraplate hotspots. Particularly voluminous intraplate island (subaerial) eruptions occurred over the Hawaii hotspot and, somewhat anomalously, along the Samoa island chain (Walker, 1990; Nunn, 1998a). The Aleutian Islands in the northernmost Pacific Ocean have volcanic bases of Tertiary age but these are almost wholly obscured at the ground surface by the products of voluminous Quaternary volcanism, perhaps greatest during the Holocene.

9.12   REVIEW

The record of environmental change in the Pacific Basin is presently too patchy to allow any detailed coherent account for most of the early and middle Pleistocene. Despite this situation, some general points emerge.

Not surprisingly, there is general agreement that ice advance and recession, and sea-level fall and rise, are correlated with each other and with glacial–interglacial alternations throughout the Pacific Basin. The effects of an equatorward compression of climate belts during glacials and their poleward expansion during interglacials appear clear from a variety of sources.

Yet no uniform picture regarding the timing of the most extreme conditions – a useful test of regional synchroneity – emerges. The most extensive glaciation in the southern Andes occurred in the early Pleistocene, whereas in most other parts of the Pacific Basin this occurred during the Last Glacial (see Chapter 11). Indications that the earliest Pleistocene was comparatively mild are known from many parts of the Pacific Basin. The time of maximum Pleistocene warmth some 0.4 Ma is known from both Chile and Japan, and may have been a Pacific-wide phenomenon.

The control of early and middle Pleistocene tectonics and volcanism by processes of plate convergence is manifest throughout the Pacific Basin. An increase in the rate of uplift throughout the Quaternary is apparent in most places except Pacific South America where this was greatest during the middle Pleistocene. In many parts of the Pacific Rim, volcanism developed along a network of fractures produced by stress associated with plate convergence.

9.13   THE LATE PLEISTOCENE PREVIEWED

The late Pleistocene encompasses two periods – the Last Interglacial and the Last Glacial – about which much more is known than the earlier Pleistocene. Yet even within the late Pleistocene, there is a conspicuous imbalance in the amount of palaeoenvironmental information available for the Pacific Basin for the Last Interglacial and early Last Glacial compared to the middle and late Last Glacial. This observation, which is globally applicable, is largely the result of the radiocarbon dating method – the most common and inexpensive – reaching its effective limit at ~40 ka BP. To understand the various dating systems used in the rest of this book, reference should be made to the section 'Conventions for Expressing Age' preceding Chapter 1.

The following two chapters treat the Last Interglacial, 128–111 ka, and Last Glacial, 111–12 ka, separately. The start of the Last Interglacial was heralded by the end of the Penultimate Glacial, a rise in both ocean level and surface temperatures, and a consequent change in vegetation patterns and the distribution of animals (including humans).

CHAPTER 10

The Last Interglacial (128–111 ka) in the Pacific Basin

10.1 THE COURSE OF THE LAST INTERGLACIAL

When the late Quaternary was first subdivided on the basis of oxygen-isotope stages, the Last Interglacial was defined as the entire stage 5 (see Table 8.2). In this book, in common with widespread usage, the Last Interglacial is considered only as stage 5e; stages 5a–d are treated as part of the Last Glacial (Chapter 11). The chronology of the late Pleistocene adopted in this book is shown in Table 10.1.

This chapter begins with a general discussion of Last Interglacial temperatures and sea levels; ice cores are discussed in a separate section. Owing to the short time span of the late

Pleistocene compared to the periods considered in earlier chapters of this book, it should be appreciated that certain low-frequency, low-magnitude, or even spatially restricted causes of environmental change, such as much tectonism and volcanism, are less important and thus merit less of a discussion here compared to earlier chapters.

10.1.1 Oxygen Isotopes and Last Interglacial Temperatures

The Last Interglacial is defined as coincident with oxygen-isotope substage 5e, 128–111 calendar years BP. While this age range may not be globally agreed upon, the implications of such

Table 10.1 Divisions of the late Pleistocene adopted in this book. Oxygen-isotope stages are from Table 8.2

Main stage	Substage	Oxygen-isotope stage	Timing (ka)	Duration (calendar years)
Last Glacial	Late	2 (cool)	24–12	12 000
Last Glacial	(Maximum)	(2)	(22–17)	(5 000)
Last Glacial	Middle	3–4	74–24	50 000
		3 (warm)	59–24	35 000
		4 (cool)	74–59	15 000
Last Glacial	Early	5a–d	111–74	37 000
		5a (warm)	79–74	5 000
		5b (cool)	91–79	12 000
		5c (warm)	99–91	8 000
		5d (cool)	111–99	12 000
Last Interglacial		5e (warm)	128–111	17 000

disagreement are less important than that which surrounds the temperature changes which took place during this period. The question of how many temperature peaks were attained during the Last Interglacial is as controversial as that referring to sea-level maxima (see below), to which it might be expected to be related. Most oxygen-isotope records show only a single maximum. The influential SPECMAP compilation shows a single peak 122 ka (Imbrie et al., 1984); that of Labeyrie et al. (1987) has a single peak 121 ka. Other records suggest that more than one temperature peak occurred during the Last Interglacial. The orbitally tuned record of Martinson et al. (1987) showed maxima 125.2 ka, 123.8 ka and 122.6 ka. Double temperature maxima also appear 132 ka and 120.5 ka

in records from Vostok ice cores (Jouzel et al., 1987); three maxima were found in Greenland ice-core data (GRIP Members, 1993).

Temperatures in the Last Interglacial Pacific Ocean reached as much as 3.9°C higher than present (CLIMAP Project Members, 1984). The Vostok ice-core record suggests that temperatures in Antarctica were 2–3°C warmer than present during the Last Interglacial (Jouzel et al., 1987).

10.1.2 Ice-Core Evidence

In an era when it looked as though the chronology of the late Pleistocene derived from oxygen-isotope analyses of ocean-floor sediments would never be improved upon, the advent of ice-core analyses was remarkable (Dansgaard et al., 1971;

Plate 10.1   Analyses of variations through ice cores have proved to be a fruitful source of palaeoclimate information. In the particularly innovative situation illustrated, solar power was used to drill through the low-latitude ice cap at Huascarán col, 6050 m above sea level, in the Peruvian Andes
Lonnie G. Thompson, Byrd Polar Research Centre, Ohio State University

Lorius et al., 1985; Plate 10.1). Some of the most recent work has involved direct measurement of various Last Interglacial palaeoclimate parameters, such as deuterium and dust in the ice and carbon dioxide (CO_2), methane (CH_4) and the oxygen-isotope ratio ($\delta^{18}O$) of air in bubbles within ice (Jouzel et al., 1993). Measurement of cosmogenic isotopes such as ^{10}Be in the ice has allowed a precise chronology of their late Pleistocene productivity, particularly as it relates to palaeotemperature and palaeoprecipitation, to be constructed (Mazaud et al., 1994).

The most fruitful site in the southern hemisphere has been at Vostok (78°28′S, 106°48′E), a Russian research station 3490 m above sea level in East Antarctica. Ice has been accumulating at a remarkably uniform rate at Vostok for 220 000 years; data for the later part of the Penultimate Interglacial were shown in Figure 9.1.

At the end of the Penultimate Glacial – a time dubbed Termination II – temperatures rose rapidly by about 8°C at Vostok culminating in a Last Interglacial temperature maximum ~128 ka. From here there was a slow fall of around 3°C until about 120 ka, when a rapid decline began until about 109 ka. There are clear correlations between changing temperature and changing concentrations of CO_2 and CH_4 (see Figure 9.1).

The importance of the Vostok ice-core records is that they show that the major climate changes during the late Quaternary were of global extent. The implication of this is that, in a crude sense, these changes were approximately synchronous globally and that the conspicuous departures from this in places are likely to be the result of a regional overprint on global change.

Other Antarctic ice-core data, including those from Dome Circe in East Antarctica and Byrd Station in West Antarctica, have been analysed and have thrown light on the palaeoclimates of the latest Quaternary. The results of coring through ice caps in the Peruvian Andes are also discussed in the following chapters.

10.1.3 The Sea-Level Record

Empirical data from throughout the Pacific Basin have been hailed as confirming the +6 m glacial eustatic sea-level maximum during the Last Interglacial on the Huon Peninsula (Figure 10.1); examples come from Tonga (Taylor, 1978) and Isla Guadalupe, Baja California (Muhs et al.,

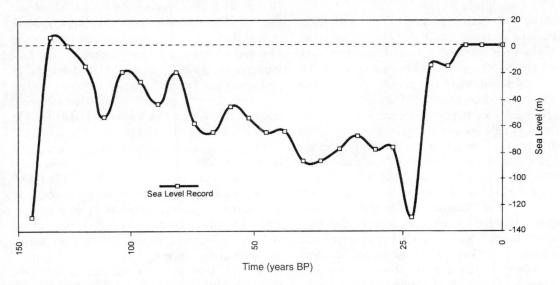

Figure 10.1   Late Quaternary sea-level change determined from ages of emerged reef terraces on the Huon Peninsula, New Guinea
Data from John Chappell (personal communication)

1994). Since melting of the West Antarctic ice sheet would cause sea level to rise about 6 m, it appears plausible to suppose that this was what happened during the Last Interglacial. Yet the ubiquity of the +6 m level has been questioned in Australia, where Murray-Wallace and Belperio (1991) found that the modal Last Interglacial maximum (125 ± 10 ka) sea level was only around 2 m above the present, and in several parts of the Pacific islands (Nunn, 1998a).

Another much-discussed issue concerning Last Interglacial sea level is whether or not there was just one maximum (also see above). It was the existence of a disconformity between emerged Reefs VIIb and VIIa on the Huon Peninsula in Papua New Guinea which led John Chappell (1974) to suggest that there had been two distinct high stands of sea level during the Last Interglacial. More recent work has found that the most reliable dates from Reef VII corals cluster around 134 and 118 ka, times which may mark culminations of two periods of rapid sea-level rise (Stein et al., 1993). More recent dates show that sea level around the Huon Peninsula reached 7 m above present mean sea level 124.5 ka then probably declined (or fell then rose again) to present mean sea level (0) 118.5 ka (see Figure 10.1).

Also influential in favour of a double-maximum scenario was the work of Ku et al. (1974) on the emerged reefs of Oahu; this work has been both supported (Sherman et al., 1993) and contradicted (Muhs and Szabo, 1994) by more recent work. A compilation of uranium-series ages of emerged Last Interglacial coral reefs showed only a single sea-level maximum 123.5 ka although the uncertainty of this (±12.5 ka) suggests that this may disguise a double maximum 129 ka and 123 ka (Smart and Richards, 1992).

Calculations of the effects of isostatic uplift resulting from changes in ice-sheet loading suggest that the volume of ocean water was constant from 135 ka until 120 ka (Lambeck and Nakada, 1992). This would mean that any sea-level maxima within that period would have been the result of local or regional lithospheric deformation and would not be expected to have been coincident across the Pacific Basin.

10.2 LAST INTERGLACIAL ENVIRONMENTS IN THE PACIFIC BASIN

A pragmatic reason for understanding the history of the Last Interglacial in the Pacific Basin is to aid the understanding of present conditions, which are thought to be analogous in many ways, and thereby to help us manage the region's future effectively. Various scenarios which could radically affect our future, such as the possibility of the West Antarctic ice sheet melting, can be evaluated by studying what happened during the Last Interglacial.

This section looks at each component of the region in turn and then produces a synthesis in which various region-wide events are recognized.

10.2.1 Last Interglacial Environments of Antarctica

Compared to its situation during the Penultimate Glacial, the lateral extent of the Antarctic ice sheet was reduced during the Last Interglacial through the combined effect of high sea levels and high temperatures. Areas like the presently ice-filled George VI Sound between Alexander Island and the Antarctic Peninsula may have been ice-free around the time of maximum warmth during the Last Interglacial. Such records confirm that Last Interglacial warmth affected high latitudes as profoundly as low and middle latitudes. A late Last Interglacial expansion of the East Antarctic ice sheet through the Dry Valleys was dated to 113 ± 45 ka (Brook et al., 1993).

Maximum temperatures during the Last Interglacial are generally thought to have been higher than the Holocene maximum (CLIMAP Project Members, 1984). Various estimates have been given, varying from just 0.5°C to a high 7–10°C. The latter may have caused collapse of the West Antarctic ice sheet, which could have caused sea level to rise some 6 m.

In 1968, John Mercer noted that the fringe of the West Antarctic ice sheet, which is below sea level, is buttressed by fringing ice shelves. He suggested that warming could have caused these ice shelves to melt and the ice sheet to have collapsed. More recently, evidence has been amassed in support of this scenario (Stuiver et al., 1981; Denton et al., 1991). In particular, the suggestion that a rapid transgression around 120 ka in the tropical western Pacific was caused by a surge of the West Antarctic ice sheet (Aharon et al., 1980) appears increasingly secure in the light of later research into ice surges (see Chapter 11).

10.2.2 Last Interglacial Environments of South America

Along the Pacific Rim in South America, the effect of repeated Pleistocene glaciations has been to destroy, confuse or obscure the legacy of intervening interglacial (and interstadial) periods. This makes the study of Last Interglacial environmental change somewhat frustrating here.

Emerged shorelines of Last Interglacial age have been mapped at several places along the rapidly rising parts of the South American coast. For example, the 68 m shoreline at Chala Bay in southern Peru is believed to have formed 125 ka; it is the only Last Interglacial emerged shoreline present at this location (Goy et al., 1992). The 20–30 m Talinay IV emerged shoreline in northern Chile is of the same age (Ota et al., 1995). The study of these shorelines has revealed more about regional Quaternary tectonic history, discussed in Chapter 9, than palaeoenvironments.

The best clues to the nature of Last Interglacial climates come from pollen preserved in the deep sedimentary basins of the Colombian Cordillera and the Central Valley of southern Chile. In their innovative work, van der Hammen and Gonzalez (1960) found that the Last Interglacial was represented by such a short interval at Sabana de Bogotá in Colombia that they suspected the lake in which most of the sediments there had been deposited had dried up at this time. This situation is consistent with an increase in temperature of 2°C (relative to the present) and afforestation of the catchment. Grassland pollen in sediments sandwiched between glacial tills at Puerto Varas in the Central Valley of southern Chile led

Heusser (1984b) to suppose that the Last Interglacial climate of this area was also warmer and drier than today.

10.2.3 Last Interglacial Environments of Central America

Deglaciation of some central Mexican volcanoes during the Last Interglacial, notably Iztaccíhuatl, is marked by strongly developed alfisol soils, probably indicating increased precipitation levels compared to earlier interglacials (White, 1986). A record of a Last Interglacial sea level 5–6 m higher than the present was reported from the Yucatan Peninsula, a little outside the study area, by Szabo et al. (1979).

10.2.4 Last Interglacial Environments of North America

Like that in South America, much of the Pacific Rim in North America was comprehensively glaciated during the Last Glacial so palaeo-environmental information about the Last Interglacial is difficult to locate. Yet, like Central America, the southern part of the Pacific Rim in North America was not glaciated during the Last Glacial so holds a legacy of late Pleistocene environmental change which provides the best keys to what probably happened throughout much of this region.

Work on emerged shorelines along the Pacific coast of North America has yielded considerable information about late Pleistocene uplift rates and sea-level changes (Muhs et al., 1990) but comparatively little about contemporary shoreline environments. Warm-water fossils found on Last Interglacial terraces in the western Los Angeles Basin, for example, confirm that contemporary shoreline environments were generally similar to those of today (Dupré et al., 1991). Oxygen-isotope data from nearby terraces suggest that temperatures offshore at the same time were cooler – by 2.3–2.6°C – than today, probably as a result of enhanced upwelling (Muhs et al., 1992b). Farther south, away from the influence of upwelling, there are clear signs that offshore waters were slightly warmer (Kennedy et al., 1992).

Interglacial sediments in the Canadian Cordillera, mostly in southern British Columbia, have been assumed to have formed during the Last Interglacial but may be older (Clague, 1991).

10.2.5 Last Interglacial Environments of Beringia

A date of 135 ± 5 ka for the widespread Old Crow tephra helped Hamilton and Brigham-Grette (1991) identify Last Interglacial deposits throughout Alaska. Macrofossils and pollen indicate the Last Interglacial climate was at least as warm as the present, perhaps warmer, as suggested by an expanded spruce forest belt. Shallow-water marine fossils from the Alaskan coast suggest that Last Interglacial sea-surface temperatures were slightly higher than today and that Arctic sea ice did not extend south of the Bering Strait during this period, as it does today.

Palaeobotanical research in northeast Siberia found that Last Interglacial temperatures were generally higher than today, as represented by 'range extensions of up to 1000 km for the primary tree species' (Lozhkin and Anderson, 1995: 147).

10.2.6 Last Interglacial Environments of East Asia

Palaeomonsoonal studies provide an important key to understanding both the nature and the periodicity of Quaternary climate changes in East Asia. During the Last Interglacial, the summer monsoon, which brings warm wet winds onshore from the ocean, was stronger than it was during the Last Glacial (although perhaps not today) resulting in a more extensive forest cover and a smaller area of desert than during the Last Glacial. A summary for the last 130 000 years is shown in Figure 10.2. As measured by aeolian quartz flux – a proxy for wind intensity – the Asian winter monsoon strengthened 145–125 ka and weakened 125–73 ka (Xiao et al., 1997a).

Analyses of pollen from emerged marine sediments on central Hokkaido in Japan show that the peak warmth of the Last Interglacial has never been equalled since (Igarashi, 1994). The extension of the Lake Biwa palaeoclimate record back to the Last Interglacial recently has also proved significant in understanding contemporary environments in this region. Of especial note is the biogenic silica flux which reflects changes in diatom productivity in the lake. For the time 141–123 ka, conditions were dry and cold; 123–115 ka the climate became warmer and wetter than today (Xiao et al., 1997b).

Emerged marine terraces of Last Interglacial age associated with transgressive deposits are found in several places in this region (Ota and Odagiri, 1994).

10.2.7 Last Interglacial Environments of Australasia

The possibility of a massive influx of cold water from an Antarctic ice surge (see Chapter 11) reaching New Guinea around 120 ka (Aharon et al., 1980) would have undoubtedly affected contemporary environments but details of these have yet to be uncovered.

Studies of palaeolakes in Australia have proved very revealing about the late Quaternary climates of this part of the Pacific Rim. The Willandra Lakes are now dry, but once covered an area of 1000 km^2; the earliest recorded period of wetness during the Last Interglacial was more than 100 000 years ago (Bowler, 1971). The suggested increase in precipitation ~126–115 ka at Lynch's Crater is based on an abrupt decline of sclerophyll taxa (Figure 10.3).

From 128 to 68 ka in New Zealand, a thick soil developed on the Mamaku ignimbrite plateau signifying a stable land surface and a continuous vegetation cover. These conditions have been interpreted as a response to a warmer, wetter climate during the Last Interglacial compared to the present (Kennedy, 1994). Pollen records of New Zealand vegetation during much of the late Quaternary were shown in Figure 9.5. The expansion of forests is the most marked characteristic of interglacial and interstadial times while herblands expanded at

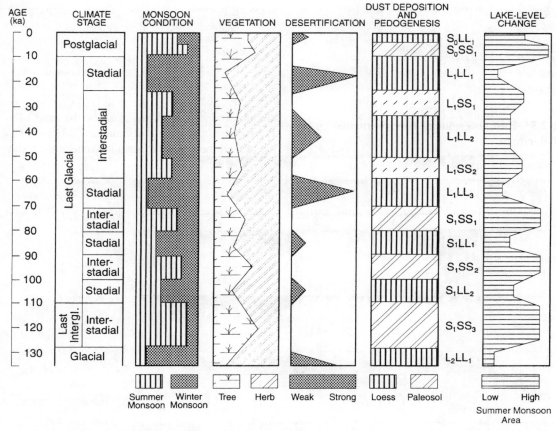

Figure 10.2   Environmental changes during the late Pleistocene and Holocene in central and eastern China
After An et al. (1991)

the expense of forest during colder times. Note that low values of *Nothofagus* and *Phyllocladus* in Figure 9.5 during the Last Interglacial are explicable by the contemporary dominance of *Podocarpus* and do not invalidate the principal conclusion.

10.2.8   Last Interglacial Environments of the Pacific Ocean and Islands

Much information about Last Interglacial environments of the Pacific Basin has been gleaned from studies of the sedimentary record on the Pacific Ocean floor, alluded to above. The only record of corresponding detail known from Pacific islands is from Hawai'i where the ice cap

which developed during the Penultimate Glacial receded (Porter, 1979).

10.3   SYNTHESIS

In a crude sense, Pacific Basin environments of the Last Interglacial were similar to those of the last 12 000 years, and there is no doubt that study of one can aid understanding of the other. Yet there are persistent signs that Last Interglacial environments experienced more extreme conditions than those of the Holocene. Particularly compelling are $\delta^{18}O$ data from ocean-floor records which show that temperatures were higher and/or ice volumes were less during the Last Interglacial. Whether this could mean that the West Antarctic

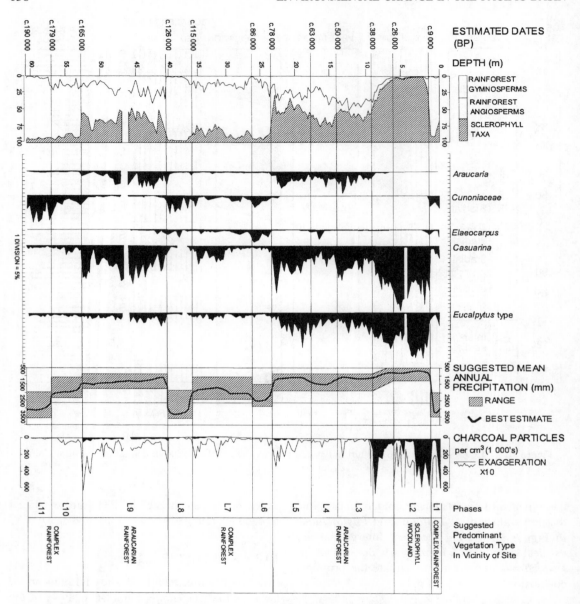

Figure 10.3   Selected parts of the pollen diagram for Lynch's Crater, Atherton Tableland, northeast Australia. Figure reproduced with modifications from Bishop, P. (1994) 'Late Quaternary environments in Australia', ESCAP Atlas of Stratigraphy XIII, Bangkok, 9–27

ice sheet melted (at least partly) during the Last Interglacial is still debated; yet this scenario might be required to explain the widespread evidence for

a Last Interglacial sea-level high stand exceeding its Holocene equivalent by 3–5 m.

CHAPTER 11

The Last Glacial (111–12 ka) in the Pacific Basin

11.1 INTRODUCTION

The Last Glacial was the last major glacial cycle to affect our planet. There is ample evidence to suggest it was one of the most enduring, perhaps the most globally extensive of the late Cenozoic glaciations. Its environmental legacy is still manifest in most parts of the modern Pacific Basin.

The Last Glacial is divided into three separate parts – early, middle and late – largely for ease of discussion (see Table 10.1). In this chapter, each of the three divisions of the Last Glacial is discussed separately for each constituent region of the Pacific Basin.

Owing largely to the widespread use and availability of radiocarbon (^{14}C) dating, the middle and late Last Glacial have become much better known than earlier times for which no comparable technique is available. Yet, while late Quaternary environmental changes have been illuminated by radiocarbon dating, there are problems reconciling the radiocarbon and calendar timescales. Recent work suggests that differences of several thousand years may exist between the two timescales for the middle Last Glacial (Kitagawa and van der Plicht, 1998). Although efforts are made to reconcile events dated by different methods in the rest of this book, these are necessarily imperfect and will require additional attention in the future.

11.2 EARLY LAST GLACIAL (111–74 ka)

This period is coincident with oxygen-isotope stages 5a–d and comprises alternating cool and warm periods. The earliest cool period (stage 5d: 111–99 ka) marks the rapid cooling at the end of the Last Interglacial and was accompanied by sea-level fall. The succeeding warm period (stage 5c: 99–91 ka) saw a rise of sea level to a maximum some 27 m below present mean sea level. Evidence that the termination of this warm period may have been caused by a surge of the East Antarctic ice sheet was assembled by Hollin (1980) and is discussed further below. The second cool period (stage 5b: 91–79 ka) was followed by the second warm period (stage 5a: 79–74 ka), the termination of which may have been an outcome of the Toba (Sumatra) eruption – by far the most voluminous on Earth within the late Quaternary (Rampino and Self, 1993).

11.3 MIDDLE LAST GLACIAL (74–24 ka)

The middle Last Glacial comprised two periods. The earlier (stage 4: 74–59 ka) was cool and saw sea level fall well below its present level. The later (stage 3: 59–24 ka) was warmer and represents the most enduring interstadial within the entire Last Glacial. Sea level attained two maxima during this period – the Göttweig (~40 ka) and Paudorf (~30 ka).

11.4 LATE LAST GLACIAL (24–12 ka)

The third part of the Last Glacial is defined as oxygen-isotope stage 2 and includes the Last Glacial Maximum (22–17 ka), the time during which most palaeoenvironmental records show that the greatest coldness and lowest sea level of the entire Last Glacial were achieved.

The later part of the late Last Glacial was marked by deglaciation throughout the Pacific Basin. Isotope data show that deglaciation occurred in two steps. The earlier – Termination IA – occurred ~14.5–11 ka BP and was characterized by rapid land-ice melt and sea-level rise. The end of Termination IA was marked by the abrupt onset of colder conditions lasting for about a millennium. This period is known as the Younger Dryas and its effects in the Pacific Basin are reviewed in a separate section below. The end of the Younger Dryas was marked by the second step in deglaciation – Termination IB – which is taken as the beginning of the Holocene. This is dated – rather unsatisfactorily on account of the presence of a radiocarbon 'plateau' – to around 10 000 radiocarbon years before present (10 ka BP in the notation of this book) which is taken as approximately equivalent to 12 000 calendar years ago (12 ka).

11.5 LAST GLACIAL ENVIRONMENTS OF ANTARCTICA

11.5.1 Early and Middle Last Glacial Environments

The ice-free valleys around the edge of the Ross Ice Shelf (see Plate 9.1) have long been a fertile source of information about Last Glacial palaeoclimate change in Antarctica. An early Last Glacial ice advance is recorded by carbonates (the Bonney drift) dammed 98–75 ka in the middle Taylor valley by an expanded Taylor Glacier. The Taylor valley is one of the few valleys bordering the Ross Embayment which can fluctuate independently of the Ross Ice Shelf. Since this advance did not occur in most of these other valleys, a likely explanation is that it represents a period of locally increased precipitation rather than lowered temperature (Hendy et al., 1979; Denton et al., 1989a).

11.5.2 Late Last Glacial Environments

Convincing reconstructions of West Antarctica during the Last Glacial Maximum were presented by Stuiver et al. (1981) and Denton et al. (1991). On account of lower sea level, Antarctic ice extended across the continental shelf and thickened around much of its periphery. Many of what had been floating ice shelves during the Last Interglacial became grounded, their bases resting on the ocean floor. Temperatures may have been as much as 10°C lower than today.

The various drifts mapped along one side of the Beardmore Glacier are shown in Figure 11.1A. The Beardmore Drift is that which formed when the glacier thickened during the Last Glacial Maximum in response to grounding of the Ross Ice Shelf (see Chapter 9; Denton et al., 1989b). The drifts along the side of the Hatherton Glacier, some 500 km away from the Beardmore, are shown in Figure 11.1B. The Britannia II Drift is that which formed during the Last Glacial Maximum (Bockheim et al., 1989). Correlation between the different valley drifts was accomplished using various properties of the soils developed on them (Table 11.1).

During the latest part of the Last Glacial, proglacial lakes formed in ice-free valleys around the Ross Sea as a result of the penetration therein by Ross Sea ice lobes. Chronologies of former lake levels obtained from algal deposits confirm the contemporaneity of ice dams and lacustrine deltas in these valleys (Stuiver et al., 1981; Denton et al., 1991).

Fragments of the mollusc *Hiatella solida*, indicative of warmer conditions than at present, found in drift in south Alexander Island, indicate that the area was ice-free 30–32 ka BP, probably an indication of stage 3 interstadial conditions. Around the Last Glacial Maximum, ice caps developed on the Antarctic Peninsula and Alexander Islands. They merged in George VI Sound discharging northwards. The varied evidence of this – from glacial striations, erratics and moraines – was assembled by Clapperton and Sugden (1982). Exposure of land above 700 m indicates significant thinning of ice sheets in this area after 18 ka BP. A 'valley till' interval some 15–14 ka BP indicates the extension of valley glaciers into George VI Sound prior to the isostatic recovery of the area around 13–12 ka BP (Clapperton, 1990b).

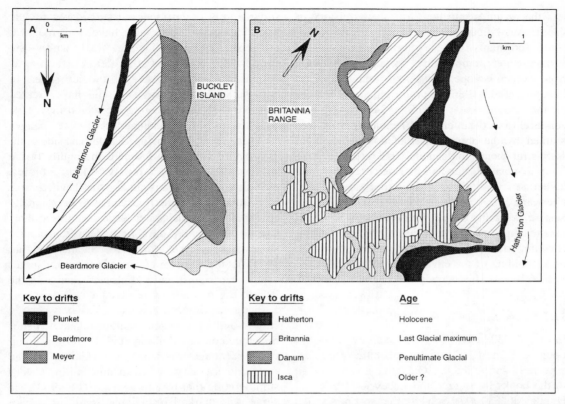

Figure 11.1   Moraines of various ages along the sides of (A) the Beardmore Glacier and (B) the Hatherton Glacier
are thought to have been deposited at times of low sea level when the ice shelf in the adjacent Ross Sea was
grounded and the glaciers thickened in response. See Table 11.1 for suggested ages of the moraines
After (A) Denton et al. (1989b) and (B) Bockheim et al. (1989)

Table 11.1   Soil properties of drifts in the Beardmore, Hatherton and Taylor valleys used to correlate drifts of
particular ages
After Denton et al. (1989a)

Drift (valley)	Staining	Coherence	Matrix salts	Suggested age
Britannia II (Hatherton)	5cd	5b	0c	Last Glacial Maximum
Beardmore (Beardmore)	6c	6b	0c	Last Glacial Maximum
Bonney (Taylor)	8b	9b	0c	Early–Middle Last Glacial
Danum (Hatherton)	10b	26a	8b	Penultimate Glaciation
Meyer (Beardmore)	10b	22ab	9ab	Penultimate Glaciation
Taylor III (Taylor)	22a	29a	12a	(Isotope stage 7)

Other insights into the Last Glacial history of Pacific Antarctica have been acquired from ocean-floor sediments. For example, estimates of surface-water temperature variations obtained from oxygen-isotope analyses of foraminifera led Labeyrie et al. (1986) to propose that the surface of the southernmost Pacific 35–17 ka BP was composed of cold meltwater indicating that large numbers of icebergs had been calving from peripheral ice shelves, a process likely to have been associated with warmer temperatures and higher sea level compared to times before and afterwards.

The possibility of the (West) Antarctic ice sheet having surged periodically during the Last Glacial is an important issue to address, particularly for an understanding of the environmental history of the Pacific Basin. This is reviewed below.

11.5.3 Ice Surges in the Pacific?

In the past 30 years or so, abundant evidence has been obtained of abrupt climate changes, especially during the late Quaternary, summarized in the books by Berger and Labeyrie (1987) and Huntley et al. (1997). Some of the most persuasive evidence comes from ice cores where rapid changes in parameters such as $\delta^{18}O$, dust, acidity and various trace elements have been interpreted as marking major environmental changes perhaps taking only a few decades to accomplish (Dansgaard, 1987; GRIP Members, 1993). These changes may have been associated with rapid changes in the extent of sea ice associated with rapid temperature changes.

At present, the lateral extent of the terrestrial Antarctic ice sheet is constrained by the sea surface. The same situation undoubtedly prevailed during the Last Glacial; since the sea level was much lower, it would be expected that the ice sheet extended across a much greater area than it does today. As sea level rose during the late Last Glacial, the ice closest to it would gradually have become detached from the land and begun to float. That ice which had existed upslope of the old shoreline would then no longer be buttressed by ice below, so might have

slid abruptly downslope and surged out to sea. Such ice surges might also have occurred as a consequence of sea-level fall when the supporting effect of the sea surface was withdrawn from the base of the ice sheet. In both situations, ice surges would have created massive shelves of floating ice which may have been causes of major and abrupt climate changes (Wilson, 1969; Hollin, 1972, 1980).

Most researchers still regard the reality of ice surges somewhat cautiously. Many studies, including some in the southernmost Pacific, have found that the Southern Ocean was occasionally covered by a 'meltwater lid' during the Last Glacial which may have been derived from ice surges (Labeyrie et al., 1986; Shemesh et al., 1994) but the question of whether or not ice surges contributed significantly to climate change in the Pacific Basin is not yet resolved.

The related question about a link between ice surges and abrupt sea-level changes is better documented, with Aharon et al. (1980) outlining the evidence from Papua New Guinea for an Antarctic ice surge ~120 ka and Hollin (1980) summarizing that for one ~95 ka. The ~95 ka ice surge was marked by a rapid sea-level rise of around 16 m in many parts of the world which is difficult to explain any other way. As with the huge outbursts of glacial meltwater from proglacial lakes during deglaciation, such as that which may ultimately have instigated Younger Dryas cooling, increasing recognition of the role of ice surges in climate change is inclining many researchers towards a more catastrophist view of environmental change than that which has dominated this subject for the last hundred years.

Comparable ice-surge events to those envisaged for Antarctica have been documented in the North Atlantic. It has been suggested that these surges may have reduced the North Atlantic thermohaline circulation triggering abrupt responses elsewhere (Broecker, 1997). Yet, perhaps because the Pacific Basin is influenced to a lesser extent by thermohaline circulation changes, the effects of these events have not proved widely apparent therein.

11.6 LAST GLACIAL ENVIRONMENTS OF SOUTH AMERICA

11.6.1 Early and Middle Last Glacial Environments

Little information is available for the early Last Glacial in South America. A major glacial advance, named Llanquihue I, occurred some 70–65 ka, most of the evidence for which is found in the Chilean Lakes region of the southern Andes (see Figure 9.2). The stage 3 interstadial during the middle Last Glacial in Pacific South America is represented by peats, formed ~43–33 ka in the Ecuadorian Andes, and others, formed >57–30 ka in the Chilean Lakes region (Clapperton, 1993a).

Evidence of changing environments from lake levels – one of the clearest indicators of changing precipitation – is summarized in Table 11.2. In addition to a dry period around 33 ka BP, these data give clear support to the idea of a wetter period around 27 ka BP which resulted in the appearance of many high-altitude lakes including palaeolake Minchin (Wirrmann and Mourguiart, 1995).

The suggestion by Colinvaux and Liu (1987) that temperatures may have reached their Last Glacial minimum in South America by some 26 ka BP is consistent with many other data. Glaciers in

the north and central Andes had reached their maximum Last Glacial extent by around 27 ka BP. Although these low temperatures may have been maintained throughout the Last Glacial Maximum (22–17 ka BP), a reduction in precipitation caused glacier recession.

11.6.2 Late Last Glacial Environments

There is increasing evidence (see above) that maximum ice extent associated with the Last Glacial Maximum in Pacific South America was widely attained during the latest part of the middle Last Glacial. A lack of precipitation during the Last Glacial Maximum caused glaciers to recede irrespective of falling temperatures.

Exceptions to this situation occurred in the southern Andes where precipitation is received in much greater quantities (and far more regularly) from westerlies compared to areas farther north. Only in this area within Pacific South America is there evidence that the maximum ice advance coincided with the time of maximum cooling (Clapperton, 1993b). The best example is the Llanquihue II glacial advance 20–19 ka BP in the Chilean Lakes region (see Figure 9.2). Modelling of the situation here suggests that the Last Glacial Maximum was marked by an overall fall in

Table 11.2 Late Last Glacial precipitation variation in the Andes indicated by lake-level fluctuations

Lake	Times of high lake level (ka BP)	Times of low lake level (ka BP)	Source of information
Palaeolake Chiuchiu-Calama, Chile	~27.4		Ochsenius (1974)
Lago de Junin, Peru		39–24	Hansen et al. (1994)
Laguna Kollpa Kkota, Bolivia	14–12.6	>14	Seltzer (1994)
Lago Llanquihue, Chile	15–13		Porter (1981)
Lago Minchin, Bolivia	29.1–24.8[1]		Servant and Fontes (1978)
Laguna de Tagua Tagua, Chile[2]	28.5–14.5	~33	Heusser (1990)
Lago Tauca, Bolivia	28–27 12.5–11[3]		Servant and Fontes (1978)

[1] Dates calculated by Clapperton (1993a)
[2] 34°30′S within zone of westerlies and thus much wetter than elsewhere in South America during the Last Glacial maximum
[3] May be erroneous owing to contamination; should probably coincide with dates from Laguna Kollpa Kkota (Seltzer, 1994)

temperature of ~3°C which combined with northward migration of precipitation belts by some 5° to allow the Patagonian ice cap, for example, to extend as much as 560 m lower than today in places (Hulton et al., 1994). Along the central Pacific coast of South America, intensified ocean–atmosphere interactions may have caused expansion of the dry region and the enhancement of its aridity during the Last Glacial Maximum (Williams et al., 1993).

Last Glacial climates in the arid northern Chilean Andes were investigated by Veit (1993). In this part of the Andes, glacial moraines indicate relatively humid phases, the two most conspicuous of which probably occurred either side of the Last Glacial Maximum during which temperatures in the area fell 6–8°C. The earlier humid phase was probably coincident with high lake levels elsewhere in the Andes (see Table 11.2). The later humid period was marked by lake formation elsewhere in the region around 15 ka BP, as well as a glacier advance in southern Chile.

Other Last Glacial ice fluctuations have also been explained by precipitation changes. The widespread Llanquihue III advance occurred 14.8–12.8 ka BP (see Figure 9.2) and – like the three late Last Glacial Huacané readvances proposed as a result of analyses of ice cores at Quelccaya (Mercer and Palacios, 1977) – probably marked a rise in precipitation superimposed on a warming trend rather than a discrete cooling event.

One of the most successful ways in which environments of the middle and late Last Glacial in Pacific South America have been investigated is by looking at the types of pollen blown and/or washed off the western slopes of the Andes and incorporated into Pacific ocean-floor sediments. For example, pollen analysis of sediments from a core off the Ecuador coast showed that around the Last Glacial Maximum, grassland expanded and rainforest areas contracted as a result of the decreased temperature and precipitation (Heusser and Shackleton, 1994). A good example comes from podocarp forest in Ecuador, the lower limit of which 33–26 ka BP fell at least 700 m, suggesting a temperature fall of at least 4.5°C, probably closer to 6°C (Colinvaux and Liu, 1987).

Lake-level data (see Table 11.2) indicate a wet period 14 ka BP which, combined with evidence for similar conditions around 27 ka BP (see above), suggest that the Last Glacial Maximum was much drier. The case for a truly arid Last Glacial Maximum in South America is reviewed in the following section.

After 14–12 ka BP, regional climates became more distinct in Pacific South America, perhaps reflecting weakening of glacial forcing, with temperatures in many places reaching near-modern levels around 12 ka BP (Markgraf, 1989).

Initial deglaciation in the Cordillera Oriental of northern Peru occurred 13.5–12.1 ka BP, followed by a period of glacier expansion 12.1–10.3 ka BP since which time all cirques in the area have been ice-free (Rodbell, 1993). In southernmost South America, initial deglaciation occurred around 18 ka BP, followed by a readvance of glaciers 15–14 ka BP, and subsequent retreat thereafter (Porter et al., 1992). The 15–14 ka BP readvance was regarded as a global montane phenomenon by Broecker and Denton (1990).

A reminder of the role of catastrophic events in environmental change is provided by the massive landslide which occurred 13.8 ± 2.7 ka BP along the north Peru coast and produced a proportionately large tsunami (Bourgois et al., 1993).

11.6.3   Aridity Around the Last Glacial Maximum in South America

The best evidence for aridity around the Last Glacial Maximum in Pacific South America is provided by reduced glacier extent in the northern and central Andes. Maximum extent was reached on either side of the Last Glacial Maximum when precipitation was sufficiently high to allow glaciers to advance despite the effect of higher temperatures (see above). Cooling around the Last Glacial Maximum 'apparently reduced the flux of moisture to the tropical and subtropical Andes, resulting in a fall in lake levels and recession of glaciers from limits reached a few millennia earlier' (Clapperton, 1993b: 201).

There is some debate about the cause of aridity during the Last Glacial Maximum in Pacific South America. Some authors regard aridity as having been endemic to the tropics during Quaternary glacials. Others regard Last Glacial aridity here as having been an outcome of the northward migration of the zone of heaviest precipitation, as inferred from both palaeoecological and geomorphological studies (Heusser, 1990; Clapperton, 1993b). Others regard the evidence as better explained by a poleward movement of this zone (Markgraf, 1989, 1993). One way of resolving such conflicting views is to suppose that aridity during the Last Glacial Maximum in South America was – as it is today – highly localized and consequently difficult to explain in terms of regional climate controls (Colinvaux and Liu, 1987). A good example is provided by Lake Kollpa Kkota in the Bolivian Andes where aridity during the Last Glacial Maximum may have been only seasonal and not much different from the modern situation (Seltzer, 1994).

Sea-level change may also have contributed to aridity around the Last Glacial Maximum. Lowering of sea level caused many lowland rivers to incise their channels and water tables to fall. The consequence was a drying-out of the soil surface, and a reduction in its vegetation cover. In turn, these conditions made the landscape more susceptible to mass movements and to wind erosion, one effect of which in Pacific South America was to cause dust levels during the Last Glacial to be as much as 200 times greater than today (Thompson et al., 1995).

11.7 LAST GLACIAL ENVIRONMENTS OF CENTRAL AMERICA

Aridity associated with the Last Glacial Maximum commenced around 30 ka BP in the Zacapu Lake basin of Mexico, and around 27–25 ka BP in the Cuenca de Mexico. The tree line in the mountains of Central America was 600–800 m lower than it is today around the Last Glacial Maximum and temperatures were perhaps 4°C lower (Bradbury, 1989; Markgraf, 1989). Recent work suggests that the magnitude of temperature fall in montane

Costa Rica and Guatemala was 7–8°C (Islebe and Hooghiemstra, 1997).

In her review of late Last Glacial palaeoclimates of Central America, Vera Markgraf (1993) found evidence to suggest that Last Glacial Maximum climates were also drier than at present. Recent data from Chalco Lake exemplify this. Around the Last Glacial Maximum, Chalco was a swamp with fluctuating amounts of inflow. A period of variable yet higher humidity, as suggested by increased proportions of *Fragilaria* diatoms, occurred around 13–10 ka BP (Urrutia-Fucugauchi et al., 1995); evidence for this period of wetness is also found elsewhere in Central America (Markgraf, 1989).

Fluctuations in the extent of ice around some Mexican volcanoes during the Last Glacial were analysed by White (1986). Traces of an early or middle Last Glacial ice advance were found on Ajusco, Iztaccíhuatl and Malinche, where it was dated to 36–32 ka BP. Traces of a late Last Glacial advance about 12 ka BP have also been found here (Heine, 1978).

11.8 LAST GLACIAL ENVIRONMENTS OF NORTH AMERICA

11.8.1 Early Last Glacial Environments

An early Last Glacial palaeoclimate record from an emerged marine terrace on the Oregon coast formed 85–80 ka shows temperature and sea-level changes consistent with those deduced from the oxygen-isotope record (Klein et al., 1997). Among these is the stage 5a sea-level maximum (~80 ka), emerged shorelines of which have also been dated at several places farther south along this coast (Muhs et al., 1994).

11.8.2 Middle Last Glacial Environments

A glaciation in the Canadian Rockies is thought to have occurred during the early–middle Last Glacial (>59 ka) but evidence is poorly exposed and dates cannot be interpreted unambiguously. Yet this glaciation was more extensive than the last one in parts of the central Yukon and the foothills of the

Rocky Mountains where ice began advancing slowly about 30–25 ka BP (Clague, 1991).

The Queen Charlotte Islands, off the west Canada coast, were never covered by the Cordilleran ice sheet but rather developed their own ice cap. The earliest known Last Glacial ice advance occurred here >51 ka. The Olympia Nonglacial Interval (~>59–25 ka), also known throughout much of mainland Pacific North America, is represented on the Queen Charlotte Islands by a landscape essentially similar to the present. The main Last Glacial cold event in the Queen Charlotte Islands began 30–25 ka BP (Clague, 1989a).

11.8.3   Late Last Glacial Environments

Much is known about conditions during the late Last Glacial in the unglaciated parts of Pacific North America. Sea-level fall around the start of this period, marking the transition from isotope stage 3 to stage 2, caused dunes to develop along the coast of south-central California as a result of exposure of the sand-covered continental shelf (Orme, 1992).

Pollen from Clear Lake in northern California indicates that junipers and pines grew in the catchment during the Last Glacial Maximum rather than the modern oak-dominated vegetation, probably in response to temperatures 7–8°C lower and precipitation as much as 350% greater than today. The tree line in the Olympic Mountains in the northwest United States was some 1000 m lower than today; temperatures were about 6°C lower (Thompson et al., 1993).

One result of the compression of mid- and low-latitude zonal climate belts at this time was the southward movement of the polar jet stream along the western seaboard of North America during the period of increasing cold and its movement in the other direction when temperatures began rising at the end of the Last Glacial Maximum. The level of Mono Lake in California was shown to have been highest 18 ka BP and 13.1 ka BP when the polar jet stream was overhead (Benson et al., 1998).

Around the Last Glacial Maximum in the western United States, mountain glaciers –

unconnected to the Cordilleran ice sheet – reached their maximum extent some 20 ka BP (Porter et al., 1983; Thompson et al., 1993). Tundra bordered the edge of the Laurentide ice sheet here in a zone 100–200 km broad. The Cordilleran ice sheet extended across much of Pacific North America during the late Last Glacial (Figure 11.2). In the Rocky Mountains of Canada, maximum ice extent was not attained until 14.5–14 ka BP (Clague, 1991), a situation consistent with aridity during the Last Glacial Maximum inhibiting ice-sheet growth despite lower temperatures.

Vegetation change, lake desiccation and glacier recession provide some of the clearest evidence of the course of Last Glacial deglaciation in the western United States. In the southern Sierra Nevada of California, juniper and pine persisted until around 12.5 ka BP when an abrupt increase in oak pollen suggests the onset of near-modern climate conditions (Adam and West, 1983). Further north, deglaciation was well underway by 14 ka BP and essentially complete in most areas by 12 ka BP (Porter et al., 1983).

A comparable picture has emerged for glaciers fringing the Great Basin, where high lake levels 16–13 ka BP suggest a period of marked expansion associated with increased precipitation, not necessarily experienced along the coast (Hostetler and Bartlein, 1990). A similar record from Death Valley has been interpreted to mean that lakes existed throughout the later part of the Last Glacial (35–10 ka BP) without drying out (Li et al., 1996).

The importance of recognizing significant spatial variations in palaeoprecipitation levels was underlined by the study of Smith and Street-Perrott (1983) who found that, for most of the late Last Glacial, 'pluvial' lakes closest to the Pacific remained full whilst those farther inland exhibited more variation; a conspicuous feature of the latter is the progressive reduction in lake level 13–10 ka BP. An excellent example is that of Tule Lake in the northern Great Basin (Bradbury, 1992). Changes in the proportions of benthic and surface-dwelling diatoms, in particular *Aulacoseira ambigua*, indicate arid conditions around the Last Glacial Maximum (~25–16 ka BP

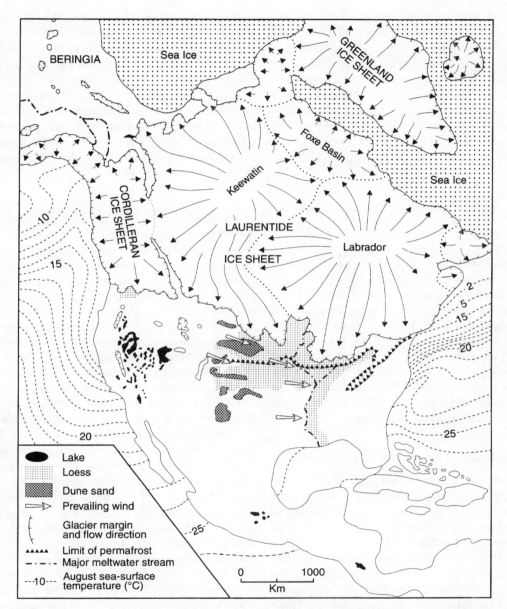

Figure 11.2   Last Glacial extent of Laurentide ice and associated environments in North America
After Porter (1988)

here). A sudden increase in diatom productivity by 15 ka BP marks the onset of warmer, wetter conditions.

Deglaciation of the Cordillera occurred 14–10 ka BP and is represented by a rise in tree lines and the slow replacement of lowland tundra or steppe by forest (Thompson et al., 1993). Remarkable analogues of the precise nature of deglaciation – involving thinning, stagnation and frontal retreat – have been obtained by the study of historic recession of glaciers, particularly those in the Saint Elias Mountains in Alaska and northwest Canada. Rates of ice loss of as much as 300 km^2 in the last 200 years – more than 50% of the mass of the Grand Pacific–Melbern Glacier system – are thought comparable to rates in Cordilleran valleys around the end of the Last Glacial (Clague and Evans, 1993).

11.9   LAST GLACIAL ENVIRONMENTS OF BERINGIA

11.9.1   Early Last Glacial Environments

The early Last Glacial (stage 5b) cold period was marked by ice advances from mountains throughout Beringia. Tephras interbedded with various periglacial and glacial deposits in the Yukon Basin provide the clearest dates for this cold period, which may have begun around 90 ka and ended perhaps 80 ka here (Hopkins, 1982). Glacial erratics in ocean-floor deposits of the north Pacific show that during the early Last Glacial (~80 ka), icebergs were calving from glaciers along the south coasts of Beringia, including the Aleutian Islands (Kent et al., 1971), suggesting the onset of warm stage 5a conditions.

11.9.2   Middle Last Glacial Environments

Following the cool and/or wet stage 4, there was a relatively mild, mesic, though dry compared to today, interstadial (stage 3) throughout Beringia in which maximum warmth was attained 35–33 ka BP (Anderson and Lozhkin, 1999).

Controversially, it has been suggested that sea level reached close to its present level at the beginning of this interstadial, as evinced by erratics within the glaciomarine Flaxman Formation in northern Alaska (Hopkins, 1979). At Bristol Bay, interstadial conditions >33 ka are indicated by an abundance of shrubs compared to the time of the Last Glacial Maximum; elsewhere in Alaska, forested conditions 70–30 ka also indicate interstadial conditions which may have been slightly warmer than those of today (Ager, 1982; Hopkins, 1982).

Palaeoenvironmental data for western Beringia were summarized by Lozhkin (1993) who recognized the effects of two cold periods within warm stage 3. The first occurred 45–39 ka BP, the second 33–30 ka BP. Most information is from dating of the bones of fauna thought to be diagnostic of particular conditions although some pollen in buried soils was also used. The warm conditions of stage 3 in western Beringia were marked by larch forests in the north, with birch–alder forest farther south.

11.9.3   Late Last Glacial Environments

The mild stage 3 interstadial was followed by stage 2 which included the Last Glacial Maximum. During this period, most of north and south Alaska was covered with ice but the central part, around the Yukon and Kuskokwim valleys, remained ice-free. Most of western Beringia remained ice-free at this time (Bespaly et al., 1993), as did the north Pacific Ocean (Mann and Hamilton, 1995). Snowlines during the Last Glacial Maximum were depressed 300–700 m in southwest Alaska (Mann and Peteet, 1994).

Climates of the Last Glacial Maximum in Beringia were also very dry, an important reason for the lack of widespread ice cover in this region. By analogy with the preceding interstadial, when climate was also drier than at present, the enhancement of aridity during the Last Glacial may have been the outcome of lowered sea level which caused emergence of the floor of the Bering Strait and much of the adjacent Arctic Ocean (Figure 11.3). The exposure of so much land imposed an element of continentality on the climate of Beringia which is absent today. It may therefore be that, at a crude level, Last Glacial aridity in this region was not an expression of a global trend but – as might be the case for Pacific South America – of only local conditions.

The biotic effects of the onset of Last Glacial Maximum coldness in Beringia are clear from records of Last Glacial insect and large-mammal faunas, many of which were conspicuously absent from the region during the Last Glacial Maximum

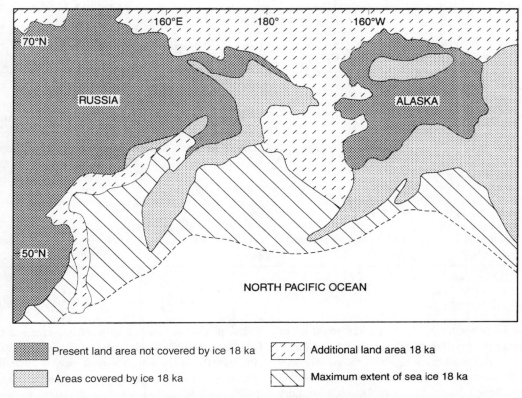

Figure 11.3   Extent of land and sea ice around the Last Glacial Maximum in Beringia
After Starratt (1993) with permission of Gebrüder Borntraeger Verlagsbuchhandlung

(Table 11.3). Other plants and animals evidently 'wintered' through the Last Glacial Maximum in the region; examples include brown bears and other large mammals in the Alexander archipelago (Heaton et al., 1996), dwarf birch (*Betula nana*) in the Bering land bridge (Elias et al., 1997), and a range of species on Kodiak Island (Karlstrom and Ball, 1969).

In those parts of eastern Beringia not covered with ice during the Last Glacial Maximum, herb-tundra pollen dominated although there are indications that spruce forests were present in places (Ager, 1982; Anderson and Brubaker, 1994). The issue as to whether the ice-free landscape of Alaska around the Last Glacial Maximum was treeless or marked by scattered groves is still controversial. Some of the sediments in which tree pollen is found may have been reworked; the tree

pollen may thus post-date the Last Glacial Maximum. Yet the evidence seems to favour the presence of some trees, not least because birch and poplar pollen becomes more abundant so rapidly in Beringian sediments associated with deglaciation 14–11 ka BP that trees must have existed in refugia within this region rather than only far to the south in central Canada as once supposed (Ager, 1982; Edwards and Barker, 1994).

Tundra and forest-tundra spread through most of western Beringia around the Last Glacial Maximum, the end of which was marked by the rapid replacement of tundra with birch and larch forest by 12.5–12 ka BP in most places (Lozhkin, 1993). In the Aleutians, deglaciation occurred 12–10 ka BP (Black, 1980), an event which also left its mark in diatom assemblages in the Bering Sea (Starratt, 1993).

Table 11.3   Well dated faunas selected to show species variations in Beringia during the Last Glacial
After Schweger et al. (1982)

Species	Middle Last Glacial[1] (mild conditions)	Last Glacial Maximum[2] (cold dry conditions)	Late Last Glacial[3] (mild conditions)
Alces alces (moose)	Present	Absent	Present
Alopex lagopus (arctic fox)	Present	Absent	Present
Dicrostonyx torquatus (collared lemming)	Present	Absent	Present
Lemmus sibiricus (brown lemming)	Present	Absent	Present
Lepus arcticus (arctic hare)	Present	Absent	Present
Lepus americanus (snowshoe hare)	Present	Absent	Present
Rangifer tarandus (caribou/reindeer)	Large numbers present	Small numbers present	Large numbers present
Saiga spp. (saiga)	Absent	Present	Absent

[1] Boutellier Interval
[2] Duvanny Yar Interval
[3] The 'birch invasion'

The well documented and rapid 'invasion' of Beringia by birch, particularly the dwarf birch shrub, around the beginning of deglaciation some 14 ka BP has been explained by flooding of what is now the Bering Strait and the replacement of a dry continental climate with moister maritime climates throughout Beringia (Young, 1982; Anderson and Brubaker, 1994). This idea was questioned by Elias et al. (1997) on the grounds that the land bridge was not submerged until after the spread of birch. In fact, they suggest that the land bridge served as a refugium for birch during the Last Glacial Maximum and that its outward spread from there during the late Last Glacial was a result of the start of postglacial warming in the region.

The step-like nature of deglaciation is conspicuous in parts of the Pacific Rim. Good evidence comes from oxygen-isotope analysis of foraminifera in ocean-floor sediments just west of the Kamchatka Peninsula. Rapid falls 14–13 ka BP, 11–10 ka BP and 7.5–5.5 ka BP are clear (Figure 11.4).

Late Last Glacial environmental change on the Bering land bridge, which may have been uncovered and available for migration by terrestrial animals (including humans) until about 11 ka BP (see Figure 11.3), was documented by Elias et al. (1997). Their main conclusion was that 'the land bridge, even though it was not ideally suited to grazing animals, provided sufficient grassy vegetation to allow safe passage from Siberia to Alaska and *vice versa*' (Elias et al., 1997: 305). Yet, if lowered sea level allowed the modern Bering Strait to be crossed on foot, it was the mountain and piedmont glaciers to the east which impeded further movement in this direction. The question of the precise timing of initial human colonization of North America therefore hinges on the opening of an 'ice-free corridor' (see Chapter 12) through these mountains, not simply on the existence of a land connection between Russia and Alaska.

A related point which has worried some scientists for a long time is that some evidence points to an initial west–east crossing of Beringia by humans some time during the Last Glacial Maximum when temperatures are thought to have reached their Last Glacial minimum. This scenario requires us to believe that these humans either knew no alternative to enduring this extreme cold and/or were so dependent on the herds of large animals which they were following that they could

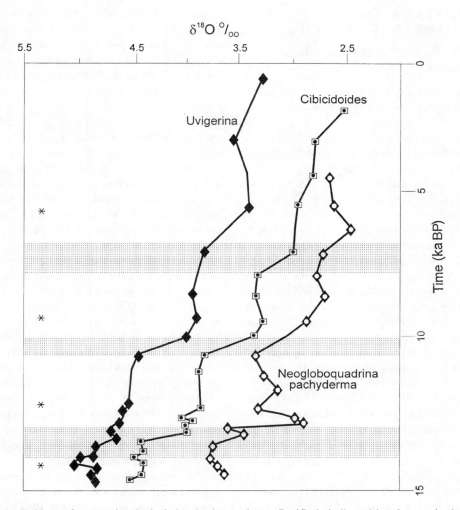

Figure 11.4 Evidence for stepwise deglaciation in the northwest Pacific is indicated by changes in the oxygen-isotope ratios of the benthic foraminifera *Uvigerina* and *Cibicidoides*, and the planktonic foraminifera *Neogloboquadrina pachyderma*

Reprinted from *Earth and Planetary Science Letters*, 111, Keigwin, L.D., Jones, G.A. and Froelich, P.N. 'A 15 000 year palaeoenvironmental record from Meiji Seamount, far northwestern Pacific', 425–440. Copyright 1992, with permission from Elsevier Science

not turn back to warmer regions. Both options could be true but have a ring of implausibility about them. This is overcome by Charles Repenning (1993) who has suggested, from his studies of Beringian fossil faunas, that the Last Glacial Maximum may have been much warmer than most scientists think and that it was additional precipitation rather than lower temperatures which fuelled ice expansion at the time. While this suggestion conflicts with many models for global climate change during the late Quaternary, it cannot be denied that much of the evidence described in this chapter concerning terrestrial conditions during the Last Glacial Maximum can be explained largely as the result of precipitation, not temperature changes.

11.10   LAST GLACIAL ENVIRONMENTS OF EAST ASIA

11.10.1   Early and Middle Last Glacial Environments

One of the only early Last Glacial climate records in this region comes from Korea where Park (1992) found evidence for cooler conditions ~86 ka, then an interstadial 78–37 ka. Changes in the Japan Sea throughout the Last Glacial are discussed in a separate section below.

Various analyses on cores from Lake Biwa bear out this broad picture of early and middle Last Glacial climate change in Japan (Xiao et al., 1997a, b). Of particular interest in assessing the strength of the winter monsoon is aeolian quartz flux, which was higher when this monsoon was stronger. A strengthening ~73–53 ka BP coincides with stage 4 cooling during which conditions here and on the adjacent continent were drier; this situation, combined with a greater wind intensity, resulted in a greater influx of aeolian material to Lake Biwa. Weakening of the winter monsoon ~53–20 ka BP marked stage 3 warming associated with moister conditions throughout this region and a reduction in aeolian quartz input to Lake Biwa.

Dry conditions 70–50 ka in Japan are also indicated by pollen data and have been explained by a weakening of the southwest monsoon. Stage 3 warming was accompanied by wetter conditions which were replaced by drier ones around 33 ka marking the start of stage 2 cooling (Heusser and Morley, 1990).

As in Beringia (see Table 11.3), the appearance and disappearance of large mammals has proved an invaluable aid in understanding middle Last Glacial climate change in East Asia. In the northern part of this continental region, for example, a cold interval 35–36 ka BP was identified by these means, particularly the abundance of woolly mammoth and woolly rhinoceros.

11.10.2   Late Last Glacial Environments

Around the Last Glacial Maximum in the northeast of continental East Asia, temperatures were some 8–10°C lower than today (Winkler and Wang, 1993). The climate of this region was uniformly drier, probably because of increased continentality associated with lowered sea level (Wang, 1984). Vegetation zones in eastern China were displaced farther south than might be expected because the comparatively subdued topography of the inland region – unlike most parts of the Pacific Rim – allowed cold continental air to reach the coast (Winkler and Wang, 1993). A strengthening of the Asian (winter) monsoon accompanied cooling during the Last Glacial Maximum, a phenomenon manifested by threefold increases in dust flux over Korea and Japan compared to interstadial or interglacial times (see Table 15.6).

Changes in the mammal fauna of northern East Asia have also been the basis for identifying the Last Glacial Maximum here 23–12 ka BP. Mammal fossils, particularly those of the woolly mammoth (*Mammuthus primigenius*), in Japan have been used to infer that there was a land connection exposed at this time between mainland East Asia and Hokkaido. Mammoth remains on the floor of the Japan Sea south of Hokkaido once led researchers to suggest that land bridges had connected Hokkaido to the rest of Japan at this time but it has since been demonstrated that these remains were carried out from the Asian mainland by the Huanghe River (Kamei, 1989).

The course of deglaciation is recorded in the vegetational history. In northern China, for example, a cold steppe existed 18–13 ka BP; 13–11.6 ka BP, pollen from boreal conifers became abundant; after 11.6 ka BP, hardwood forest dominated by basswood (*Tilia*) is a sign of increasing warmth (Liu, 1988). An 'invasion' of birch comparable to that in Beringia (see above) occurred in the Beijing lowlands during the latest Last Glacial, some 12 ka BP, indicating temperatures 4–5°C lower than today (Chen, 1979).

The observation that abundances ~14–13 ka BP of the planktonic foraminifera *Globigerinoides sacculifera* and *Pulleniatina obliquiloculata* increased rapidly in a deep-sea core from the South China Sea (Figure 11.5) suggests that the

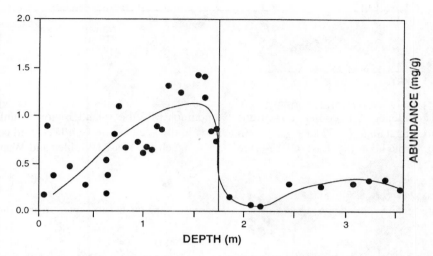

Figure 11.5   Abundance of the shells of the planktonic foraminifera *Pulleniatina obliquiloculata* as a function of depth in a core from the South China Sea. The rapid increase in abundance about 1.75 m depth has been dated to ~13.5 ka BP and may indicate the rapid afforestation of the nearby continent associated with a correspondingly rapid increase in precipitation and temperature levels

Reprinted with permission from *Nature* (Broecker et al., New evidence from the South China Sea for an abrupt termination of the last glacial period, volume 333, 156–158). Copyright (1988) Macmillan Magazines Limited

end of the Last Glacial Maximum in this area was quite abrupt. Broecker et al. (1988) argue that such observations reflect changes in ocean-surface productivity, which directly manifest the change from high levels of sediment being discharged from the continent under dry conditions (when savanna existed) to lower levels when the climate became wetter and rainforest returned.

Cold and dry conditions in the Japanese islands during the Last Glacial Maximum can be largely explained by their land connection with the mainland which increased the continentality of their climate. Forest dominated by spruce and fir persisted through most of Japan at this time, suggesting temperatures about 8°C lower than today and precipitation perhaps one-third of present values (Tsukada, 1983). Owing to sea-level rise and the rapid re-establishment of maritime climates, the later warming 15–10 ka BP was more rapid in Japan than on what is now the East Asia continental margin. In Japan, terminal Last Glacial warming is marked by changes in forest composition: birch, alder and oak became common (Igarashi, 1994).

Analyses of lake cores from Taiwan island, which was also connected to the East Asian mainland during the Last Glacial Maximum, imply that cool, moderately dry conditions prevailed at this time. Rapid warming marked deglaciation 14–12 ka BP (Tsukada, 1967, 1983).

11.10.3   The Japan Sea During the Last Glacial

Ocean palaeotemperatures for the late Last Glacial in the Japan Sea have proved instructive concerning the nature of environmental change in East Asia at this time. The present current system is shown in Figure 11.6 together with the modern depths of the various straits connecting the Japan Sea to the rest of the Pacific.

Between 85 and 27 ka, the warm Tsushima Current did not enter the Japan Sea, perhaps in part because it had been weakened, and perhaps also because of the shallowing of the Tsushima Strait. Indeed it has been suggested that humans and other animals crossed it routinely during the Last Glacial (Reynolds and Kaner, 1990). Between 27 and 20 ka

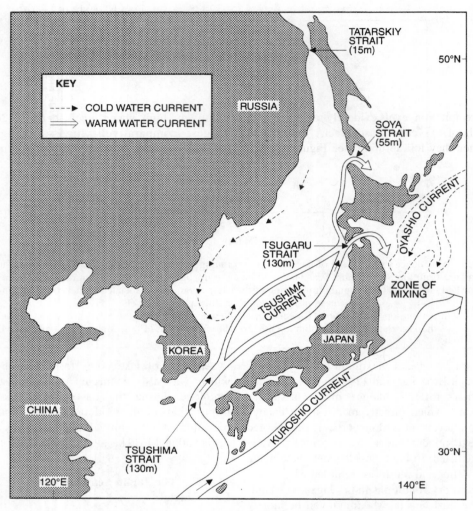

Figure 11.6   Ocean currents and maximum depths of straits in the Japan region
After Chinzei et al. (1987) and Oba et al. (1991)

BP, fresh water from the Huanghe River, reaching the sea on the west side of the Korean Peninsula, dominated inputs through the Tsushima Strait (possibly because the climate was wetter). This fresh water caused the water column to become stratified, and led to the extinction of deep-sea organisms. In the late Last Glacial, 20–10 ka BP, the cold Oyashio Current penetrated the Japan Sea through the Tsugaru Strait, facilitating deep-water ventilation and allowing a benthic biota to be re-established. The reintroduction of the warm

Tsushima Current through the Tsushima Strait about 10 ka BP heralded the establishment of modern conditions (Oba et al., 1991).

11.11   LAST GLACIAL ENVIRONMENTS OF AUSTRALASIA

11.11.1   Early Last Glacial Environments

The early Last Glacial in Australia was associated with a period of high precipitation some 110–

80 ka, indicated by high lake levels and increased fluvial activity. Between ~80 ka and the Last Glacial Maximum, the climate became progressively drier, as shown by falling lake levels and decreasing fluvial activity. A good example is given by the record from Lynch's Crater on the Atherton Tableland in northeast Queensland where precipitation levels evidently remained high for much of the early Last Glacial until about 78 ka when they fell abruptly (see Figure 10.3)

Vegetation changes in New Zealand indicate the occurrence of an interstadial centred on 80 ka (McGlone and Topping, 1983).

11.11.2 Middle Last Glacial Environments

Cooling >50 ka BP in New Guinea is marked in the central highlands by the lowest Last Glacial altitudinal occurrence of cold-adapted taxa (Haberle, 1998). An expansion of tropical forest

~32–28 ka BP signals a slight warming, which may have coincided with a sea-level maximum on the Huon Peninsula (Chappell and Veeh, 1978).

Lake Carpentaria existed in the modern Gulf of Carpentaria when this was above sea level 35–12 ka BP (Torgersen at al., 1988; Figure 11.7). Relatively high water levels here 35–26 ka BP suggest comparatively high (although substantially lower than present) precipitation levels.

The Early Margaret glacial deposits of Tasmania were produced during isotope stage 4 (Clapperton, 1990a) which was probably also characterized by dry conditions throughout Pacific Australia. Subsequent wetness 50–35 ka in southeast Australia is indicated by high lake levels and renewed flow in some palaeochannels (Kershaw and Nanson, 1993). Fossil pollen in Tasmania indicates interstadial conditions 48.7–

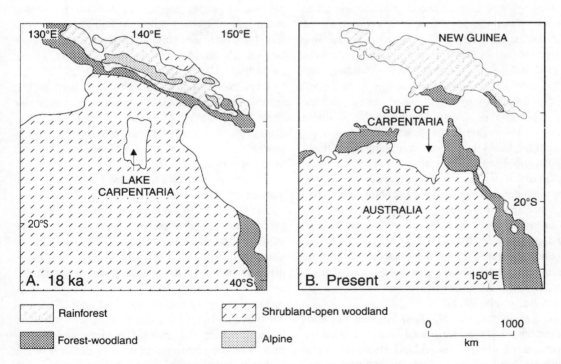

Figure 11.7   Shorelines, lakes and vegetation pattern in tropical Australasia around (A) the Last Glacial Maximum compared to (B) present
After Markgraf et al. (1992) and Kershaw et al. (1994)

25 ka BP, during which a cool, moist climate persisted, in contrast to the cold alpine conditions of the Last Glacial Maximum (Clapperton, 1990a). Aridity persisted throughout the middle and late Pleistocene at Lynch's Crater with a pronounced fall of precipitation ~38 ka BP, deduced from a change from araucarian vineforest to savanna woodland, and a lowering of lake levels here (see Figure 10.3).

Palaeovegetation records from New Zealand indicate a second Last Glacial interstadial ~50–25 ka (McGlone and Topping, 1983). Conclusions from pollen data supported by dendro-chronological analysis suggest that the warmest time of this interstadial occurred ~41–34 ka, when a diverse forest dominated by the diagnostic kauri (*Agathis australis*) existed in northernmost North Island (Ogden et al., 1993).

11.11.3 Late Last Glacial Environments

Cooling in New Guinea lasted from around 28 ka BP throughout the Last Glacial Maximum, some 18.5–16 ka BP. During this time, the tree line descended nearly 1000 m and temperatures were 7–11°C below their present levels (Swadling and Hope, 1992). The highest mountains were topped by small ice caps which spawned valley glaciers. An expansion of forest dominated by southern beech (*Nothofagus*) in the New Guinea highlands around the Last Glacial Maximum appears to have been a response to temperature fall yet also signals adequate moisture, perhaps because of the development of a persistent cloud belt rather than high precipitation levels. Ice began receding 15–14 ka BP and most had disappeared by the end of the Last Glacial (Bishop, 1994). Rainforest began expanding around 13 ka BP, reaching its present limits ~9 ka BP; there is a suggestion of significantly reduced precipitation 13–11.5 ka BP (Markgraf et al., 1992).

A combination of uplift and lowered sea level during the middle and late Last Glacial caused the diversion of the Fly–Strickland river system in southern New Guinea. This event, which had ramifications throughout northern Australasia, is discussed in the following section.

The Last Glacial Maximum was unequivocally drier than today throughout Pacific Australia. A major line of evidence is the increase in the extent of sand dunes which Bishop (1994) suggested to have resulted from both increased aridity and increased windiness, the former explicable largely by the increased continentality – a result of lowered sea level – particularly in the northeast.

Temperatures in Australia around the Last Glacial Maximum were 4–7°C lower, precipitation as much as 50% lower, and there was increased wind transport into adjacent oceans (Bowler et al., 1976; Thiede, 1979; Colhoun, 1991). This time was also marked by weakening of both the northwest and southwest monsoons.

In addition to increased aridity, an increased seasonality of precipitation around the Last Glacial Maximum is indicated by the Lake Carpentaria record. On the basis of the extent of savanna vegetation in the surrounding catchment ~23–15 ka BP, Torgersen et al. (1988) inferred that the contemporary runoff–evaporation ratio was about half of today's value. Maximum dryness at Lynch's Crater was achieved shortly after the Last Glacial Maximum, perhaps around 15–17 ka BP (see Figure 10.3).

Evidence of vegetation consistent with aridity around the Last Glacial Maximum was reported from Tasmania by Markgraf et al. (1986). The timing of the Last Glacial Maximum here is bracketed by outwash at Lake Margaret ~21–19 ka BP (Clapperton, 1990a).

Deglaciation during the latest Last Glacial was marked by lake-level rise ~12 ka BP and forest expansion, both indicators of increasing precipitation. The occurrence of this 10–9 ka BP in southeastern Australia is thought to herald the northward migration of the subtropical high-pressure belt into its present position (Harrison, 1993).

Of all the areas constituting the Pacific Rim, it is Australia and New Guinea (and nearby offshore islands) where the clearest evidence of human impact on the environment prior to the start of the Holocene is found. Evidence of shifting cultivation involving forest clearance for planting is known around 35 ka BP from the rainforests of New

Plate 11.1   The upland *maquis*, Plaines des Lacs, 230 m above sea level in southeast New Caledonia. Pollen from such *maquis* is found to replace that from burned *Nothofagus*-dominated forests at lower elevations on several occasions during the Last Glacial. Had humans been present on New Caledonia at this time, forest burning might be attributed to their impact. Yet, since humans were not present here until about 3100 BP, these forests must have burned for other reasons. It follows that even in places where humans were demonstrably present at this (and later) times, it cannot be automatically assumed that they were responsible for forest burning
Geoff Hope

Britain[1] (Pavlides and Gosden, 1994). Charcoal began to be produced in the Kosipe area of the New Guinea highlands around 30 ka BP. This has been interpreted as an outcome of the burning of the upper forest fringe by people who were already established users of higher-elevation subalpine grasslands and who wished to increase pandanus crops (Hope and Golson, 1995). Most other reviews of vegetational history during the Last Glacial Maximum regard aboriginal disturbance as a likely explanation for the replacement of particular assemblages, especially through the use of fire (Kershaw and Nanson, 1993; Haberle,

1998). Yet in New Caledonia there is evidence of disturbances of a kind which have been attributed to humans in New Guinea and Australia yet which must have occurred here long before humans set foot on this island (Hope, 1996). The upland *maquis* on the main island (Plate 11.1) spread on several occasions during the period ~40–26 ka BP (and later) into forested areas following their burning, which – given that the island was settled by humans only about 3.1 ka BP – could only have been a response to a natural fire regime. This emphasizes the need to be cautious when assigning particular late Quaternary environmental changes to humans, a theme explored more in Part E.

Deforestation around the Last Glacial Maximum in New Zealand, indicated clearly in pollen

---

[1] Strictly part of the 'Pacific Islands' in this book

data (McGlone, 1988), is also suggested by the rise in silt sedimentation at this time on the surrounding ocean floors (Thiede, 1979; Nelson et al., 1993), and the development of sand dunes (Neall, 1975). The depression of the snowline in southern New Zealand by some 800–830 m during the Last Glacial Maximum indicates a fall of temperature relative to the present of around 5°C yet the vegetational response here was more severe than would be expected from a temperature fall of this magnitude alone. McGlone (1988) argued that an additional cause of contemporary deforestation was a northward movement of cold subpolar air masses at the time.

Deglaciation in northern New Zealand is marked by afforestation 14.5–13 ka BP. In colder South Island there was no corresponding change, probably because of the continued proximity of subpolar air masses. Instead grassland was replaced by shrubland after 12 ka BP (McGlone, 1988). 'The speed and the size of these geo-morphic and biotic changes were astounding. At 15,000 BP the New Zealand mainland was one large island approximately 50 per cent larger than now, but with less than 10 per cent of its land area in forest. In less than 4000 years its size had dwindled by 30 per cent, it had split into three main islands, and tall, evergreen vegetation had re-occupied approximately 90 per cent of the land area' (Anderson and McGlone, 1992: 205–206).

11.11.4   The Diversion of the Fly–Strickland River System, New Guinea

The huge Fly–Strickland river system in southern New Guinea presently flows into the Gulf of Papua; both rivers turn abruptly southeastward in their middle courses (Figure 11.8). Prior to ~35 ka BP, these rivers flowed southward into a large estuarine embayment on the floor of the modern Carpentaria Basin. Uplift of the Oriomo Plateau occurred around 27 ka BP causing them to be diverted southeast. The lower parts of the Bain and Merauke rivers are probably the beheaded courses of the middle Last Glacial Fly and Strickland rivers respectively (Swadling and Hope, 1992).

As a result of sea-level fall, this embayment became the closed Lake Carpentaria (see Figure 11.7) which existed ~35–12 ka BP (Torgersen et al., 1988). The diversion of the Fly–Strickland system would have reduced water discharge into Lake Carpentaria substantially. This would have influenced human settlement in this area and also in the modern lower valley of the Fly–Strickland where diversion had caused a 'change from a gently-sloping plain to a dissected landscape dominated by narrow ridges and valleys' (Swadling and Hope, 1992: 21).

11.12   LAST GLACIAL ENVIRONMENTS OF THE PACIFIC OCEAN AND ISLANDS

Of the few terrestrial pollen records extending from the Holocene to the Pleistocene in this region, only those on Easter Island (Rapanui) and the Galápagos Islands have been able to provide information about Last Glacial palaeo-environments (Flenley and King, 1984; Colinvaux, 1984).

The earliest palaeoclimate reconstruction on Easter Island is of a period cooler and drier than at present >32.5 ka BP. This was followed by an interstadial of variable length – 32.5–26 ka BP at the Rano Aroi site – indicated by the presence of the tree *Sophora toromiro*, now extinct on the island. The Last Glacial Maximum, around 21–12 ka BP, was cooler and/or drier than at present.

Work at the El Junco crater lake on San Cristóbal island in the Galápagos Islands has illuminated Holocene environmental and climate change (see Figure 13.4) but all that can be said about the Last Glacial is that there was no lake some 30–10 ka BP, from which it can be inferred that the climate was too dry for one to form.

Some corroborative evidence of widespread aridity during the Last Glacial in the Pacific, at least in its tropical parts, is available elsewhere. For example, Matthew Spriggs (1986) found that aggradational river terraces had accumulated around 23 ka BP on Aneityum island in Vanuatu, perhaps under the influence of dry conditions

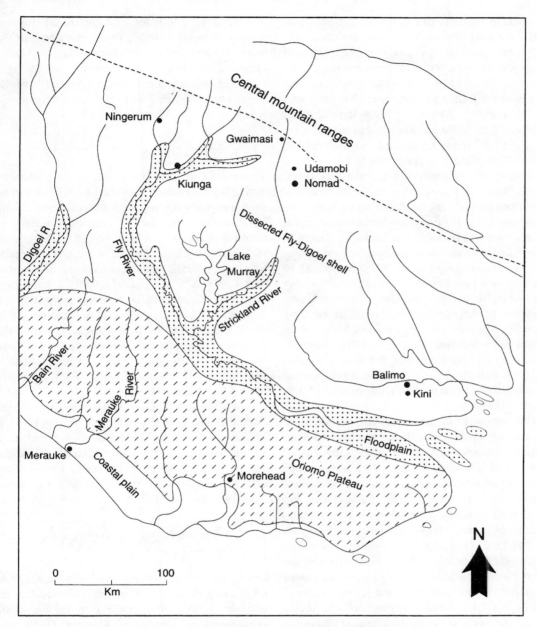

Figure 11.8 The Fly-Strickland area of southern New Guinea
Compiled from various sources

during which sediment was mobilized only occasionally by torrential rain associated with tropical cyclones. In New Caledonia, an expansion of conifers around the Last Glacial Maximum is also consistent with cooler and drier conditions (Méon and Pannetier, 1994).

An apparent exception occurs on the island Hawai'i, where Stephen Porter (1979) was forced to conclude that wetter conditions had prevailed during the Last Glacial Maximum. He concluded this because, from his study of Last Glacial snowlines, he believed that temperatures in the area must have been some 5°C lower. Yet the authoritative CLIMAP estimates for the area (CLIMAP Project Members, 1976) showed a fall of only 1–2°C, so Porter inferred that the extra ice must have been produced by increased precipitation. The inference was justified given the paucity of comparative data from the central Pacific Basin. Yet some other CLIMAP estimates for temperatures during the Last Glacial Maximum elsewhere in the region were also underestimated, and it now seems probable that Hawai'i – like most other Pacific islands – experienced drier conditions at this time (see review of CLIMAP below). More recent work focused on the loess of Hawai'i shows that, since it derived mostly from pyroclastic eruptions on the island, it is the timing of these rather than the prevailing climate which is the main control of loess deposition (Porter, 1997). The supplementary conclusion is that, from consideration of the orientation of loess deposits, it is clear that Hawai'i has remained within the tradewind belt since ~50 ka BP.

The advance of polar fronts towards the Equator during the Last Glacial compressed other climate zones and steepened meridional temperature gradients. There is ample evidence from the Pacific Ocean to support the idea, deduced from the study of the Pacific Rim and associated marginal seas, that the ocean–atmosphere system intensified during cool periods of the Last Glacial. This resulted in faster tradewinds – manifested by increased aeolian quartz flux (Molina-Cruz, 1977) – which increased equatorial current velocities and led to enhanced upwelling and hence higher productivity (Pedersen, 1983). It has been calculated that ocean productivity fell by a factor of ~1.7 at the time of the last glacial–interglacial transition in the Pacific (Herguera and Berger, 1994).

The chronology of deglaciation has been determined from a variety of ocean-floor data, but has proved least ambiguous in $\delta^{18}O$ data (see Figure 11.4). Recent work in the North Pacific

(Gorbarenko, 1996), for example, dated the two periods of rapid warming found elsewhere in the world to 12.5 ka BP (Termination IA) and 9.3 ka BP (Termination IB). It is becoming increasingly clear that a severe cold period occurred between these two. The evidence for this is reviewed below.

11.13   SYNTHESIS

Throughout the Pacific Basin, climate changes during the cold intervals of the Last Glacial can be explained satisfactorily in a general sense by an equatorward movement of climate zones. Good examples from around the Last Glacial Maximum include the shift of the Subtropical Divergence between Australia and New Zealand north from ~30°S to ~26°S (Martínez, 1994), a northward migration of precipitation belts in southern South America (Hulton et al., 1994), and the southerly displacement of storm tracks originating from the Aleutian low-pressure centre across Owens Lake in California (Bradbury, 1997). The situation is less clear in the northwest Pacific Basin (COHMAP Members, 1988).

For almost every constituent region of the Pacific Basin, it is also necessary to invoke local influences in order to explain observed Last Glacial climate changes. In montane parts of the Pacific Rim, for instance, a lowering of altitudinal climate zones was superimposed on gross latitudinal movements.

11.14   THE YOUNGER DRYAS IN THE PACIFIC BASIN

One of the major challenges in understanding late Quaternary climate change is why short-lived yet extreme events, of opposite sign to long-term, orbitally forced trends, should occur. One of the best studied of these events is the Younger Dryas Climatic Reversal, which occurred some 11–10 ka BP, and caused temperatures in northwest Europe, where it has been studied most intensively, to fall to almost full-glacial levels. The global extent of the Younger Dryas is controversial. This section reviews the evidence from the Pacific Basin.

Owing to the existence of a radiocarbon plateau at the time of the Younger Dryas, there is a problem in making a meaningful comparison of ages obtained using radiocarbon. While it is considered that all the evidence given below refers to the same event, there is presently no way of demonstrating this unequivocally to be the case.

11.14.1 The Younger Dryas in Antarctica

No direct evidence for the Younger Dryas event is known from the Pacific Rim in Antarctica although the prominent two-step pattern of deglaciation inferred from ice-core records may be found to delimit the Younger Dryas cooling event once these have been securely dated. Clapperton (1990b) suggested that moraines in the Ablation Massif of Alexander Island, on the western side of the Antarctic Peninsula formed during a Younger Dryas advance (see Plate 13.2).

11.14.2 The Younger Dryas in South America

In southernmost South America, an abrupt shift in a forest–tundra boundary occurred at the same time as the Younger Dryas in Europe, and was considered contemporaneous by Heusser and Rabassa (1987). Yet for much of Pacific South America, the latest Pleistocene and earliest Holocene glacial advances occurred 13–12 ka BP and 10–9 ka BP and both are likely to manifest increased precipitation rather than rapidly decreased temperature (Heine, 1993). Palaeo-botanical data from the Pacific Basin in northern and central South America have yielded 'no strong indication of a warm to cold climate reversal' (Hansen et al., 1994: 273) in the late Last Glacial yet many writers admit the evidence is inconclusive. Other workers are more hopeful: Seltzer (1993: 134) wrote of 'increasing evidence for climate reversals and glacial advances' 12–10 ka BP, one of which was subsequently elucidated from the Ecuadorian Cordillera by Clapperton et al. (1997).

11.14.3 The Younger Dryas in Central America

It is unfortunate that lake levels in Mexico – potentially one of the most illuminating sources of information about Last Glacial palaeoclimates in Central America – fell so low after 16 ka BP in many places that they can no longer be used as recorders of subsequent climate change (Markgraf, 1989). Yet, the dominance of fresh-water marshes at Chapultepec 14–9 ka BP may reflect increased precipitation in Mexico and may correlate in part with readvances of mountain ice, but it seems clear that these advances were short-lived and of only local extent (Bradbury, 1989). Clearer evidence of the effect of Younger Dryas cooling in Central America was reported by Leyden (1995); interestingly, while pollen data from Costa Rica indicate 1.5–2.5°C cooling, those from Panama show no cooling suggesting that local monsoonal influences may have obscured the effect of cooling here.

11.14.4 The Younger Dryas in North America

A widespread drought 11.3–10.9 ka BP in the western USA, coincident with the Younger Dryas, may have contributed to the contemporary decimation of megafauna (Haynes, 1991), particularly the mammoth (*Mammuthus columbi*), an event which has popularly and long been ascribed to the arrival of human hunters in the region at this time (see below). Evidence from both fossil pollen and foraminifera for a Younger Dryas event on the coast of British Columbia was gathered by Mathewes et al. (1993). A change from forest to open herb-rich vegetation in the Queen Charlotte Islands occurred some 11.1 ka BP and lasted about 1000 years. Peak abundances of the cold-water indicator foraminifer *Cassidulina reniforme* were found in nearby deep-sea sediments deposited 11–10.2 ka BP. Late Last Glacial readvances of ice, possibly contemporaneous with the Younger Dryas, have been documented in the mountains of western North America. In the Rocky Mountains, advances coincident with the last major extension of Lake Bonneville occurred 13–11 ka BP; in the northern Cascades, 12–11 ka BP; in southwest Canada, 11.8–11.4 ka BP (Heusser et al., 1985).

11.14.5   The Younger Dryas in Beringia

Younger Dryas ice advances occurred in the Brooks Range of northern Alaska 12.8–12.5 ka BP, and in southern Alaska 12.8–11.8 ka BP (Heusser et al., 1985). A vegetational record of a possible Younger Dryas event in Beringia was obtained by Engstrom et al. (1990) from southeast Alaska. Pollen showing an abrupt change from a pine parkland to shrub- and herb-dominated tundra was obtained from lacustrine sediments dated 10.8 to ~9.8 ka BP.

11.14.6   The Younger Dryas in East Asia

In the Beijing lowlands of continental East Asia, coring of peat has shown that from about 11.4–11 ka BP, there was a conspicuously dry climate – possibly a Younger Dryas correlative – indicated by increases in *Artemisia* pollen, after which time conditions became warmer and wetter (Liu, 1988). An abrupt cooling event occurred around the same time on the Loess Plateau (An et al., 1993).

A conspicuous fall of sea level, which interrupted the postglacial rise around Japan, occurred 11–10 ka BP (Umitsu, 1991). Terrestrial evidence for Younger Dryas cooling here comes from the identification of pollen assemblages ~12.4–11.8 ka BP in Hokkaido which indicate vegetation similar to that of modern Sakhalin Island to the north. This relationship led Igarashi (1994) to suppose that Younger Dryas climate in Japan involved temperatures 7–9°C lower than today and annual precipitation levels at least 735 mm lower. Movements of the cold Oyashio ocean current traced in ocean-floor cores in the northwest Pacific show that there was a southward advance 11–10 ka BP thought by Chinzei et al. (1987) to manifest the Younger Dryas event. Additional support for the occurrence of this event was found in the same part of the Pacific Ocean floor by Heusser and Morley (1990).

11.14.7   The Younger Dryas in Australasia

For Australasia, the inferred drop in the $^{14}C/^{12}C$ ratio in the ocean around the Huon Peninsula, Papua New Guinea, some 12–10 ka BP has been linked to Younger Dryas ocean cooling (Edwards et al., 1993).

A period of dune emplacement 11.2 ka BP in the Cape York area may have been facilitated by a rapid sea-level fall associated with cooling (Lees et al., 1993). The occurrence of a dust peak in northeast Australian sediments 11.4–10.4 ka BP led DeDeckker et al. (1991) to suggest this to be good evidence for the Younger Dryas cooling having also affected this area. Other evidence in Australia and New Guinea has not been forthcoming despite many studies where such an event might be expected to have been noticed.

In New Zealand, a well documented advance of the Franz Josef Glacier around 11.05 ka BP has been linked to the Younger Dryas event (Denton and Hendy, 1994) but otherwise evidence in these islands is absent (McGlone, 1995).

11.14.8   The Younger Dryas in the Pacific Ocean and Islands

In the north Pacific Ocean, as in the north Atlantic, there is clear evidence from foraminiferal indicators that the Younger Dryas cooling affected the region, culminating around 10.55 ka BP (Duplessy et al., 1989). The record of sea-level rise across the Younger Dryas in Papua New Guinea (Edwards et al., 1993) was used by Fletcher and Sherman (1995) as a basis for assigning a Younger Dryas age to the submerged (−56–67 m) Penguin Bank shoreline complex off Oahu island in Hawaii. An analysis of Sr/Ca ratios in fossil corals from Vanuatu showed that the sea-surface temperature in the area 10.2 ka BP was about 5°C cooler than today, possibly a Younger Dryas signal (Beck et al., 1992). Elsewhere in the Pacific, especially the central part, there is limited evidence for the Younger Dryas; however, since it is well marked in the north and it seems to have been explicit along most of the Pacific Rim, it is probably only a matter of time before evidence is obtained from here, including perhaps some islands, perhaps similar to the abrupt Younger Dryas change in upwelling character around some Caribbean islands (Overpeck et al., 1989).

11.14.9 Heinrich Events in the Pacific Basin

If Younger Dryas cooling was induced by meltwater flow into the North Atlantic (Broecker et al., 1989), then precise information about the strength and the timing of this event in the various parts of the Pacific could be used to calibrate the subsequent dispersal of this meltwater and, ultimately, to test the idea.

The rapid, short-lived Younger Dryas cooling event could be regarded as the most recent in a series of such events – termed Heinrich events – which have occurred throughout the Last Glacial (Dowdeswell and White, 1995). The effects of Heinrich events are best marked by massive ice-rafting in the North Atlantic and it seems possible that, through the consequent disruption of the thermohaline circulation (Broecker, 1997), the influence of these events was felt as far afield as the Pacific Basin. The evidence is not widespread here and potential correlations with events in the North Atlantic suffer from problems with precise dating. Despite these, correlatives of particular Heinrich events are known from California (Benson et al., 1998) and the Loess Plateau in China (Porter and An, 1995).

11.15 SELECTED ISSUES CONCERNING THE LAST GLACIAL

There are two issues of importance to an understanding of Last Glacial environments in the Pacific Basin which merit separate and fuller discussion than was appropriate above. The first concerns the results of the innovative and influential CLIMAP project in the light of more recent research. The second is the question of mammalian extinctions during the late Last Glacial and their relationship to contemporary climate changes and human activities.

11.15.1 The CLIMAP Initiative Revisited

One of the most innovative approaches to the study of late Pleistocene climate change in the 1970s and early 1980s was the CLIMAP (Climate: Long-Range Investigation, Mapping, and Prediction) project, one aim of which was to reconstruct global conditions around the Last Glacial Maximum (CLIMAP Project Members, 1976). The CLIMAP map of sea-surface temperatures at this time (18 ± 3 ka BP) proved to be one of the project's most influential conclusions but the marked disparity between ocean and land temperature differences provoked comment. CLIMAP found that the tropical Pacific had surface temperatures during the Last Glacial Maximum similar to those it has today and – most surprisingly – that Last Glacial Maximum temperatures were even greater at certain times of year around the centres of the two Pacific gyres than today.

Inevitably the CLIMAP reconstructions attracted criticism, related especially to the model's inability to portray Last Glacial Maximum ocean circulation realistically and to the likely incorrectness of some of the transfer functions used to derive sea-surface temperatures from particular assemblages of marine micro-organisms. Rebuttals of CLIMAP's estimates of Last Glacial Maximum temperatures, particularly in the tropical Pacific, came from high-elevation tropical palaeotemperature interpretations of snowline lowering (Rind and Peteet, 1985) and ice accumulation (Thompson et al., 1995) at Pacific Rim sites which indicate a much greater degree of cooling during the Last Glacial Maximum.

The CLIMAP reconstruction has been defended by others who point out that most of the terrestrial data used to criticize it derive from tropical highlands, and note that palaeo-temperatures derived from tropical lowland pollen blown offshore and accumulated in ocean-floor sediments actually agree much better with the CLIMAP estimates (Van Campo et al., 1990). Recently, it has been suggested that the disparities between surface palaeotemperature estimates of marine and terrestrial derivation for the Last Glacial Maximum may be due to the physiological effects of Last Glacial conditions on the organisms used for study. In support of this view, the UK_{37} (unsaturation degree of long-chain alkenones) method of estimating palaeotemperatures from a group of planktonic algae occupying a possibly less vulnerable ecological niche than most other organisms used for palaeotemperature estimation has yielded results for the western tropical Pacific very similar to those of CLIMAP (Ohkouchi et al., 1994). Similar confirmation of the overall CLIMAP estimates of palaeotemperature in this area has also been obtained using methods other than those used originally (Thunell et al., 1994).

A recent simulation of Last Glacial Maximum conditions using a more realistic model than was available to the CLIMAP group has produced results for the low-latitude Pacific Ocean that are in better accord with the estimates of cooling derived from terrestrial studies around the Pacific Rim (Bush and Philander, 1998). This model estimates that sea-surface temperatures during the Last Glacial Maximum in the tropical Pacific fell generally by 3–6°C. The lowest temperature fall was just 1°C south of latitudes 20–22° in the southwest Pacific.

11.15.2   Mammal Extinctions in the Terminal Pleistocene

Around the end of the Last Glacial, there were massive extinctions of (mostly) large-bodied mammals. In North America, around 11 ka BP, 91% of mammals weighing more than 5 kg including 73% of the megafauna (those weighing more than 44 kg) became extinct perhaps within just 3000 years. In South America, 80% of megafaunal species disappeared. In Australia, 86% of the megafauna vanished 25–11 ka BP. The megafauna of eastern Japan disappeared after 15 ka BP. Yet there was no marked change in the rate of marine organic extinctions during this time (Lundelius, 1983; Barnosky, 1989; Keally, 1991).

Although nowhere near as severe as the end-Cretaceous (K–T) extinctions (see Chapter 5), the terminal Pleistocene mammalian extinctions caused major ecosystem changes in many parts of the Pacific Rim. There have been two explanations proposed for these extinctions: one that they resulted from climate changes, the other that they were caused by sudden increases in human demands (overkill). Both explanations have their strong and their weak points, and the truth may involve elements of both.

The climate change hypothesis supposes that increased aridity, warming and increased seasonality, exacerbated by abrupt and extreme cold events like the Younger Dryas, all contributed to a reduction in vegetation heterogeneity during the late Pleistocene and led to a loss of animal habitats. Why similar extinctions do not mark the end of every Pleistocene glacial period is less clear, and is a weakness of the climate-change hypothesis to which there is no clear answer.

The overkill hypothesis finds its greatest support in North America, where the beginning of the Clovis culture some 12 ka BP is marked by the development of stones with fluted points for use as spearheads – the most effective way of despatching big game at the time – and a rapid decline in the numbers of slow-moving megafaunal species with long gestation periods. To some, a weakness of this hypothesis is that extinctions occurred during the Last Glacial in other parts of the Pacific Basin where humans had coexisted with these large mammals for thousands of years. Other weaknesses were outlined in a thoughtful paper by Grayson (1988). Among these is the absence of stones with fluted points from Clovis-age archaeological records in much of North America (see Figure 12.2). The absence of such weapons – in the Great Basin, for example – suggests that the people there at the time were not hunting megafauna, at least not effectively. Grayson (1988: 116) regards the problems with the overkill hypothesis as 'overwhelming'. Others regard it quite differently, arguing, for example, that the coincidence of the presence of Clovis hunters with megafaunal extinction 'is beyond the probability of chance' (Agenbroad, 1988: 72).

It has been suggested that water tables were

falling and surface moisture was less abundant during the latest Pleistocene in North America (Haynes, 1991). Water-table fall would probably have been associated with the rapid sea-level fall which marked the Younger Dryas period at the end of the Pleistocene (see above). These conditions would have led to both humans and megafauna (among others) congregating around the remaining water sources, encouraging the localized extermination of one by the other. Yet many other megafauna may have died at this time simply because there was no available water. In this scenario, humans accounted for only a small part of the total extinction of large-bodied mammals there; climate change led to the demise of the greater number.

Elsewhere along the Pacific Rim, any connection between human activities and megafaunal extinction is less secure. In Australia, for example, fossils of megafauna are concentrated in formerly wooded areas, which has been interpreted as indicating that Late Glacial aridity forced these large animals to remain close to water holes, where they soon ate up the remaining vegetation (Horton, 1984). The combination of a long history of coexistence of humans and megafauna and an absence of 'kill sites' in Australia has led researchers to regard the overkill hypothesis as implausible here (Gorecki et al., 1984) although it has its persuasive advocates (Flannery, 1994). A similar story appears true of western Beringia and Japan where there is no suggestion of any massively increased human predation of megafauna during the late Last Glacial (Sher, 1997; Keally, 1991).

Although this debate is relevant to the whole question of human–environment relations in the Pacific Basin, it is more so to areas just outside. In North America, for example, most megafauna lived in places east of the Rockies – in what is now the United States – which Clovis people entered from the north via an ice-free corridor (see Chapter 12).

11.16 THE HOLOCENE PREVIEWED

Although humans reached the Pacific Rim long before the start of the Holocene, it was only during this epoch that they spread throughout it and that their impacts – on account of rising population densities and increasingly sophisticated lifestyles – became more than merely negligible. So the story of the Holocene is the story of environmental change induced both by non-human factors – as in earlier times – and by human factors, the influence of which is felt increasingly throughout the epoch.

Chapter 12 focuses on humans, the history of their development, their Pleistocene colonization of the Pacific Basin, and their dispersion and environmental interactions here by the beginning of the Holocene. The discussion of Holocene environmental changes in the Pacific Basin is divided into three parts: the early Holocene (12–6 ka) is discussed in Chapter 13, the middle Holocene (6–3 ka) in Chapter 14, and the late Holocene (3–0 ka) in Chapter 15.

PART E

Holocene Environments of the Pacific Basin

CHAPTER 12

Human Arrival in the Pacific Basin

In those cultures lacking unfamiliar challenges, external or internal, where fundamental change is unneeded, novel ideas need not be encouraged ... But under varied and changing environmental or biological or political circumstances, simply copying the old ways no longer works. Then, a premium awaits those who, instead of blandly following tradition, or trying to foist their preferences onto the physical or social Universe, are open to what the Universe teaches. Each society must decide where in the continuum between openness and rigidity safety lies.

Carl Sagan, *The Demon-Haunted World* (1996)

12.1 INTRODUCTION

In many accounts of environmental change, particularly during the late Quaternary, little effort is made to separate non-human from human impacts. In a few texts, environmental change is even equated uncritically with human impact. Personal experiences suggest that a whole generation of students of recent Earth history are being misinformed as to the proper context of human impact.

It appears increasingly convenient for many authors to treat all changes during the most recent period of human settlement of a particular region as being primarily a product of human behaviour. Not only is this convenient but it is also perhaps more palatable to a human readership who – be it to apothesize or demonize – are concerned mostly with their own rather than any other species. In this book, the balance is redressed by placing human impact in the Pacific Basin in a long-term context and, in the following chapters, showing how the relationship between humans and the environments they occupied was not all one-way traffic. Undeniably humans have had a massive influence on their environments but the difference between downtown Los Angeles or Tokyo today and rainforest clearance in New Britain around 35 ka BP is one only of degree. And just because modern urban dwellers in the Pacific Basin may feel that their lives are not influenced by their environments,

the historical record – of even a few decades – is proof of the contrary. No part of the Pacific Basin has been rendered immune from earthquakes or El Niño simply by reason of the degree of human impact. And we need take our histories back only a few centuries or millennia further to be able to demonstrate how non-human agents had even greater environmental influence. Without a doubt, changes in the natural (rather than human) environment have influenced human behaviour. To deny this – as has been increasingly *de rigueur* since the demise of the paradigm of environmental determinism a few decades ago – appears ludicrous if at the same time we admit that our earliest human ancestors adapted their lifestyles to the varied environments they encountered.

This chapter deals exclusively with the origins, arrival, spread and impact of early, largely pre-Holocene, humans in the Pacific Basin. It stops short of detailed discussions of long-resident human groups in the various subregions of the Pacific Basin; such discussions are incorporated into the accounts of Holocene environmental changes in Chapters 13–15.

12.2 THE ANCESTORS OF THE EARLIEST PACIFIC PEOPLE

The earliest known bipedal human ancestors are represented by *Australopithecus afarensis* who existed in parts of East Africa 4.0–2.5 Ma. The first stone tools were made by *Australopithecus africanus* (3.0–1.0 Ma) and *Homo habilis* (2.0–1.5 Ma), both of whom may have been confined to Africa. The later *Homo erectus* (1.8–0.2 Ma) migrated throughout Africa, into Europe and much of southern and eastern Asia including Java (1.8 Ma), China (1.7 Ma) and possibly the Philippines. *Homo erectus* may also have been distinguished by an ability to use fire.

Owing to its great age, a record of stone tools from Lena River terraces 140 km upstream of Yakutsk, possibly dating from 3.4–1.8 Ma (Hall, 1992), poses a challenge for students of human history in the Pacific Basin. A better constrained record of the earliest human presence in this region – indeed the first permanent occupation of a non-tropical region – is that of *Homo erectus* more than 1 Ma in the Loess Plateau of China (Wang et al., 1997). It is plausible to suppose that *Homo erectus* reached the Pacific Ocean although this is likely to have been perceived as a barrier to movement rather than an adaptational opportunity.

An important site in the western Pacific is at Zhoukoutien, close to Beijing, and only about 100 km from the ocean today where *Homo erectus* was present perhaps 0.7–0.2 Ma. A noteworthy change in the stratigraphy is represented by faunal remains. In the earlier parts of the sequence, forest faunal remains dominate; in the younger part, most remains are of xerophytic grassland herbivores found in association with comparatively intricate tools. The change, dated to around 370 ka, has been interpreted as signalling a change in human behaviour from resource gathering to hunting (Ho and Li, 1987).

Homo erectus also reached southeast Asia around 1.8 Ma and may have persisted there – long after *Homo erectus* had vanished elsewhere – until at least 0.1 Ma, where their presence is represented by the poorly dated Ngandong (Java) material (Santa Luca, 1980). Although there is no indication that humans slaughtered all the animals (numbering 25 000) whose bones were found at the Ngandong site, it could be noted that the remains of just 11 human skulls are associated with them. Other hominid remains in Java, most

notably at Mojokerto and Sangiran, have been dogged by classificatory and dating controversy.

The possible presence of stone tool-using human ancestors in Japan >0.3 Ma has been reported by Kamata et al. (1993) although a general lack of stone artefacts from East Asia may be because bamboo (which does not preserve so well) was used for toolmaking (Stringer and McKie, 1996).

Although some workers have favoured the idea that *Homo erectus* evolved into *Homo sapiens* in China (Yinyun, 1991) as well as in Africa and possibly elsewhere, most biomolecular and other evidence suggests that this 'multiregional' model of human development is less plausible than the 'single origin' model which sees *Homo sapiens* as the product of a single ancestral population in Africa, which differentiated and then migrated out of Africa around 200 ka (Stringer and Andrews, 1988; Cann, 1996; Stringer and McKie, 1997). If this is the case, then early human (*Homo erectus*) impacts on the Pacific environment would have become overprinted by those associated with the later arrival of the anatomically modern *Homo sapiens*, a more sophisticated user of the environment. The implications of the alternative 'multiregional' model, well articulated by Pope (1992) who argued that its unpopularity stems largely from an ignorance of East Asian human palaeontology, would clearly be quite different; a gradual rather than abrupt increase in the amount of environmental change would be expected. As yet, the environmental record is silent on this issue.

There is little evidence with which to challenge the view that, because of their low numbers and a presumed nomadic habit, early humans had negligible impacts on the environments they inhabited. Early humans certainly seem to have had their lifestyles influenced far more by environmental changes than to have been causes of such change (Foley, 1987). Nevertheless the organization of groups for hunting, the use of fire coupled with a low population density probably meant that early humans had tangible effects on environments locally but that these effects were transient and superficial compared to what is usually classed as 'human impact' today.

12.3 THE ARRIVAL OF *HOMO SAPIENS* IN THE PACIFIC BASIN

Conventional wisdom holds that modern humans (*Homo sapiens*) moved out of Africa within the past 200 000 years. They had reached Israel at least by 90 ka, which suggests that the rate of this movement was extremely slow. Yet there is some evidence that they moved faster into south and east Asia than they appear to have done into the Near East and Europe (Stringer and Andrews, 1988).

There are indications that pre-modern *Homo sapiens* were present in East Asia earlier than would have been possible with this 'out-of-Africa' single-origin model. The best evidence is the date of 0.2 Ma for a cranium from Dali in China. At the site of Jinniushan in northeast China, pre-modern *Homo sapiens* may also have been present by 280 ka. Despite its near-coastal location, there is no indication that the Jinniushan humans interacted with the ocean. At Zhoukoutien, deer teeth have been used to date the first signs of (pre-modern) *Homo sapiens* to no more than 171 ka.

Anatomically modern humans appeared in China perhaps at 108 ka but do not show up at Zhoukoutien until <34 ka BP. Their association with tropical animals such as cheetahs and hyenas here suggests the prevalence of stage 3 warm conditions (see Chapter 11). The presence of the bones of teleost fish together with worked marine shells in association with the remains of these humans are the first clear indications of human–ocean interaction in the Pacific Basin.

In southeast Asia, *Homo sapiens* is known to have been living at Niah Cave in Borneo around 40 ka (Harrisson, 1967). This cave is 16 km from the sea and full of bones of animals assumed to have been eaten by the resident humans. These animals include orang-utan, pangolin, porcupine, wild cat, tiger, tapir, pig, deer, rhinoceros and crocodile (Hooijer, 1963). What is of interest is that, of all these animals, only the giant pangolin (*Manis palaeojavanica*) is now extinct, so it might be concluded that humans were not present here at this time in such numbers that they had a major impact on the fauna. The Niah people also ate fish, shellfish and birds. Their diet was clearly non-selective, but they were distinguished from *Homo erectus* in this region by their use of marine resources.

Although there is vigorous debate about the time of arrival of *Homo sapiens* in Australasia (see below), it is important not to allow this to overwhelm other implications of the move by humans from one to the other. Low sea level during Pleistocene glacials would have facilitated the initial crossings of ocean gaps between southeast Asia and Australasia; the shortest gap is about 20 km between Bali and Lombok islands but some scientists have also favoured a crossing between Borneo and Sulawesi (Celebes). Once in Lombok or Sulawesi, humans would have had to make only a few more comparatively short hops between islands to reach the limits of the glacial Australasian continent (Figure 12.1). But they would already have arrived in a completely different world, a 'naive' land. For Australasia had been cut off from other continents for more than 50 million years and a distinctive biota had evolved there, in the absence of large predators such as humans. The encounters between modern humans and these naive biotas may have led to a 'great leap forward' in human development, as discussed in Chapter 2.

On the New Guinea coast, stone blades which were hafted to make axes are associated with 40 ka old tephra on the Huon Peninsula (Groube et al., 1986; see also Plate 2.1). It is plausible to suppose that such advanced tool technology may have arisen from a desire to clear forest. The rainforests of New Guinea (and elsewhere in the Pacific Basin) cannot support high densities of hunter-gatherers so one option, once population densities began rising, would have been to open the canopy through clearance to allow for the management of food plants like *Pandanus* and *taro* (*Colocasia* sp.). At present, the earliest known evidence for forest clearance in Papua New Guinea occurs 35 ka BP on the offshore island of New Britain (Pavlides and Gosden, 1994; see also Chapter 11).

Pleistocene human-occupation sites in the New Guinea highlands contain plentiful bones of animals, some of which are now extinct. Humans

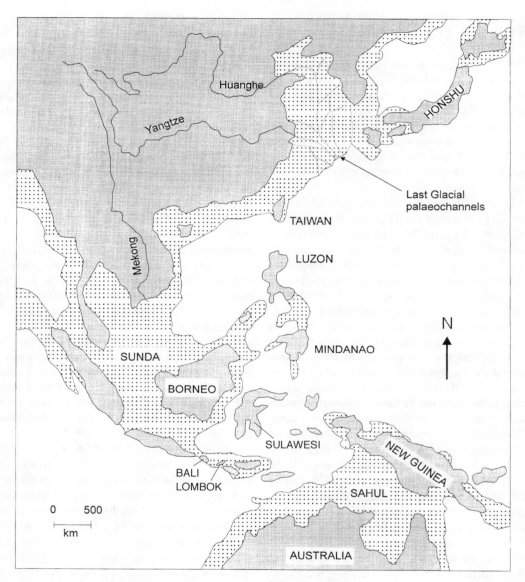

Figure 12.1   Part of the eastern Pacific Rim showing the modern limits of land areas and the limits which existed around the Last Glacial Maximum when sea level was ~130 m lower than today. Note the broad Sahul Shelf connecting Australia and New Guinea, and the Sunda Shelf connecting many of the islands in southeast Asia Compiled from various sources

undoubtedly had a role in this process but, like many comparable situations, it is not yet possible to distil the effects of human impact from those of non-human impacts, particularly latest Pleistocene climate changes.

12.3.1   Homo Sapiens in Australia

The first humans are generally thought to have arrived in Australia 50–20 ka BP. The classic sites around Lake Mungo in New South Wales date from around 32 ka BP (Barbetti and Allen, 1972). Artefact-bearing gravels in the Cranebrook Terrace near Sydney have been dated by thermoluminescence to as much as 47 ± 5.2 ka BP (Nanson et al., 1987). The earliest traces of settlement in Tasmania are from around 30 ka BP (Cosgrove, 1989).

The suggestion by Singh et al. (1981) that a rise in the amount of charcoal in sediments of Lake George near Canberra 75–128 ka might indicate an initial human presence was dismissed by most workers, not least because this was a time of higher sea level which presumably would have made the movement of humans from southeast Asia to Australasia more difficult than when the sea level was low. Yet the idea of such an early human presence has been revived recently by Kershaw et al. (1993) who argued that charcoal and pollen evidence from an ODP ocean core off the Queensland coast point to a human presence on the continent 140 ka. This suggestion has been rebutted by several archaeologists who stress that there is no unimpeachable evidence of a human presence in Australia so early (Anderson, 1994; White, 1994). Further, evidence of fires is not evidence that humans lit them. Although humans and their ancestors had been using fire for many millennia, fires can obviously start and spread for reasons completely unrelated to humans.

For most of the time that they have occupied the Australian environment, humans have had little impact compared to their contemporaries in other Pacific environments, because there was no widespread need (as there was in New Guinea perhaps 40 ka – see above) to change from the dominantly hunting and gathering lifestyle of the earliest arrivals. Yet conversely, it is wrong to equate apparent economic simplicity with low or no environmental impact. Aboriginal Australian hunter-gatherers used environmental management techniques to maximize food acquisition and some of these – firestick farming for instance – may have contributed to persistent vegetation and landscape changes, the cumulative effects of which have been identified in the Australian environment, at least locally. Yet, were it to be demonstrated conclusively that humans had reached Australia nearly

100 000 years earlier than conventionally thought, this would mean that these effects may have accumulated over a much longer time span and therefore ideas about late Quaternary human impact here would have to be revised.

12.3.2 Homo Sapiens in China

Modern *Homo sapiens* may have appeared first in eastern China during the early Last Glacial, perhaps descended from earlier human occupants of this region, perhaps newly arrived from Africa (see above). Around the time of the Last Glacial Maximum, anatomically modern *Homo sapiens* were scattered throughout China (Chen and Olsen, 1990). Some classic sites close to the Pacific coast have been dated to this time.

Sites along the Huanghe River are all several hundred kilometres upstream from the coast, much closer to the Last Interglacial shoreline than to the modern shoreline. They include numerous river-terrace sites at Xiachuan for which dates of human occupation of 26.2–15.9 ka BP have been obtained (Wang et al., 1978). Sites along the Yangtze River farther south are closer to the modern shoreline. All are suspected to be of latest Pleistocene age. The upper cave at the classic site Zhoukoutien near Beijing was first occupied by modern *Homo sapiens* <34 ka BP.

Modern humans reached many parts of the Pacific coast in China during the latest Pleistocene and, to judge from most dates of the earliest human presence in North America (see below), had begun dispersing along the Pacific Rim before the start of the Holocene. Yet this idea is being seen increasingly as conservative, not just because of a number of dates for human occupation of the Americas from before the Last Glacial Maximum, but also because of the evidence for a much earlier occupation of western Beringia than traditionally supposed.

12.3.3 Out from China

Just as there is controversy about the timing of initial human settlement of Australasia so too is there controversy about that in the Americas. The situation is comparable. The most widely accepted view holds that low sea levels during the later part of the Last Glacial exposed the Bering Strait in the northernmost Pacific allowing humans to traverse it and thus enter the largest unpeopled continents of the age (see Chapter 11). Yet there is some evidence, reviewed opposite, that humans crossed into the Americas much earlier.

According to linguistic evidence, most Pacific islands were colonized from southern China and Taiwan only within the last few thousand years (Kirch, 1997). This extraordinary feat was accomplished only once humans had developed sailing and navigation techniques of appropriate sophistication. It is probable that the first Pacific islanders traversed the Pacific Ocean several centuries before European explorers first saw it.

Both the settlement of the Americas and that of the Pacific islands is believed to have begun from southern China and Taiwan. It is plausible to assume that modern humans had settled the Pacific coast of China during the late Last Glacial and slowly adapted their lifestyles to their new situation. It is likely that ocean foods came to play an increasingly important part in the diets of these coastal dwellers, in the same way as specialized technology for coastal living would gradually have displaced that acquired specifically for inland living (Table 12.1). Within a few generations this change may have restricted the migration options for these people. In future their migratory routes would be controlled by proximity to the ocean.

It is therefore reasonable to suppose that the coastal people of China began to disperse along the western Pacific Rim towards the end of the Pleistocene. Sea level was still low enough to allow access to many of what are now offshore islands, including those of the Japanese archipelago.

12.3.4 A Japanese digression

Of particular interest is the settlement of Japan. It is now clear that the Tsushima Strait between western Japan and Korea, and the Tsugaru Strait between Honshu and Hokkaido remained open during the late Pleistocene. Yet the island Hokkaido was connected to mainland East Asia by a land bridge and it is this which the earliest human inhabitants of Japan probably used to reach the islands. Not that minor sea crossings were impossible for these early settlers: obsidian on islands south of Tokyo, which were never connected by dry land to the main islands, was being exploited by humans some 30 ka BP (Hashiguchi, 1994).

A date of around 35 ka BP seems likely for the earliest presence of *Homo sapiens* in Japan. As the Last Glacial Maximum approached, so the forests in the northern (parts of) these islands became less hospitable to humans and other animals, so they moved south where they could. Many of the

Table 12.1 Some Proto Malayo-Polynesian maritime words intended to demonstrate that the earliest settlers of most Pacific islands and parts of southeast Asia, including parts of the Philippines and Papua New Guinea, had a long history of interaction with the ocean which facilitated their long-distance voyaging at a time when most humans elsewhere in the world were able to sail only comparatively short distances. While these words can be traced throughout most modern languages in this region, they cannot be demonstrated to have derived from the aboriginal inhabitants of southern China and coastal Taiwan (because their maritime vocabulary has been lost) although there are other linguistic trails which suggest that this was the case. (PAN = Proto-Austronesian, including the aboriginal Taiwanese language)
Table provided by Paul Geraghty

waŋkaŋ	canoe
qabaŋ	canoe (PAN)
baŋkaq	canoe, boat
para[qh]u	boat
katiR	small outrigger canoe or canoe hull
tekén	punting pole
láyaR	sail
quliŋ	rudder, steering oar, steer
beRsay	paddle, oar
paluja	paddle [verb]
limás	bail, bailer
seŋkar	cross-seat in boat, thwart
laŋen	roller for beaching canoe
saRman	outrigger float
kawíl	fishhook (possibly PAN)
saruk	fishnet

earliest settlement sites are in the south, along coastal plains, parts of which are today under water.

12.3.5 Across Beringia into America

A land bridge probably existed across Beringia at some time during every Pleistocene glaciation. Plants and animals exploited the connection to move from Asia to America throughout the Last Glacial, but most humans crossed only after the Last Glacial Maximum about 12 ka (Bonnichsen and Turnmire, 1991; Plate 12.1). This model has undergone some important 'fine tuning' recently.

It is now apparent that Beringia was not open for biotic migration for the whole of the Last Glacial for, despite lowered sea level, it was periodically covered by ice flowing south from the Arctic continental shelf. Like other species of animal, humans would have had to exploit 'windows of opportunity' to move from Asia to America (Hughes and Hughes, 1994). An increasing number of secure dates from before the Last Glacial Maximum for a human presence

in Beringia and the Americas is paving the way for a revised model of human occupation.

In northeast Asia, adjacent to western Beringia, the Diuktai Cave site was occupied as early as 35 ka (Mochanov, 1977), presumably by *Homo sapiens*, and has since been revealed as just one of a number of sites dating from the middle Last Glacial which were central rather than marginal to contemporary human development (Goebel and Aksenov, 1995). At the Bluefish Caves site in northwest Canada, worked fragments of caribou and mammoth bone have been dated to around 24 ka BP (Cinq-Mars, 1990) and indicate that some humans entered North America before the Last Glacial Maximum.

The major barrier to human colonization of most of North America was not an open Bering Strait but the presence of ice covering the mountains in eastern Alaska (see Chapter 11). An ice-free corridor permitting the movement of humans and other animals is postulated to have opened only about 12 ka BP, suggesting that reports of earlier human occupation of the Americas are incorrect, or that the route by which

Plate 12.1   Hunters followed mammoths and other animals across a region of shrub birch and sedge in Beringia to enter the Americas, apparently for the first time about 35 000 years ago. Painting by Carl Buell to illustrate the reconstruction of the Bering land bridge by Scott Elias and colleagues

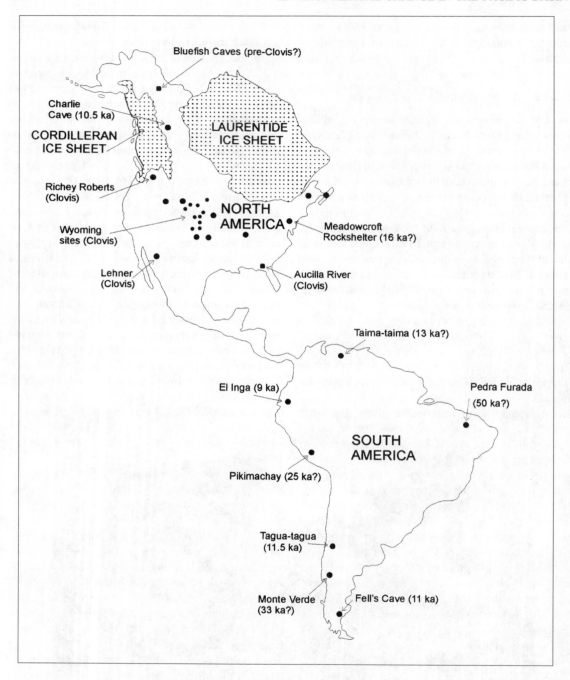

Figure 12.2   Location of late Pleistocene and early Holocene settlements in the Americas. The North American ice sheets are shown at their approximate terminal Last Glacial extent. Clovis sites are mostly late Last Glacial or early Holocene in age. Far older sites apparently occur in South America

Compiled from various sources

the 'first' Americans arrived did not involve this ice-free corridor (Dixon, 1993).

American sites of possibly considerable antiquity to the south include the Meadowcroft rockshelter in Pennsylvania where humans may have lived 16 000 years ago (Adovasio et al., 1988); the Monte Verde site in central Chile, occupied perhaps as much as 33 000 years ago (Dillehay, 1989); and the Pedra Furada rockshelter in northeast Brazil where it has been suggested that people lived continuously 50 000–55 000 years ago (Bahn, 1993). Sites are shown in Figure 12.2[1].

The situation in the Americas is comparable to that in Australia where the possibility of human occupation some 100 000 years earlier than generally thought will, if demonstrated to be correct, require the pace of associated human impact to be rethought (see above).

12.3.6 The Pattern of Initial Colonization of the Americas

To the 'uninformed' observer, the insistence on an initial entry by humans into the Americas through Beringia may seem improbable given that what are apparently the earliest known settlement sites – at Pedra Furada and Monte Verde – are both in South America. It may be that the first humans moved quickly through North America leaving hardly a trace behind; it may be that the earliest sites in North America have not yet been discovered (which makes the discovery of those in South America doubly surprising); it may be that indications of pre-Clovis humans in North America have been routinely dismissed as erroneous since they conflicted with a widely accepted orthodoxy; it may be that humans reached the Americas by another route (Dixon, 1993). There is no direct evidence yet available of any such explanations.

---

[1] It must be noted that controversy surrounds pre-Clovis ages from these and other American sites. It may be that future re-evaluations of the human record here will bring their ages into line with the model of humans entering the greater part of the Americas only after the ice-free corridor in the north opened

At present then, most scientists believe that the first humans reached the Americas some time during the Last Glacial across the Beringian land bridge. They then spread southwards, either along the coast and/or along an ice-free corridor between the Cordilleran and Keewatin ice sheets (see Figure 11.2) before fanning out through what is now the United States and thence across the central American isthmus into South America.

12.3.7 The Peopling of the Pacific Islands

At least 40 000 years ago, but perhaps much earlier, *Homo sapiens* moved out of southeast Asia across the ocean into Australasia. From New Guinea, they moved east to the other islands of what we now call Papua New Guinea and Solomon Islands by around 25 000 years ago. The initial human settlement of the vast numbers of small islands north, south and east of here was not apparently accomplished until much later (see Table 12.2) for reasons which have been widely discussed (Irwin, 1992; Kirch, 1997). It is possible that the earliest known dates are not in fact those of the earliest human settlement. Earlier settlements may have been exclusively coastal and their remains have since been drowned by rising sea level, buried by sediment, or even overgrown with coral reef (Gibbons and Clunie, 1986; Nunn, 1994c).

Yet it is also possible that the first inhabitants of Papua New Guinea and Solomon Islands were constrained within that area by an inability to make successful sea voyages over longer distances (Wickler and Spriggs, 1988). It was left to later arrivals (the 'Lapita' people) – equipped with vessels and navigational skills for long-distance voyaging – to make the first successful forays in search of other islands in the South Pacific.

The Lapita people had their origins in southern China and Taiwan and can be traced by their languages across the Pacific (see Table 12.1). Their initial migration led them to Halmahera island in the Molucca Sea, thence to the Bismarck Archipelago in the outer islands of Papua New Guinea. From here they colonized most other Pacific islands – Vanuatu, Fiji, Tonga and Samoa – very rapidly; then, after a pause, their

Table 12.2   Earliest known settlement dates of selected Pacific islands
From Spriggs and Anderson (1993), Nunn (1997c) and sources therein

Island (group)	Site	Date (years BP)
Melanesia		
Papua New Guinea	Huon Peninsula	>40 000
New Ireland	Matenkupkum cave	32 700 ± 1550
Solomon Islands	Kilu	~28 000
New Britain	Misisil cave	11 400 ± 1200
Fiji	Natunuku, Vitilevu	3240 ± 100
New Caledonia	Naia	3165 ± 120
Vanuatu	Malo	3150 ± 70
Polynesia		
Tonga	Tongatapu	3540 ± 70
Western Samoa	Ferry Berth, Upolu	3251 ± 155
Marquesas	Hane, Ua Huka	~1650
Hawaii	Wai'ahukini, Hawai'i	~1300
Easter Island	Rano Kao	1180 ± 230
Society Islands	Huahine	<1200
Cook Islands	Ureia, Aitutaki	~1140
New Zealand	Wairau Bar, North Island	~825
Henderson Island		790 ± 110
Mangareva		760 ± 80
Micronesia		
Marianas Islands	Chalan Piao, Saipan	3479 ± 300
Caroline Islands	Bolipi, Lamotrek	3310 ± 85

descendants moved on to the Marquesas and other island groups in the south-central Pacific, finally colonizing the remote outposts of Hawaii, Easter Island and New Zealand (Table 12.2).

Experiments involving the viability of coconuts in seawater suggest that they could not have reached the Panama coast (where they were recorded in AD 1514 by Gonzalo de Badajox) unaided by humans, implying that Pacific islanders had traversed the entire Pacific Ocean from west to east before it was known to Europeans (Ward and Brookfield, 1992).

12.4   THE IMPACT OF PRE-HOLOCENE HUMANS ON THE PACIFIC ENVIRONMENT

The discussion in Chapters 9–11 involved explaining observed environmental changes largely without reference to humans, even though

humans and their ancestors were present in the Pacific Basin. This is justified on the grounds that population densities were much less (compared to later times) and because the potential impact of an individual on the environment was so little (compared to a modern human). For ease of discussion, it is assumed that the beginning of the Holocene marks the time when human impact typically began to be registered in settled environments of the Pacific Basin.

It is difficult to quantify the subsequent rise of human impact for particular parts of the Pacific Basin because insufficient is yet known about the chronology of human settlement. It is often necessary to argue from the specific to the general, from the evidence of a single site to an entire region – a procedure which may produce different results 20 years from now when additional data are available.

There are also many methodological and philosophical problems involved in reconstructing human impact. For example, natural events often produce similar results to those produced by an abrupt human impact. The tendency to mistake the former for the latter may be exacerbated by the propensity, neatly expressed by Lamb (1977: 243), that 'because man is so adaptable and so apt to personalize the blame for his misfortunes by finding scapegoats and making victims of his fellows, the evidence of climatic causes is commonly obscured'. These issues are discussed in more detail in the following chapters with reference to specific instances.

That most pre-Holocene humans in the Pacific Basin practised only or largely hunting and gathering is also an important reason to suppose that their environmental impact was small, highly localized and often transient. Exceptions occur where humans developed agriculture, burning the forest to assist growth of food crops. Of those human groups who acquired a lifestyle linked to ocean resources, terrestrial environmental impacts may have been much less. Such groups would undoubtedly have had impacts on the surrounding environment, particularly as a result of a need for wood with which to build houses and boats and to burn for domestic purposes which, in the latest Pleistocene in Japan, evidently included the firing of clay pots (Imamura, 1996; Plate 12.2).

12.5 HUMANS IN THE PACIFIC AT THE START OF THE HOLOCENE

By the start of the Holocene, humans occupied all parts of the Pacific Rim excluding Antarctica.

Plate 12.2   Nail-impressed pottery reconstructed from fragments found at Shimachi, Iwate, Japan, and dating from the Incipient Jomon period 12.3–10.1 ka BP. This represents one of the earliest types of pottery to have been made in the Pacific Basin
Tadahiro Ogawa

Their presence on Pacific Islands was confined to those of western Melanesia. Generally low population densities at this time were probably associated with only small environmental impacts, probably overwhelmed in most places to the point of invisibility by the effects of late Last Glacial climate changes (Chapter 11).

Most human groups at this time relied mainly on hunting and gathering although there is evidence that agriculture was established in some highland areas of Papua New Guinea. There is a tendency among many writers to assume that the beginning of agriculture in a particular society represents a threshold in a developmental continuum. This view was challenged by Jared Diamond (1992) who showed – as is argued in the present book – that the development of agriculture was a response to the constraints of a particular environment and that early human communities would not necessarily have viewed it as a desirable alternative to a solely hunter-gatherer lifestyle where this remained viable.

Since it is a region that abuts the ocean, many human groups in the Pacific Basin at the start of the Holocene utilized ocean resources. It is probable that the continuing interaction with the ocean throughout the late Last Glacial and early Holocene was accompanied by an increasing ability to sail long distances. This ability would have extended human settlement of the region, not just in the islands, but also along parts of the rim. It is reasonable to suppose that this process would have been accompanied by an increase in the movement throughout the region of those plants and animals which were important to humans. The deliberate introduction of these plants and animals to 'naive' environments set the stage for the massive biotic transformation of Pacific environments during the Holocene, which is continuing still.

CHAPTER 13

The Early Holocene (12–6 ka) in the Pacific Basin

... so the wat'ry throng,
Wave rolling after wave, where way they found -
If steep, with torrent rapture, if through plain,
Soft-ebbing; nor withstood them rock or hill;
But they, or underground, or circuit wide
With serpent error wandering, found their way,
And on the washy ooze deep channels wore:
Easy, ere God had bid the ground be dry,
All but within those banks where rivers now
Stream, and perpetual draw their humid train.
John Milton, *Paradise Lost*, VII

13.1 DIVIDING THE HOLOCENE

The Holocene Epoch – the last 12 000 calendar years of the Earth's history – can be divided in different ways using a variety of criteria. The system adopted here – largely for convenience of discussion – is loosely based on both climatic and sea-level criteria, as discussed in the text. The early Holocene (12–6 ka) was a time of overall warming and sea-level rise, the middle Holocene (6–3 ka) was a time of comparatively high temperature and sea levels in many (though not all) parts of the Pacific Basin, and the late Holocene (3 ka to the present) was a time of mostly falling sea level and cooling climate during which human impacts on Pacific environments increased massively.

Most books dealing with the Holocene focus, either explicitly or implicitly, on human impact and the development of our species' comparatively sophisticated lifestyles throughout this epoch. Since the details of this are best known in northwest Europe and – to a lesser extent – North America and Australia, it is on these areas that most such books concentrate.

This book strives to be different: firstly, by maintaining its regional focus irrespective of the interesting things humans may have been doing elsewhere; and secondly, by concentrating on environmental change rather than human-induced environmental change. The importance of making this distinction will become clear later in Part E where it is argued that many environmental changes once routinely ascribed to 'human impact' might in fact be more realistically explained by largely non-human impact. This is not to say that humans had no significant impact on Pacific Basin environments, only to insist that their role here (as elsewhere in the world) needs to be placed in careful perspective.

In Chapters 13–15 in this book, there is an increasing preoccupation with short-lived events which manifests the increasing precision with which environmental change can be detected and measured throughout the Holocene. Many more sources of information become available, and more techniques can be used to synthesize and help interpret the greater mass of data which has survived from this epoch compared to earlier times. In contrast, the effects of volcanism and tectonics on Holocene environments of the Pacific Basin become less important in explaining regional change, for which reason they are reviewed in a separate section for the entire Holocene in Chapter 15.

This chapter considers only the early Holocene. Following a general statement concerning early Holocene environmental change in the Pacific Basin, each constituent region is discussed separately.

13.2   OVERVIEW OF THE EARLY HOLOCENE

The beginning of the Holocene is marked by the end of the Younger Dryas cooling event (Termination IB) about 12 ka (or about 10 ka BP). Ever since this time, in most middle and low latitude areas, climate has been warmer and sea level has been higher, for which reasons the Holocene is regarded as an interglacial interval, still in progress. Most of the ice formed only during the Last Glacial Maximum had melted by ~9 ka BP, although remnants of the Laurentide ice sheet in central north America persisted until 7–6 ka BP. Information about the early Holocene is quite sparse in most parts of the world, including the Pacific Basin; few studies have focused explicitly on this time.

In her survey of the origins of modern agriculture, Mannion (1991) recognized the importance of the period 10–7 ka BP when climate amelioration worldwide was creating conditions suited increasingly to the domestication of plants and animals by humans. Yet with that in mind, it is likewise clear from environmental records that changing human lifestyles during the early Holocene generally had little impact on contemporary Pacific environments, probably because population densities were low and/or the environments under pressure were resilient enough to withstand this. The first signs of significant human impact in most parts of the region are detected only in the middle Holocene (see Chapter 14).

Many plants and animals were domesticated first in the Pacific Basin. The first record of gourd/ squash (*Cucurbita* spp.) cultigens dates from 10.7–9.8 ka BP in Central America. In the same area, capsicum (*Capsicum annuum*), cotton (*Gossypium* spp.), certain beans and maize (or corn – *Zea mays*) – which was to come to dominate tropical American agroecosystems – were domesticated during the early Holocene. In South America, potato (*Solanum tuberosum*), sweet potato (*Ipomoea batatas*) and manioc (or cassava – *Manihot esculenta*) were domesticated during the early Holocene. Llama and alpaca were domesticated as beasts of burden in the Andean *puna* (grassland) around 6 ka BP. Dogs and cattle were domesticated in North America by 9 ka BP. Less information is available for the 'cradles of agriculture' in the west Pacific Basin although it seems probable that rice (*Oryza sativa*) was domesticated in China at least by 7 ka BP, soon followed by soybeans (*Soja max*) and foxtail millet (*Setaria italica*). Breadfruit (Plate 13.1) was an early Holocene domesticate in the western Pacific.

13.3   EARLY HOLOCENE ENVIRONMENTS OF ANTARCTICA

As for the Last Glacial (Chapter 11), the fringes of the Ross Ice Shelf have proved a fertile area for reconstructing Holocene climate changes. The recession of grounded ice from the western part of the Ross Embayment was underway during the early Holocene. A date of 7020 ± 60 BP for shells of *Adamussium colbecki* in a diagnostic moraine at Terra Nova Bay exemplifies the evidence of this process although most dates from the area fall within the middle Holocene, when ice recession had ended. The withdrawal of grounded ice from this area also led to isostatic uplift; emerged beaches testify to the progress of this during the early Holocene. Around the mouth of Taylor Valley, where beaches are absent, an emerged delta testifies to the time of ice recession here; dates from *Adamussium colbecki* bracket this event between 8340 BP and 6670 BP (Denton et al., 1989a).

In contrast to the situation in the Ross Embayment, few changes are known to have occurred in the adjoining Transantarctic Mountains during

Plate 13.1   The breadfruit (*Artocarpus altilis*) tree may have been an early Holocene domesticate in the western Pacific. Although never rivalling *taro* or yams as a staple food, it still forms an important element of arboriculture on many Pacific Islands. The belief that breadfruit would be a cheap food source for slaves in the West Indies led the British government to dispatch William Bligh in 1788 to transport enough from Tahiti to the West Indies to establish it there. Bligh's crew mutinied and sailed the *Bounty* to Pitcairn Island where their descendants live to this day. Bligh sailed the *Bounty's* longboat across the Pacific to Timor, returning eventually to Tahiti in 1792 to complete his original commission. Breadfruit did not prove popular in the West Indies
From the narrative of the *Voyage of the Beagle*

the early Holocene. This suggests that the principal cause of the changes in the Ross Embayment was probably not climatic (temperature or precipitation related) but associated with sea-level change. As sea level rose during the

early Holocene, ice which had previously been grounded on the floor of the Ross Embayment would have been released and begun to float, destined perhaps to drift offshore and break up. Ice upslope, which had previously been buttressed by

Plate 13.2   View along the east coast of Alexander Island where glaciers meet the George VI ice shelf. Moraines such as that between the ice edge and talus apron around Ablation Point (left middleground) formed some 6.5 ka BP when sea level was higher than today
Chalmers Clapperton

the grounded ice, may have surged downslope and also floated away from the shore. Thus did the Antarctic ice-sheet margins recede.

Evidence for the time of maximum Holocene warmth on the Antarctic Peninsula comes from a date of 6500 BP obtained from fragments of the barnacle *Bathylasma corolliforme* in a moraine on Alexander Island. This date was interpreted by Clapperton (1990b) as being that of the time when sea level in the nearby George VI Sound was higher so that the ice shelf was pushed farther onshore than today (Plate 13.2).

13.4   EARLY HOLOCENE ENVIRONMENTS OF SOUTH AMERICA

Pollen data indicate that the warmest time of the entire Holocene in southern Chile occurred 8.5–6.5 ka BP (Heusser, 1974). Later work in the Chilean Lake region allowed the situation to be defined more precisely. The progressive decline of *Podocarpus andinus* during the early Holocene tracks this warming but other pollen types indicate that the climate was still cool compared to today. Around 8350 BP *Nothofagus* forests came to dominate this region implying cooler and wetter conditions, especially after 6960 BP (Heusser, 1984a). The coincidence of this with glacier extension in Patagonia after 6850 BP (Mercer, 1982) suggests that the event was of considerable extent. Pollen records confirm this general picture for Tierra del Fuego and Isla de Chiloé (Clapperton, 1993a).

In central Chile, the wet conditions of the late Last Glacial were succeeded by dry conditions in its final part and into the early Holocene, as attested to by an abundance of salt-flat vegetation in pollen records, 'the almost complete absence of arboreal and aquatic taxa, and a general decrease in the diversity of the semiarid shrubland indicators' (Villagrán and Varela, 1990: 205). A study of lake sediments in the northern Chilean Andes found evidence for greater humidity 9.5–7.5 ka BP (Veit, 1993). Lacustrine sediment records indicate that glaciers receded rapidly in the Bolivian Andes 10.7–9.7 ka BP (Abbott et al., 1997).

A similar picture is available for the central Peruvian Andes where the start of the Holocene is marked by abrupt increases in nettles (*Urticales*) and other montane pollen types – indicative of warmer, wetter conditions – and by abrupt decreases in trees, shrubs and herbs (*Polylepis-Acaena* and *Compositae*) – indicators of cooler,

drier conditions (Hansen et al., 1994). Ice-core records from Huascarán col in the north-central Peruvian Andes suggest that maximum Holocene warmth occurred 8400–5200 BP (Thompson et al., 1995: see Plate 10.1).

Some investigators have detected rapid, comparatively short-lived departures from prevailing climate conditions during the early Holocene. For example, pollen data for sites in Andean Peru show an abrupt expansion of grasses 7.6–7 ka BP suggesting cold and dry conditions (Graf, 1981) which were undoubtedly linked to the 54 m lowering of nearby Lake Titicaca 7.7–7 ka BP (Figure 13.1). As elsewhere in the Pacific Basin, lake-level fluctuations here have proved a potent source of palaeoclimate information yet their interpretation is not straightforward. The curve shown in Figure 13.1 is a product of not only

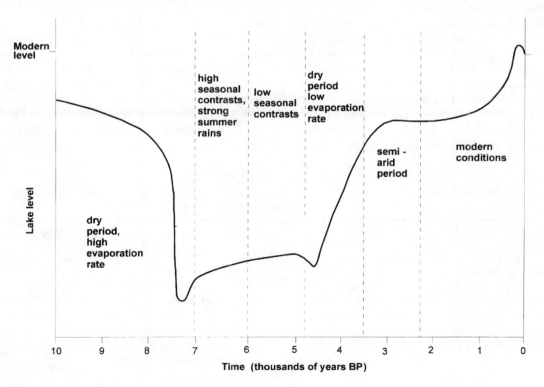

Figure 13.1   Holocene changes in the surface level of Lake Titicaca, Bolivia and Peru, and their climatic interpretation. Note the rapid fall in lake level of perhaps 54 m some 7700–7000 BP
Reprinted from Wirrmann et al. (1988) in Rabassa, J. (ed.), *Quaternary of South America and Antarctic Peninsula. Volume 6* 318 pp. Hfl. 140/- US$70.00. A.A. Balkema, PO Box 1675, Rotterdam, Netherlands

interannual changes in precipitation and temperature (as a proxy for evaporation) but also of the changing seasonality of these variables.

Some of the interesting work on the coast of Peru has focused on the response of its human occupants to early Holocene sea-level rise. Middens associated with lowland sites studied by Sandweiss (1996) exhibited increasing amounts of marine molluscan shells throughout the early Holocene, interpreted as a response to the progressively shorter travelling time needed to reach the shoreline as sea level rose during this period. Investigations by Lisa Wells (1992) showed that sea level rose during the early Holocene until 7–6 ka BP after which time shoreline progradation became dominant and coastal environments in which humans could flourish were created (see Figure 15.4). In comparison, changes in early Holocene environments in the corresponding parts of the Andes have proved harder to recognize because the overprint of deglaciation is so strong. One of the few studies (Miller et al., 1993) showed that these landscapes were unexpectedly stable during most of the Holocene. Following early Holocene deglaciation, gullying affected most valley-side slopes but this was followed by a blanketing of loess across the landscape, which led to its

stabilization, soil development, and colonization by vegetation.

It seems clear that glacial conditions during the earlier part of the early Holocene in much of Pacific South America were similar to those of today. The Quelccaya ice cap in the Peruvian Andes, for example, was the same size 10 ka BP as today (Mercer and Palacios, 1977). Falling precipitation levels during the early Holocene led to glacier and ice-cap recession and heralded a period of widespread aridity.

A summary of environmental conditions at key sites in Pacific South America during the Holocene is given in Table 13.1 using only the best-constrained records. In most extratropical areas, the time of maximum warmth during the Holocene was in its early part; the record in tropical areas is ambiguous. The pattern of changing precipitation during the early Holocene in Pacific South America is simpler to explain in general terms by the northward movement (and/or expansion) of zonal storm belts.

Although humans were present in Pacific South America during the early Holocene, only very little information about their lifestyles has been forthcoming compared to those of later times. While lowland peoples depended largely on marine and river-valley food sources, highland

Table 13.1   Summary of environmental change at key sites in Pacific South America during the Holocene
Adapted largely from Clapperton (1993a)

Location	Early Holocene (10–6 ka)	Middle Holocene (6–3 ka)	Late Holocene (3–0 ka)
Tierra del Fuego	10-5 ka, warm, drier	5 ka, cold	
Isla de Chiloé	12–8.5 ka, warm, wetter 8.5–5 ka, warm, drier	5 ka, cold, wetter	
Chilean Lakes	10–8.4 ka, warm, drier >7 ka, maximum warmth 7 ka, cool, wetter		
Peru			<3 ka, cold
Bolivia	9.5–7.6 ka, becoming humid 7.6–7 ka, dry interval	5.5–3 ka, humid, cooler	1.5 ka, humid, cooler

dwellers hunted deer and wild camelids to supplement what could be gathered in this generally less productive region. Indeed the poverty of potential wild food sources may have encouraged experiments in agriculture in the highlands before the need for this became apparent in most lowland areas. By way of example, it has been suggested that maize was domesticated by the close of the early Holocene in Andean Peru independently from Mexico (Bonavia and Grobman, 1989). A more detailed example of human–environment relationships, discussed further in Chapter 14, is provided by a record from the Atacama Desert where for many years it had been assumed by archaeologists that hostile environments had been rendered even more hostile to humans by enhanced aridity during the early Holocene (Grosjean et al., 1997).

13.5 EARLY HOLOCENE ENVIRONMENTS OF CENTRAL AMERICA

The progress from late Last Glacial coolness to early Holocene warmth is generally clear in the available pollen and related records for Central America (Bradbury, 1982). A temporary reversion to cooler conditions occurred about 7300 BP in Panama (Bartlett and Barghoorn, 1973) as in parts of South America (see above).

Recent work at Cobweb Swamp in Belize has shown that the entire early Holocene here was characterized by accumulation of a basal clay indicative of drier climate conditions which persisted to at least 6400 BP when, in response to rising precipitation levels, the site was flooded for the first time (Jacob and Hallmark, 1996). At Chalco Lake, in central southern Mexico, an abrupt change in limnological conditions around 9000 BP signals its transformation from a shallow fresh-water lake with low productivity to a saline marsh; this situation is attributed to the onset of higher temperatures which persisted to ~6500 BP when wetter conditions ensued (Lozano-Garciá et al., 1993). Similar variations were not detected in a core from the highlands of Costa Rica, which show that the site has been surrounded by treeless

páramo (grass and shrubs) since deglaciation 10 ka BP (Horn, 1993). As for South America, such data suggest that early Holocene climate variations in the tropics were comparatively small or non-existent in contrast to those outside the tropics which register the effects of changes in the positions of zonal climate belts associated with the transition from glacial to interglacial conditions.

Although most available data from Central America and Mexico indicate that the early Holocene was generally dry, glacial advances occurred on the mountains Ajusco, Iztaccíhuatl and Malinche in southern Mexico 10–8.5 ka BP (White, 1986). Since they lack correlates elsewhere in this region, these may represent a climate fluctuation of only local influence or extent.

The earliest known humans in this region reached Panama around 9000 BC settling the Pacific foothills where forest was easier to clear than on the coastal plain. Although hunting and gathering remained the more important part of their lifestyle, they also 'tended trees such as palms and may have grown arrowroot and other native plants in small hillslope gardens' (Cooke et al., 1996: 120). The development of more sophisticated agriculture 7–6 ka BP in Central America is attested to by the discovery of maize phytoliths and pollen. The change from a saline to a fresh-water marsh at Lake Chalco some 6000 BP may have been linked to an attempt by humans to control water level (Bradbury, 1989).

13.6 EARLY HOLOCENE ENVIRONMENTS OF NORTH AMERICA

13.6.1 The Early Holocene in the United States

The dominance of sagebrush (Artemisia) on the western slopes of the Sierra Nevada during the early Holocene is only one of many indications of contemporary aridity in the area until about 7 ka BP. Studies of the bristlecone pine (Pinus longaeva) in the nearby White Mountains of California have revealed much about Holocene palaeoclimates here. This pine is the longest-living tree species on Earth, some living almost 4000

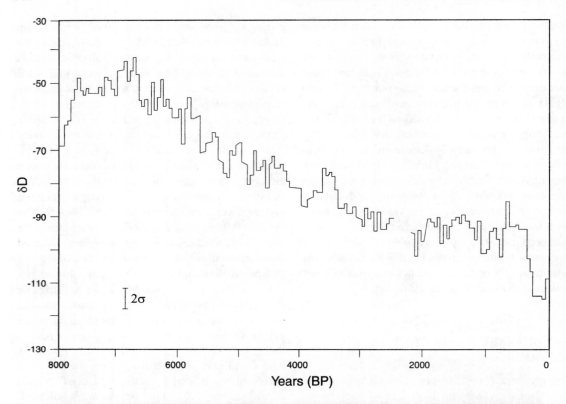

Figure 13.2   Temperatures during the last 8000 years for the White Mountains of California. The hydrogen-isotope ratio (δD) values were obtained from three long-lived bristlecone pine trees
After Feng and Epstein (1994)

years. Its dendrochronology was used by Feng and Epstein (1994) as a baseline for determining palaeotemperatures using stable hydrogen isotope composition (Figure 13.2). Warming is indicated from 8 ka BP (when the record begins) to 6.8 ka BP – the warmest time of the entire Holocene – and cooling since then.

A similar conclusion for the same area was obtained from the study of plant remains in packrat middens. The climate was similar to the present 9–7.5 ka BP; thereafter, the replacement of a dominantly pinyon forest by a pinyon–juniper forest in the subalpine zone indicates warming (Jennings and Elliott-Fisk, 1993).

Various records from the Great Basin suggest that the early Holocene was cooler and/or drier until 7–6.5 ka BP (Thompson et al., 1993). The dominance of Douglas fir (*Pseudotsuga menziesii*)

on the Olympic Peninsula 9 ka BP testifies to drier conditions during the early Holocene in the northwest United States (Heusser, 1977). A combination of grassland and low lake levels elsewhere in the northwest United States suggests very dry conditions around this time (Davis and Sellers, 1987).

The Hilgard glacier advance in the Sierra Nevada has been dated, albeit somewhat uncertainly, to the early Holocene. It has been suggested that an advance in the Cascade Mountains is of the same age, certainly older than the Mazama tephra (6.6 ka BP) which covers it in places (Burke and Birkeland, 1984). These data suggest that warming during the later part of the early Holocene was accompanied by increasing precipitation in parts of the westernmost United States.

Changes of this kind are important in helping explain changes in the distribution of human settlement sites in Pacific North America during the early Holocene. Many such explanations have been premised on the idea that early Holocene human groups in the region exploited only a part of the potential range of food resources, a view challenged in more recent studies. Willig (1996: 250), for example, concluded that 'economic strategies did *not* involve a specialized focus on *either* hunting large terrestrial game *or* foraging in lake-marsh habitats. Rather, there was a generalized, broad spectrum exploitation of a wide variety of habitats "tethered" within mesic catchment areas.'

As in Central America, the effects of early Holocene sea-level rise on the environments of western North America and their human occupants are poorly known. Indeed, Holocene coastal studies along the western seaboard of the United States 'have been greatly neglected' (Fairbridge, 1992: 14). The course of the Early Holocene transgression is known only in outline and depends for its validation on models of sea-level change founded on assumed lithospheric response to deglaciation.

13.6.2 The Early Holocene in Canada

From studies of glacier margins and fossil pollen, the early Holocene in Pacific Canada is inferred to have been warmer and drier than today (Clague, 1991; Burn, 1997; Plate 13.3). Progressive warming of the western part was tracked by the northward movement of the southern limit of boreal spruce (*Picea glauca*) forest 11–7 ka BP (Ritchie, 1984).

A recent study from the Queen Charlotte Islands, which were not covered by Cordilleran ice during the Last Glacial and which therefore

Plate 13.3 Pine log dated to 9120 ± 120 BP found in sediments well above the timberline, Castle Peak cirque, British Columbia, shows that trees grew higher during the early Holocene than they do today, probably because the climate of the area was warmer
John Clague

hold a longer record of postglacial climate change than the mainland, has shown that conditions ~9.6–6.6 ka BP were warmer and drier than today (Pellatt and Mathewes, 1997).

Most other indications of palaeoclimate suggest that maximum Holocene warmth in this region was attained around 6 ka BP, at which time much of western Canada was laid open to the influence of dry Pacific air masses (Ritchie and Harrison, 1993).

The few sea-level records from the Pacific coast of Canada are overprinted by tectonic changes and are consequently hard to interpret correctly (see Figure 15.12). The sea-level curve developed by Friele and Hutchinson (1993) for the west coast of Vancouver Island involved a sea-level maximum beginning around 6 ka BP.

13.7 EARLY HOLOCENE ENVIRONMENTS OF BERINGIA

Pioneering work in Alaska showed that a cool moist climate prevailed 10–8 ka BP followed by a warm period 8–3.5 ka BP (Heusser, 1960). Later work showed that such conditions occurred only in southwest Alaska. In the rest of Alaska, early Holocene climate was probably warmer and wetter than today, maxima occurring around 9 ka BP.

A precipitation minimum is evident in southern Alaska at 8 ka BP (Anderson and Brubaker, 1993). In northeast Alaska, the early Holocene probably included the warmest time of the entire epoch marked by expansion of *Picea glauca* in major river valleys ~9 ka BP, the spread of juniper (*Juniperus*) 10–8 ka BP, suggesting particularly warm and dry conditions, and an expansion of alder (*Alnus*) 8.5–7 ka BP indicating a shift to the moister conditions which have prevailed since (Edwards and Barker, 1994). Early Holocene vegetation in northwest Alaska was typically *Betula–Alnus* tundra, the modern *Picea* forest–shrub tundra being established only around 6 ka BP (Anderson et al., 1994).

One of the few early Holocene lacustrine records from western Beringia shows that warmer winters some 9 ka BP led to the arrival of *Pinus pumila* and the establishment of modern

vegetation. Continued warming during the early Holocene is indicated by the expansion of tree *Betula* and large *Alnus* shrubs north to the shores of the East Siberian Sea (Lozhkin et al., 1993). The Holocene thermal maximum in this area occurred 9.5–8 ka BP and was marked by the spread of woody plants into modern barren-ground tundra (Lozhkin, 1993). A date of 7.9 ka BP for a settlement site on Zhokov island in the modern Arctic Ocean indicates that early Holocene conditions there were equable enough to attract humans and the animals they hunted, probably more so than today (Pitulíko, 1993).

Deglaciation on the Aleutian islands[1] was almost completed during the earliest early Holocene but the period 7.5–5.5 ka BP was marked by renewed glacier advances signalling increased moisture availability throughout the islands (Black, 1980). Pollen data led Heusser (1978) to conclude that Aleutian climate was cool and moist 10–8.5 ka BP and warm and dry 8.5–3 ka BP. A conspicuous 'step' indicating rapid warming 7.5–5.5 ka BP was found in ocean-floor sediments just east of the Kamchatka Peninsula (see Figure 11.4).

13.8 EARLY HOLOCENE ENVIRONMENTS OF EAST ASIA

13.8.1 The Early Holocene in China

Pollen data from eastern China indicate a rise of temperature ~10–9 ka BP followed by a moist and warm climate optimum ~9–3 ka BP within which there was a short-lived cold episode 7.3 ka BP. Within this climate optimum, the warmest, wettest conditions occurred 7.2–6 ka BP; temperatures were 3–4°C higher than they are today in northeast China. Desert areas became less extensive; the spread of human settlement into areas which are now uninhabitable is a notable feature of the period 7.2–6 ka BP (Feng et al., 1993; Shi et al., 1993; Winkler and Wang, 1993).

An example of the detail which is becoming available for Holocene palaeoclimates is given in

---

[1] Strictly part of the Pacific Islands in this book

Table 13.2 for part of coastal China. The general pattern is of a warmer-than-present early Holocene with short-lived cooler excursions.

Sea level rose for much of the early Holocene but stabilized in a relative sense around the mouth of the Yangtze around 7.5 ka BP because of the compensatory growth of its delta. Palaeo-geographic maps of the area give a clear picture of the maximum encroachment of the rising sea on the land before this movement was reversed by delta outgrowth and, during the later Holocene, by sea-level fall (Figure 13.3). The Holocene sea-level maximum occurred around 6 ka BP along most other parts of the China coast (Winkler and Wang, 1993).

Table 13.2 Holocene temperatures at Qingfeng Brickworks, coastal China After Zhao et al. (1995)

Time (ka)	Temperature relative to present
2.1–1.2	+1.0°C (maximum)
2.4–2.1	−0.2°C (minimum)
2.9–2.4	(rise)
3.0–2.9	(fall)
6.5–3.0	+1.0°C (peaks of 1.7°C ~5.8 ka, 4.4 ka, 4.1 ka; trough 6.2–6.1 ka)
6.6–6.5	−0.4°C (minimum)
7.2–6.6	+1.3°C
7.3–7.2	−0.1°C
8.5–7.3	+1.3–1.6°C
9–8.5	−0.6–0.8°C
10–9	+0.5°C

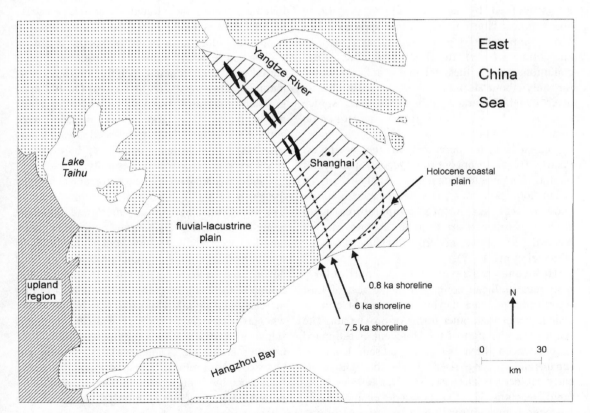

Figure 13.3 Holocene changes in the southern part of the Yangtze river delta, China, showing the shorelines at 7.5 ka BP, 6 ka BP and 0.8 ka BP
After Zheng et al. (1994) with permission of the Japanese Association for Quaternary Research

13.8.2 The Early Holocene on Offshore Islands in East Asia

On the offshore islands of Japan and Taiwan, considerable light has been shed on early Holocene environmental change from studies of fossil pollen. The Holocene temperature maximum occurred 7–4 ka BP in Japan, a time when the upper limit of beech (*Fagus*) forest was 400 m higher than today. At mountain sites in central Taiwan (~23°S), the time of maximum Holocene warmth some 8 ka BP was coincident with the maximum extent of forests dominated by *Castanopsis* (*chinquapin*) and *Trema* (Tsukada, 1967; Iwauchi, 1994).

Within this general picture, variations of interest have been recognized. An early Holocene boundary between pollen zones in Japan 8.5–8 ka BP is marked by a rapid fall of *Betula* in northernmost Honshu, and a rapid spruce decline and a complementary expansion of oak (*Quercus*) and elm (*Ulmus*) in Hokkaido. The warm conditions which these changes indicate lasted for only about a thousand years until climate deteriorated, as marked by the expansion of spruce (*Picea*) and fir (*Abies*) forest in Hokkaido (Igarashi, 1994).

Sea-surface temperatures in the northwest Pacific Ocean bear out the general pattern of onland temperature changes. Analysis of an ocean-floor core near Honshu island in Japan showed that sea-surface temperatures were slightly warmer 9 ka BP than at present, but reached ~3°C above modern values about 6 ka BP (Chinzei et al., 1987).

Holocene sea-level studies, particularly numerous in Japan compared to the rest of East Asia, often indicate the broad pattern of climate change better than other forms of proxy data. The sea level at the start of the Holocene is estimated to have been 40 m below its present level off Japan, and to have risen slowly throughout the early Holocene. The rise 8–6.5 ka BP was quite rapid, reaching 2 cm/year; sea level attained its Holocene maximum 6.5–5 ka BP. Although temporary regressions occurred along parts of the Japanese coast during the early Holocene, it is unclear which of these regressions were local – linked perhaps to uplift – or regional phenomena (Umitsu, 1991).

Compared to elsewhere in the Pacific Basin at the time, there were high human population densities in many parts of East Asia during the early Holocene, which makes the understanding of contemporary human–environment interactions, discussed in the following section, of especial interest.

13.8.3 Early Holocene Humans in East Asia

Continental East Asia is probably that part of the Pacific Basin inhabited longest by humans (see Chapter 12) so it is no surprise that it was also this area where innovations related to interactions between humans and the environments they inhabited were concentrated. East Asia was also an important source area for the dissemination of people and ideas throughout the Pacific Basin.

The nature of environmental impact by humans varied within this region depending on the type of natural environment they occupied. On the open grasslands of major river valleys, where there was no need to clear large areas, impacts were generally slighter than in forested areas where trees had to be cleared – usually by burning – if crops were to be grown successfully.

There is abundant evidence for agriculture in the Huanghe and Yangtze valleys from at least 7–8 ka BP. Crops, agricultural tools and domestic animals displayed distinctive regional characteristics which disappeared only gradually as interaction between human groups increased during the remainder of the Holocene. For example, the main crop in northern China during the early and middle Holocene was foxtail millet whereas rice (*Oryza sativa*) was the common staple in southern China (Zhimin, 1989). Rice farming along the Yangtze valley became the basis of communal living during the early Holocene; by 7 ka BP, towns like Hemudu supported large, wealthy populations (see also Chapter 14).

An uncommonly high incidence of forest fires 8.5–7 ka BP in the area around Nonbara Bog in

Japan was explained by Matsuo Tsukada (1986) as the consequence of burning by humans. If this interpretation is correct, then it would be the earliest record of human impact on vegetation in Japan. Yet the abrupt end to the Nonbara fires 7 ka BP suggests that they could also be explained as the products of unusually dry, warm periods or manifestations of natural fire regimes[2], as affected New Caledonia around this time (see Chapter 11). The first unequivocal signs of forest clearance associated with shifting agriculture in Japan date from 6.6 ka BP (Tsukada, 1986), while in Taiwan from perhaps 11 ka BP (Gorman, 1971).

Just as humans had effects on early Holocene environments of East Asia, so changes in these environments affected humans. Nowhere is this clearer than with the artefacts created by people of the Jomon culture (10–2.3 ka BP) in Japan, which show affinities with those of mainland Asia during the early part of the early Holocene and an increasing use of locally available materials thereafter, explicable by increasing insularity resulting from sea-level rise (Imamura, 1996).

Although the earliest known pottery may have been made in Japan (see Plate 12.2), it may have been invented independently in southeast China during the early Holocene. Pottery remains from this time have been found in places where the remains of deer, boar, small mammals such as monkeys and rabbits, fish, shellfish, turtles, crabs and possibly domestic pigs are also found. Cultivated root crops at these sites suggest that this time marked the transition from a mobile lifestyle based on hunting, fishing and gathering to a more settled one founded on agriculture to which food storage and processing techniques – for which pottery was needed – were central (Fung, 1994).

13.9 EARLY HOLOCENE ENVIRONMENTS OF AUSTRALASIA

13.9.1 The Early Holocene in New Guinea

During the early Holocene in the mountains of New Guinea, both ice and high-elevation (subalpine) grasslands receded as temperatures rose and forests expanded upslope. Southern beech (*Nothofagus*) pollen decreases in most early Holocene records owing to the development of more mixed forests in response to warmer, less misty conditions at lower elevations (Hope and Golson, 1995).

Vegetation belts in Papua New Guinea lay 100–200 m above their modern limits from ~9 ka BP into the late Holocene (Flenley, 1979). Temperatures rose during the early Holocene reaching a maximum of some 1–2°C above modern values in the New Guinea highlands some 6.5 ka BP (Harrison and Dodson, 1993). Wetter conditions 9.5–8.2 ka BP were inferred from the increasing water depth of Lake Wanum, followed by warmer and/or drier conditions 8.2–5 ka BP (Garrett-Jones, 1979).

Sr/Ca measurements on fossil *Porites* corals, which lived ~8.92–7.37 ka BP on a now-emerged reef on the Huon Peninsula in eastern New Guinea, showed that contemporary sea-surface temperatures were 2–3°C cooler than those of today (McCulloch et al., 1996). Since sea-surface temperatures had presumably increased significantly in this area since the Last Glacial Maximum, this suggests – albeit crudely – that these had been much lower at that time, more like those in the model of Bush and Philander (1998) than that of CLIMAP (see Chapter 11).

Evidence for early Holocene sea-level rise around New Guinea is less abundant than for much of the rest of Australasia. A 52 m core drilled through an emerged Holocene reef on the Huon Peninsula showed that while sea level was rising rapidly 11–7 ka BP, the reef was able to grow upwards at the same rate (Chappell and Polach, 1991) – an uncommon condition when the entire Pacific is considered (Nunn, 1994a).

Increased erosion and changes in secondary forest structure point to a discernible human

---

[2] These explanations do not preclude human involvement in the fires

Plate 13.4   Excavated section along the ~9000 year old drainage ditch at Kuk Swamp in the New Guinea Highlands. Note the traces of the bottom concretionary layer, containing datable organic litter, at the far end of the excavation and the remains of trees which once grew across the channel in the foreground. The lower wall on the left is part of the channel fill which was purposely left in place. This fill represents an eight-fold increase in swamp sedimentation thought to have begun only once this channel fell into disuse. The Kuk ditches represent one of the earliest known examples of deliberate environmental manipulation by humans in the Pacific Basin
Jack Golson

impact on highland vegetation ~9 ka BP in the view of most authors (e.g. Gillieson et al., 1987). The remarkable discovery of field drainage systems, excavated at Kuk Swamp by Jack Golson (1982), shows that comparatively sophisticated agriculture was underway perhaps 9 ka BP at this highland site where it came to be associated with an eight-fold increase in erosion of the catchment (Plate 13.4). This suggests that humans cleared and utilized New Guinea's forests increasingly during the early Holocene. At higher altitudes, the area of subalpine grasslands – restricted to a higher mean altitude than during the late Last Glacial when they were used more intensively – was still in use. Hope and Golson (1995) infer that hunting in these cold grasslands was carried out by transitory groups, perhaps dispatched from permanent settlements in the valleys where population levels were rising.

An important debate concerning the interaction between climate change and human innovation arose from the study of environmental change during the early Holocene in the highlands of New Guinea – arguably that area of the tropical Pacific Basin where these issues are best understood. Some researchers have suggested that the transfer of food plants (like *taro*, yams and bananas) and associated cultivation techniques from lower to higher altitudes around 9 ka BP occurred shortly after the time that highland climates had changed to allow these innovations (Golson and Hughes, 1980); in other words, humans responded rapidly and appropriately to new opportunities created by environmental change. Other writers have questioned this idea, typically arguing that the human response was less well defined and took place over a much longer time period (Swadling and Hope, 1992). Whatever the eventual resolution of this debate, it is clear that the New Guinea highlands are one place where the legacy of early Holocene agriculture was so profound that later Holocene agricultural adaptations in the area must be viewed – at least in part – as a response to earlier agricultural practice (Hope and Golson, 1995).

The incidence of periodic drought and frost in the New Guinea highlands, which had massive effects on agricultural productivity and human mortality during the late Holocene, may have been much less during the earlier Holocene owing to reduced ENSO variations (Hope and Golson, 1995).

13.9.2 The Early Holocene in Australia

Most pollen records in eastern Australia tell a tale of early Holocene climate change similar to that elsewhere in the Pacific Basin. The first part of the early Holocene was marked by rising temperatures and precipitation levels (Table 13.3). The vegetation response was particularly well marked in Tasmania by an expansion of wet eucalypt forests in the east of the island, and the spread of *Nothofagus*-dominated rainforests elsewhere. Pollen analyses from six sites in northeast Australia showed that grassy sclerophyll woodland dominated by *Eucalyptus* was replaced progressively by rainforest during the early Holocene (Walker and Chen, 1987).

Table 13.3   Climate data for the Holocene in northeast
Queensland
Kershaw (1976)

Time	Temperature (°C)	Precipitation (mm)
9 ka	18	1600
6 ka	20.5	3200
Present	20	2500

The early Holocene precipitation increase in northeast Australia has been linked to the re-establishment of the northwest monsoon (Bishop, 1994) and an equatorward shift of the subtropical high-pressure belt (Harrison, 1993).

Towards the end of the early Holocene, the climate was notably warmer and wetter than at present, a situation which persisted for most of the middle Holocene (Markgraf et al., 1986; Harrison and Dodson, 1993). One conspicuous exception is provided by pollen records from the Atherton Tableland of northeast Queensland, where temperatures were apparently lower than today 7–5 ka BP. Rather than reduced solar radiation, this may have been a function of greater cloud cover locally (Kershaw et al., 1994), also an explanation for the apparently anomalous Last Interglacial record from this site (see Chapter 10). This record is also anomalous within the early Holocene because of the delay between the onset of suitable conditions for rainforest colonization and the colonization itself, a delay attributed to both the time which it took rainforest species to migrate from source areas and the role of fire in inhibiting this process (Walker and Chen, 1987).

Lake Carpentaria, which existed in the now-submerged land area connecting north Australia and New Guinea during much of the Last Glacial (see Figure 11.7), became connected to the ocean by the start of the Holocene. Rapid sea-level rise (1–3 mm/year) was held by Torgersen et al. (1988) to account for the change from estuarine conditions about 12 ka BP (when sea level was −53 m) to shallow nearshore conditions around 8.5 ka BP (sea level was −45 m) to fully marine conditions about 7 ka BP (sea level was −10 m).

The early Holocene in north Australia, particularly the period 7–6.5 ka BP, was marked by the widespread development of mangrove swamps in consequence of a deceleration in the rate of sea-level rise and the infilling of coastal embayments which created environments suitable for mangrove colonization. Interestingly, this period of mangrove development lasted only some 1000 years because thereafter, sea level stopped rising and sediment infilling increased, reducing available mangrove habitats (Woodroffe et al., 1985).

Population densities were very low in Australia during most of the Holocene for which reason human–environment relations are difficult to recognize. Humans have been implicated in megafaunal extinctions, yet it is clear that humans and megafauna coexisted for more than 40 000 years in Australia so that models of overkill proposed for the Pacific United States, for example, may not apply here (see Chapter 11), although Flannery (1994) argued that the process was simply more prolonged and thus more difficult to pinpoint. The two representative examples of the use of fire by Aborigines during the early Holocene in Pacific Australia described by Head (1989) led to the conclusion that in neither case was this able to prevent forest expansion although elsewhere on the continent, particularly in more vulnerable (commonly drier) areas, this may not have been the case.

A good example of the effects of environmental change on humans comes from Tasmania which was effectively isolated from mainland Australia until the arrival of Europeans as a result of early Holocene sea-level rise. In consequence of this isolation, humans in Tasmania lost the ability (among others) to use boomerangs and also could not share any of the technological innovations (or dingoes) which spread through mainland Australia during the pre-European Holocene. In the case of dingoes, this also meant that Tasmania was shielded from the effects of their competition with the Tasmanian devil and Tasmanian tiger, both of which became extinct on the Australian mainland several thousand years ago while surviving into colonial times on Tasmania (Gollan, 1984).

13.9.3 The Early Holocene in New Zealand

The earliest part of the early Holocene 10–8 ka BP in New Zealand was 1–2°C warmer than at present, a time marked by the almost complete afforestation of the islands. The presence of the frost-sensitive *Ascarina* throughout western New Zealand 10–7 ka BP suggests that climates were also drier and less windy than today, a conclusion which can also be reached by considering the restricted distribution of the hardier *Nothofagus* at this time in places where they now dominate. Such data have been interpreted as meaning that the influence of westerlies on New Zealand, particularly South Island, was less than today. The existence of the mangrove *Avicennia marina* in the Poverty Bay area ~9840 BP is another clear indicator of contemporary warmer and wetter conditions; this mangrove now occurs only in the Bay of Plenty ~1° farther north (Mildenhall and Brown, 1987).

After 7.5–7 ka BP, the climate became markedly cooler, producing changes in vegetation patterns and associations. The extent of *Ascarina* diminished rapidly and cold-tolerant species such as *Nothofagus* spread, particularly in upland areas (Salinger, 1988; McGlone et al., 1993). Encouraged in particular by reduced moisture after 7 ka BP, *kauri* (*Agathis australis*) trees spread southwards across New Zealand at rates as high as 197 m/year (Ogden et al., 1992).

13.10 EARLY HOLOCENE ENVIRONMENTS OF THE PACIFIC OCEAN AND ISLANDS

A lake came into existence for the first time some 10.3 ka BP in the El Junco crater on San Cristóbal island in the Galápagos Islands. The varying extent of this lake during the Holocene was determined from variations in the proportion of pollen of the water fern *Azolla microphylla* in lake sediments (Colinvaux, 1972, Figure 13.4). Results show that an ephemeral lake existed 10.3–8.6 ka BP, indicating either lower precipitation and/or higher temperatures than today. During the remainder of the early Holocene, increased precipitation is indicated by increasing lake depth. A possible correlate was reported from Mangaia in the Cook Islands by Joanna Ellison (1994) although the interval 7–6 ka BP may have been drier here than at present (Kirch et al., 1992).

The final disappearance of Makanaka (Last Glacial Maximum) ice from Mauna Kea on the island Hawai'i probably took place during the early Holocene, as shown by a minimum age of 9.1 ka BP for the first sediments unaffected by

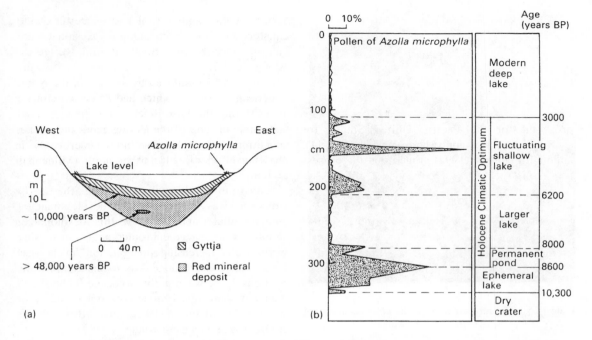

Figure 13.4   Palaeoclimatic interpretation of El Junco lake, San Cristóbal island, Galápagos. (a) Section across El Junco lake showing the likely stratification of lake sediments and the present marginal distribution of the water fern, *Azolla microphylla*, the pollen of which has been used as a palaeoclimate indicator. (b) Average proportion of *Azolla microphylla* from two cores through El Junco lake sediments. The interpretation of its earlier conditions and their chronology is also shown
From Nunn (1994a), after Colinvaux (1972)

meltwater to have been deposited in Lake Waiau (Porter, 1975).

A record of episodic sea-level rise during the early Holocene is preserved by submerged shorelines off Oahu island in Hawaii (Fletcher and Sherman, 1995; Plate 13.5). During the early Holocene, a prominent stillstand ~9–8 ka BP produced the −24–30 m Kaneohe shoreline complex, which is found at similar depths off eastern Australia and New Zealand. On other Pacific islands, early Holocene sea-level records are few, most showing or implying a continuous rise for this period.

Coral reefs became established around many Pacific islands during the early Holocene and have been classified according to their subsequent response to sea-level rise as 'keep-up', 'catch-up' or 'give-up' reefs (Neumann and MacIntyre, 1985). Few keep-up reefs occurred except in the

warmest parts of the Pacific Ocean. Most reefs appear to have been catch-up reefs, a fact that is important to an understanding of the Holocene evolution of reef-fringed Pacific coasts (Hopley, 1984; Nunn, 1994a; see Figure 14.4C). Give-up reefs are found particularly in the region between Samoa and Tuvalu which is assumed to have experienced oceanographic conditions unfavourable for coral-reef upgrowth during the early Holocene (Nunn, 1998a).

13.11   SYNTHESIS

Throughout the Pacific Basin, the early Holocene was a time when warm conditions – typical of late Quaternary interglacials – were established. The Holocene thermal maximum occurred (at least partly) within the early Holocene in most parts of the Pacific Rim.

Plate 13.5   The submerged intertidal notch 24 m off the
coast of Oahu, Hawaii
Chip Fletcher, University of Hawaii

Variations in other climate parameters can be
linked to shifts in zonal climate belts. The best
example is the changes in precipitation
experienced along the Pacific Rim in parts of

South America and Australia as the result of the
equatorward movement and/or expansion of the
belt of storm-bearing South Pacific westerlies.
Along the Pacific Rim in the North Pacific, the
situation was complicated by melting of the North
American ice sheets, which had deformed climate
zones during the Last Glacial, and the continued
(effective) closure of the Bering Strait preventing
cold Arctic water from reaching the Pacific, which
moderated the influence of polar climate in this
region.

Environmental change during the early
Holocene was critical to the development of
human culture. Yet to view this development
simply as a benevolent outcome of increasing
warmth and increasing precipitation levels – as
some authors have done – is to misinterpret the
relationship between early Holocene humans and
their environments. Humans did not suddenly or
universally abandon their past lifestyles as
environments changed. Rather they responded to
change and, where that change did not directly
threaten the continuation of their traditional ways
of living, these continued. Only where environmental change reduced the viability of former
lifestyles to a great extent were new opportunities
sought.

CHAPTER 14

The Middle Holocene (6–3 ka) in the Pacific Basin

... the bare Earth, till then
Desert and bare, unsightly, unadorned,
Brought forth the tender grass, whose verdure clad
Her universal face with pleasant green;
Then herbs of every leaf, that sudden flowered,
Opening their various colours, and made gay
Her bosom, smelling sweet; and, these scarce blown,
Forth flourished thick the clust'ring vine, forth crept
The smelling gourd, up stood the corny reed
Embattled in her field: add the humble shrub,
And bush with frizzled hair implicit:
John Milton, *Paradise Lost,* VII

14.1 THE MIDDLE HOLOCENE

In some parts of the world, the middle Holocene includes, straddles or falls within the time of maximum Holocene temperature, sea level, and often either precipitation or aridity. It is a time which has been given a number of labels (the Hypsithermal, the Holocene Climatic Optimum, the altithermal interval), the range and definition of which vary from place to place. The use of such labels is thus deliberately avoided in this book.

Although maximum Holocene temperatures were attained in some parts of the Pacific Basin during the early Holocene (Chapter 13), in some this maximum occurred in or continued into the middle Holocene. As would be expected, given that postglacial sea-level rise lagged behind temperature rise, the Holocene sea-level maximum occurred during the middle Holocene in almost every part of the Pacific Basin. Although many parts of the region were conspicuously drier than today during the middle Holocene, precipitation has generally a more variable, more localized pattern of change through time.

This chapter reviews the environmental changes that took place during the middle Holocene for each of the constituent regions of the Pacific Basin. Note that the description of environmental changes associated with volcanism and tectonics is given for the entire Holocene in a separate section in Chapter 15.

14.2 MIDDLE HOLOCENE ENVIRONMENTS OF ANTARCTICA

The only part of the Pacific Rim which humans are not known to have inhabited by the beginning of the middle Holocene is Antarctica. A recurring theme in Holocene studies in Antarctica, particularly pioneering ones, is the absence of significant variation in Holocene climate (e.g. Pickard et al., 1984). Later work, employing higher-resolution and novel investigative methods, has shown that this conclusion is certainly not as widespread in Antarctica as once implied, although diagnostic data are still lacking from most parts.

Of the few Holocene palaeoclimate studies in this region, one of the most informative concerning the middle Holocene is that by Björck et al. (1996) from James Ross Island, off the Antarctic Peninsula. They found evidence for glacier recession before 5 ka BP, owing to cold arid conditions, and the onset of humid warm conditions about 4.2 ka BP. Aridification about 3 ka BP caused glaciers to vanish from the study area. A similar conclusion was reached from the study of abandoned penguin rookeries along the Victoria Land coast (Baroni and Orombelli, 1994). The greatest diffusion of rookeries occurred 4–3 ka BP, suggesting the existence of what were probably the warmest conditions during the entire Holocene at this location. The sudden decrease ~3 ka BP implies rapid cooling.

Evidence for middle Holocene sea-level maxima in Antarctica is abundant. The compilations by Nunn (1984) and Berkman (1992) suggest that at least one maximum occurred during the middle Holocene. Identifying the elevation to which sea level rose at this time is problematical because most records are from ice-free areas around the continent's periphery which have risen isostatically in the last few millennia, raising emerged shorelines to levels well above those at which they formed. For example, the Marble Point area of McMurdo Sound is thought to have risen isostatically some 18 m since ~4.6 ka BP (Nichols, 1966). One convincing attempt to separate eustatic and tectonic effects was by Baroni and Orombelli (1991) working in Terra Nova Bay in McMurdo Sound, but they could still not identify a eustatic maximum because of fast rates of uplift (>3 mm/year during the middle Holocene).

14.3 MIDDLE HOLOCENE ENVIRONMENTS OF SOUTH AMERICA

A view of Holocene climate monotony similar to that noted for Antarctica (see above) was found in a study of marine fossils in emerged beaches along the sides of the Beagle Channel in Tierra del Fuego, in southernmost South America (Gordillo et al., 1992). In this case, comparative climate uniformity is more believable owing to the location of the field area in the path of the cool

Antarctic Circumpolar Current, which there is no reason to suppose varied significantly in either temperature or position during the Holocene. Such an interpretation does not necessarily conflict with the significant terrestrial variations found at a nearby site, Caleta Róbalo, by Heusser who interpreted the dominance of *Nothofagus* (southern beech) forest after 5 ka BP as 'an apparent response to cooler, wetter climate with increased storminess and cloud cover' (Heusser, 1989: 403).

An absence of temperature variation has been noted elsewhere in South America, at least during the (early and) middle Holocene. In the semiarid region of coastal Chile, for example, dry warm conditions prevailed for most of the Holocene (Villagrán and Varela, 1990), suggesting stability of the southeast Pacific anticyclone during this time. Data from fossil beetles, which have proved to be highly sensitive indicators of Holocene climate change elsewhere, indicate a uniform climate in part of the Chilean Andes during the middle Holocene (Heusser, 1984b).

There are a variety of indications of glacier expansion in the southern Andes 4.7–4.2 ka BP (Coltrinari, 1993). Pollen data from the Mallin Book site in the same region suggest that this expansion was a consequence of increased precipitation, a condition that lasted until about 3 ka BP (Markgraf, 1983). Humid conditions prevailed in the Bolivian Andes 5.5–1.5 ka BP (Graf, 1981).

Farther north, there is an abrupt change to this picture. Decreases in montane forest in the central Peruvian Andes about 6 ka BP denote the onset of drier conditions although the possibility of significant anthropogenic disturbance at this time is acknowledged. The increase in Chenopodiaceae–Amaranthaceae about 4 ka BP suggests that drier conditions were persisting, even strengthening, here although there are undoubted signs of human activity at this time (Hansen et al., 1994). Much of the environmental change in montane Pacific South America during the middle Holocene was associated with cooling although much of the evidence for this – such as the lowering of tree lines 5–3 ka BP – could also be ascribed to increased aridity (Markgraf, 1993).

Drier conditions in the Peruvian Andes around 4 ka BP may have led to the abandonment of the Santa Delta region in coastal Peru by humans 4–3 ka BP, a time marked by increased agricultural activity in the hinterland (Wilson, 1988; see Figure 15.4). Yet in such a low area, it is possible that there is an incomplete record of early settlement sites because of the effects of both flooding associated with large El Niño events and sea-level rise.

It is not just flooding which is associated with El Niño events in coastal Peru. Today the warm-water countercurrent along the Peruvian coast which develops during an El Niño event causes shortages of cool-water fish and other marine fauna. During an unusually severe El Niño some 4 ka BP, it is possible that the coastal people of the Santa Delta and elsewhere in this region were forced to depend more on the products of their fledgling agriculture and eventually decided to move inland where this could be practised with less disruption from flooding. Another El Niño flooding event affected coastal Peru around 3200 BP (Wells, 1990).

If the climate signals appear mixed, the sea-level record is unequivocal. As shown in Table 14.1, the maximum Holocene sea level along the Pacific coast of South America occurred during the middle Holocene. Although the effects of the transgression may have been offset by uplift in places, it undoubtedly caused problems for many coastal peoples, particularly those in tectonically quiescent, stable or subsiding parts of the continental margin. It is plausible to suppose that sea-level rise, along with water shortages and the

environmental effects of El Niño, drove many coastal dwellers inland during the middle Holocene in search of less fickle environments to occupy. The magnitude of the stress involved in coastal living under such circumstances can be measured by the fact that some of the least hospitable parts of the hinterland were settled at this time.

An excellent example comes from an eco-logical refuge in the Atacama Desert, shown recently to have been inhabited ~8.0–3.1 ka BP (Grosjean et al., 1997). Many of the settlement sites in this area are buried by debris flows presumed to have been caused by unusually heavy rainfall events recurring every 500–1200 radiocarbon years during the otherwise generally arid middle Holocene. Most of these events left a regional imprint on the landscape and remain largely unexplained at present. Much clearer are the changes in human–environment relations in the area. Earlier residents (before 5.9 ka BP) were hunters who occupied temporary campsites around wetlands formed upstream of a point where debris flows had blocked the Puripica River. Both hunting strategies and tool-making had become more sophisticated by around 5.9 ka BP. Semi-permanent settlement was associated with the domestication of animals (supplemented by hunting) by 5.1 ka BP, a number of additional debris flows having helped to maintain the wetlands. These survived until 3100 ± 70 BP after which they disappeared gradually and the people moved elsewhere.

Although maize and other cultivars had arrived in Andean communities by the start of the middle

Table 14.1   Middle Holocene sea-level maxima in Pacific South America

Location	Date of maximum (ka BP)	Height of maximum (m)	Source of information
Bahía San Sebastián, Tierra de Fuego	5.27	1.8	Vilas et al. (1987)
Estero Tongoy, north-central Chile	~6	3	Ota and Paskoff (1993)
Santa Delta, Peru	~6	?	Wells (1992)

Holocene (see Chapter 13), these were not wholly dependent on agriculture but continued to hunt (and domesticate) wild camelids, the numbers of which had increased greatly as they outcompeted other vertebrates to dominate grazing niches in most grassland areas. This process was part of a broader change favouring grazers over browsers resulting from the increase in grassland area at the expense of forest which accompanied the warmer, drier conditions of the middle Holocene in the main parts of the inhabited Andes (Hansen et al., 1994).

14.4 MIDDLE HOLOCENE ENVIRONMENTS OF CENTRAL AMERICA

Drier, perhaps cooler, conditions affected Panama 7.3–4.2 ka BP (Bartlett and Barghoorn, 1973). Maize phytoliths appear first 3910 BP, a sure sign that slash-and-burn agriculture was underway (Piperno, 1994). In contrast, the highest parts of southern Central America exhibit no changes at this time (Horn, 1993) suggesting that climate change did not affect these areas, perhaps because they were already cool and dry and, more surprisingly given the contrary situation in South America, that they were not permanently occupied by humans. The latter observation is readily explained as a consequence of low population densities at lower elevations and, later, by the efficacy of lowland agricultural intensification, which obviated the need for people to utilize the highest parts of this region.

In the central highlands of Mexico, a prolonged dry period occurred ~6–4.5 ka (Metcalfe, 1995). Chalco Lake in the central basin of Mexico reached its lowest level 6–5 ka BP and a saline swamp was established (Lozano-Garciá et al., 1993). The aridity which characterizes the modern Sonoran Desert began during the middle Holocene; conditions may even have been drier than today (Spaulding, 1991). While these conditions persisted in northern Mexico, a modest precipitation increase in central Mexico led to the establishment of the modern climate regime in this region (Bradbury, 1989).

Short-lived erosional events 3640–2890 BP are thought to be connected to the first appearance of maize in the highlands of central Mexico (O'Hara et al., 1993). Human impact was apparently so profound throughout Central America by 3.5 ka BP that subsequent palaeoclimates have proved almost impossible to reconstruct (Markgraf, 1993).

Although Central America was the principal centre of agricultural and societal development in the Americas, and many innovations originating there later spread through other parts of the continent, other innovations came from elsewhere. Pottery, for example, first began to be made in Brazil more than 7 ka BP (Roosevelt et al., 1991), reaching Panama only ~5.4 ka BP (Ranere and Hansell, 1978).

14.5 MIDDLE HOLOCENE ENVIRONMENTS OF NORTH AMERICA

14.5.1 The Middle Holocene in the United States

Palaeoclimate data from the White Mountains of California suggest that temperatures were as much as 2°C higher than today 7350–3450 BP (LaMarche, 1973). A temperature maximum ~6800 BP was followed by cooling throughout the middle Holocene (see Figure 13.2). The larger-than-present size of fossil perch (fish) at Clear Lake, California, has also been taken to indicate that summer temperatures were higher than present during the middle Holocene.

Compared to the early Holocene, there was increased moisture available during the middle Holocene in the Sierra Nevada. There are some data which suggest a precipitation maximum ~4500 BP in California, and another in the Great Basin 3800–2000 BP. The onset of cooler, wetter conditions, associated with much of the late Holocene, has been dated to 4700 BP in the Great Basin, and 3–2 ka BP in the Sierra Nevada (Thompson et al., 1993).

The record of climate change from San Joaquin Marsh in California is also discussed in Chapter 15. Suffice to note at this point that there is evidence here for a period of cooling 3800 BP (Davis, 1992). This falls within the period 5200–3250 BP identified by Cole and Liu (1994) as a time of arid climate on Santa Rosa island, California.

Evidence for middle Holocene changes in ice extent in the mountains of the Pacific United States is sparse. One of the exceptions are the 4960–4700 BP ages for wood from trees sheared off by advancing ice at Dome Peak in the Cascade Range (Miller, 1969).

The broad view of sea level around the United States reaching a middle Holocene maximum around 5700 BP then falling, perhaps in an oscillatory manner (Fairbridge, 1992), has not been clearly identified along the Pacific coast.

14.5.2 The Middle Holocene in Canada

The middle Holocene in Pacific Canada was marked by a progressive fall in temperature as the Arctic front moved south from the point it had reached (~56°N) at the time of the temperature maximum about 6 ka BP. In much of western Canada, the aridity of the early Holocene was replaced by conditions of increasing moisture during the middle Holocene (Ritchie and Harrison, 1993). In western British Columbia and the Canadian Rockies, there was a widespread glacial advance ~3100–2500 BP which may represent the culmination of a climate deterioration 5–3 ka BP (Luckman et al., 1993) although it could also have been influenced by increasing precipitation levels.

One of the few sea-level histories for Pacific Canada is for Vancouver Island (Friele and Hutchinson, 1993) where a ~3.2 m maximum occurred 6000–4800 BP.

14.6 MIDDLE HOLOCENE ENVIRONMENTS OF BERINGIA

The effect of sea-level rise during the early Holocene may have been to change the climate of Beringia from a warmer, more continental regime to one more maritime and marked – throughout the middle and late Holocene – by an overall cooling. Modern vegetation communities became established throughout Beringia by the start of the middle Holocene and the few data that show variations during this time signal minor climate changes, possibly of only local extent. Examples include the slight cooling of summer temperatures ~5 ka BP inferred from increases in *Pinus pumila* pollen near Magadan in west Beringia (Lozhkin et al., 1993). A slight precipitation maximum was thought by Patricia Anderson and Linda Brubaker (1993) to have affected vegetation in southern Alaska about 4 ka BP and may have coincided with the ~3.6 ka BP advance of the McCarty Glacier in the area (Wiles and Calkin, 1994). Most Aleutian islands were deglaciated 5500–3500 BP, a period followed by one of renewed glacial advance (Black, 1980).

Humans occupied most parts of Beringia during the middle Holocene but, as along the Pacific Rim in North America, their numbers were probably never sufficient to cause significant widespread environmental change. One of the most congenial environments in Beringia would have been in the Russian far east, where up until ~4.5–4 ka BP human impacts were restricted to burning and trampling near hunting, fishing and gathering sites. After this period, cattle breeding and agriculture began, and impacts can be assumed to have increased proportionately (Kuzmin, 1995).

14.7 MIDDLE HOLOCENE ENVIRONMENTS OF EAST ASIA

14.7.1 The Middle Holocene in China

In their review of middle Holocene environments in China, Shi et al. (1993) concluded that in sharp contrast to the early Holocene, the period 6–5 ka BP was characterized by a fluctuating climate manifested as generally adverse conditions for human cultural development. Records from the interior of China influenced the conclusion of Shi et al. (1993) that this period was not the climate optimum for China, but the assertion seems premature for its Pacific fringe where there are several indications of warmer, wetter conditions 6–5 ka. A record from Nanjing, for example, near the mouth of the Yangtze River suggests temperatures around 5200 BP were 3.6°C higher than today (Lamb, 1977). This picture is also confirmed by data from northeast China showing that middle Holocene climate was warmer by 3–5°C and, until ~5 ka BP, more humid than at present (Winkler and Wang, 1993). The synthesis of Holocene climate change in eastern China by Feng et al. (1993) showed that the middle Holocene was part of the warmest time of the entire Holocene despite the occurrence of a major cool episode 5.7 ka BP and a minor one ~4.5 ka BP.

The period 4–3 ka BP is regarded in available human records as a time of calamities, including a period of catastrophic floods in the major river lowlands which prompted King Dayu to begin projects for regulating water flow. The end of the middle Holocene (3 ka) marked an end to the period of maximum Holocene warmth in Pacific China and the start of the cooling which has largely persisted since then. Holocene sea level peaked around the beginning of the middle Holocene along the coast of most of mainland China excepting the Yangtze delta (see Chapter 13, especially Figure 13.3).

Increased warmth and moisture availability during the later part of the middle Holocene may have assisted agricultural development in parts of China, notably the valleys of the Yangtze and Huanghe. Development was rapid here compared to other parts of the world, perhaps partly because of the climate but also because the landscape was itself conducive to the spread of agriculture – extensive flat plains of fertile, well drained alluvial (loessic) soil compared, for example, to Mesopotamia where the spread of agriculture was constrained more by mountains. Most importantly though, the development and spread of agriculture in Pacific East Asia, which is such a conspicuous feature of its middle Holocene history, might be explained largely by the absence of opportunities for alternative lifestyles. The earlier inhabitants of the region must have survived largely by hunting, fishing and gathering but, as numbers grew, this lifestyle proved inadequate as a means of subsistence.

Throughout the East Asia region of the Pacific Rim during the middle Holocene, vegetation changes associated with the spread of agriculture – especially increases in cereal and chenopod pollen – make it difficult to identify the effects of subsequent climate changes (Winkler and Wang, 1993). As an illustration of how productive middle Holocene agriculture was in eastern China, Yan (1993) excavated pits at Cishan which could have held 100 tonnes of millet, and ~6000 year old rice remains at Hemudu equivalent to a crop of 120 tonnes. The environmental effects of the 'rice revolution' are discussed in more detail in a separate section below.

14.7.2 The Middle Holocene in Japan

The warmest temperatures in Japan were attained about 6–5 ka BP, since which time cooling and increased precipitation have caused vegetation change and glacier advance (Fukui, 1977; Tsukada, 1986).

Pollen analysis of sediments from Ukinuno Pond in Japan show that millet was first cultivated in the area around 3 ka BP; the intriguing find of pollen from the cantaloupe (*Cucumis melo*) – a plant endemic to Africa – at the same stratigraphic level has been used to argue for the ultimate African origin of the contemporary settlers (Tsukada, 1986). The spread of rice through East Asia is discussed in the following section.

Shell middens found up to 50 km inland on the Kanto Plains of Japan show the position of the shoreline at the sea-level maximum during the early middle Holocene (Tsukada, 1986). Studies of fossil diatom assemblages in sediments along the Kyushu coast showed that there was a highstand of sea level ~5.5 ka BP at around 1 m above its present level (Yokoyama et al., 1996). A minor sea-level fall within the period 5–4 ka BP (Umitsu, 1991) may have been associated with a short-lived cooling ~4.4–4.2 ka BP (Sakaguchi, 1983).

14.7.3 The Rice Revolution

Common rice (*Oryza sativa*) was domesticated first in East Asia, certainly by 7 ka BP in the lower Yangtze valley. The key characteristics of this

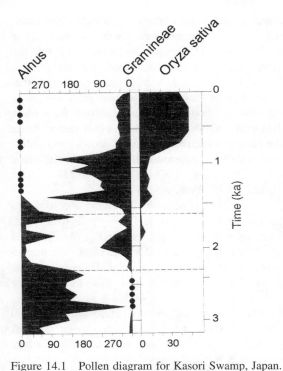

Figure 14.1 Pollen diagram for Kasori Swamp, Japan. The increase in rice and grass pollen is coincident with a decrease in alder pollen, interpreted as an indicator of forest clearance by humans, largely for rice cultivation
After Tsukada (1986) with permission of Matsuo Tsukada and Center for Japanese Studies, University of Michigan

region which made it especially suitable for rice-growing were (and are still) the dependable monsoon rains, the warm humid climate, and the presence of valleys with seasonally flooded grassland depressions. Initial gathering of wild *Oryza* spp. may have involved selection of those plants bearing larger panicles and heavier grains. Methods of shattering the panicles may have involved the accidental dropping of some, which fell and began growing close to settlements. This may have led to the purposeful dropping of seeds in such places and, as for other crops, the organically enriched soil here may have given rise to higher yields encouraging people to continue the process (Chang, 1976, 1989).

From southern China, rice-growing technology spread both inland and offshore (Chang, 1989).

Rice was being cultivated on Luzon in the Philippines 3.4 ka BP (Snow et al., 1986) employing an awe-inspiring system of terracing. In Japan, the beginning of rice cultivation around 3.2 ka BP is marked by an abrupt fall in alder (*Alnus*) pollen (Figure 14.1). Alone among the Pacific islands, rice reached the Marianas before European contact, probably ~0.9 ka BP (Hunter-Anderson et al., 1995).

Economies based on rice cultivation flourished in the lower Yangtze valley by ~5 ka BP; the Chinese term for rice began to be written around 4 ka BP (Grist, 1986). Today, rice in the Pacific Rim grows as far as 53°N in the Amur valley on the China–Russia border.

The practice of lowland rice cultivation has changed little in terms of its environmental impacts since the middle Holocene. It involves planting in waterlogged soils, which occur naturally on many parts of river floodplains or can be created in better-drained situations by purposeful flooding. In the latter case, a network of dykes to prevent water draining away may be created, as shown in Plate 14.1. The creation of these *padi* optimize the beneficial effects of purposive nutrient recharge (using manure and compost).

14.8 MIDDLE HOLOCENE ENVIRONMENTS OF AUSTRALASIA

14.8.1 The Middle Holocene in New Guinea

At Waigani Lake in lowland Papua New Guinea, discussed in more detail in Chapter 15, peat indicative of swamp forest accumulated throughout the middle Holocene from at least 4400 BP in association with a rise in water level (Osborne et al., 1993). An increase in precipitation during the late middle Holocene can also be inferred from the development of a glacier on Mount Jaya some 3500 BP (see Plate 15.7) – the first ice to develop here since the Last Glacial (Hope and Peterson, 1976). Owing to their locations, neither of these two sites were affected by people during the middle and late Holocene.

Plate 14.1   Rice farming became an important cause of lowland environmental change in East Asia during the middle and late Holocene. Here, in an 1808 illustration, rice seedlings are shown being transplanted by hand in orderly rows in flooded fields – a method still used in many parts of the western Pacific today, and one which was probably first developed several thousand years ago
Courtesy of Harvard University Press

A similar picture of middle Holocene palaeo-climate change has emerged from studies elsewhere in New Guinea, although this should be treated more cautiously because of the greater possibility of human influences. At Telefomin, for example, the period 3500–2500 BP was marked by the replacement of forest by grassland – the opposite to what might be expected had the climate been becoming wetter (see above) – yet consistent with the effects of forest clearance by humans (Hope, 1983).

14.8.2   The Middle Holocene in Australia

During the early part of the middle Holocene, temperature reached its Holocene maximum in most of Pacific Australia; precipitation was also higher than at present. These conditions caused an expansion of *Nothofagus* (southern beech) forests at Barrington Tops and alpine Mount Kosciusko in New South Wales, and at most lowland sites in eastern Australia. A reduction in effective pre-

cipitation 5.0–4.5 ka BP on the Atherton Tableland in northeast Queensland, probably marking a temperature maximum, is suggested by the development of a more open forest canopy and falling lake levels. Several indicators suggest that conditions in this region became drier after 4 ka BP. Peat development occurred during the middle Holocene at Mount Kosciusko but ended some 3.5 ka BP with the onset of cooler conditions, manifested by increased slope instability (Costin, 1972; Kershaw, 1976; Dodson et al., 1986; Hiscock and Kershaw, 1992).

A contraction of moisture-loving forests ~6 ka BP in Tasmania occurred in response to a short-lived episode of cooling and drying (Markgraf et al., 1986; Harrison and Dodson, 1993); the human response to early and middle Holocene forest expansion continued across this interval (see below).

The history of middle Holocene sea-level changes in eastern Australia has been debated vigorously over the past few decades. Debate continues, yet today most workers consider that sea level reached a Holocene maximum sometime around 5.5 ka BP (Chappell et al., 1982) and/or 3.5 ka BP (Flood and Frankel, 1989). The culmination of Holocene sea-level rise 5.2 ka BP in northeast Australia was marked by a period of dune formation (Lees et al., 1993; Plate 14.2). Disagreements about the timing of the Holocene sea-level maximum may be resolved by assuming that there were actually several sea-level maxima during the middle and late Holocene – both along the east Australian coast and elsewhere in the Pacific Basin – and that the magnitude of particular maxima may have been increased or reduced by tectonic changes (shown in Figure 15.9A). An oscillatory trend is more plausible than that involving a single maximum because of the undisputed oscillatory character of middle and late Holocene temperature changes, which are thought to have been the major control of contemporary ocean-volume changes (Fairbridge, 1992).

Despite the undeniable presence of humans in Australia during the middle Holocene, there is little evidence of any significant human impact, probably because they were nomadic rather than

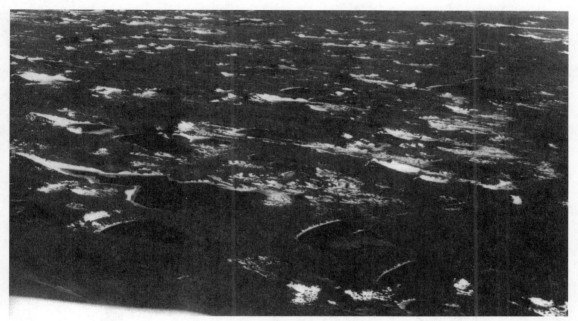

Plate 14.2   The middle Holocene dune field at Shelburne Bay, Cape York, northeast Australia, was reactivated
within the last millennium
Brian Lees

sedentary, and their lifestyle involved mainly hunting and gathering rather than agriculture. In places, environmental changes may have amplified or subdued the effects of human lifestyles but there is no unequivocal evidence that these altered fundamentally in response to particular environmental changes. For example, the change to drier conditions on the Atherton Tableland ~4.5 ka BP may have facilitated the spread of fires which, on account on the sharp increase in charcoal particles in sedimentary records there at this time, Hiscock and Kershaw (1992) contend were probably lit by humans. Yet such records do not reveal anything directly about changes in human lifestyles.

The situation is slightly different in Tasmania to which its inhabitants were confined for the greater part of the Holocene because of the submergence of the land bridge connecting the island to the Australian mainland during the Last Glacial. Throughout most of the early and middle Holocene, humans here were squeezed by expanding forests out of areas they had once

occupied and onto the coasts where they developed more sedentary lifestyles marked, for instance, by the establishment of oyster middens in the east and southeast parts of the island by around 6–5 ka BP. Following the time of peak warmth/moisture around 5–4 ka BP, conditions became cooler and drier. Rainforests began to contract and the area settled by Aborigines began to expand, often aided by fire, setting the stage for more rapid cultural development in the late Holocene (Lourandos, 1997).

14.8.3   The Middle Holocene in New Zealand

The overall picture of middle Holocene climate change for New Zealand is one of increasingly cool and dry conditions following the early Holocene temperature maximum (see Chapter 13). In their synthesis, McGlone et al. (1993) emphasize the increasing seasonality of middle Holocene precipitation, with summers becoming drier and winters becoming wetter. These changing conditions placed many elements of

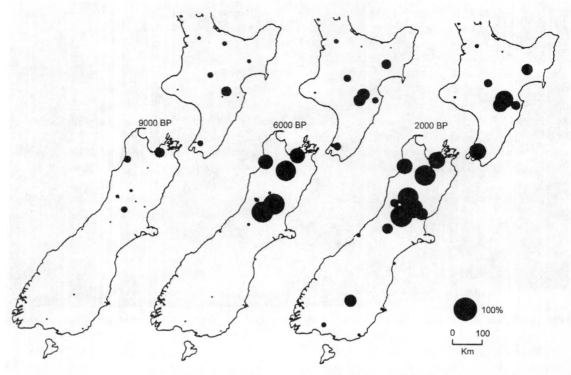

Figure 14.2   Percentages of *Nothofagus* (southern beech) pollen at various times during the Holocene in New Zealand. Areas of circles are proportional to the percentages of total terrestrial pollen. Note the sharp rise in *Nothofagus* by the start of the middle Holocene (6 000 BP) and its steady rise subsequently
After McGlone (1988) with kind permission of Kluwer Academic Publishers

early Holocene ecosystems under stress but were successfully exploited by *Nothofagus* (Figure 14.2) and *kauri* (*Agathis australis*) (Ogden et al., 1992).

Glacier fluctuations, well documented for the late Holocene (see Chapter 15), have also been recorded for the middle Holocene. In a survey of 18 glaciers, Wardle (1973) found evidence of a glacial episode before 4730 BP which bears out the general picture of a cooler, seasonally wetter, middle Holocene.

Of particular interest, and relevance to the debate about the impacts of early human settlers on the environments they inhabited, is the record from the Prospect Hill plateau in South Island, where five layers of charcoal representing burning of the surrounding vegetation were dated to 5.8 ka BP, 3.8 ka BP, 3.5 ka BP, 2.6 ka BP and 0.86 ka BP

(Burrows and Russell, 1990). The first four episodes of burning occurred long before people arrived in New Zealand ~800 BP. Charcoal layers of this kind were probably associated with minor climate changes which made New Zealand's forests highly vulnerable to fires ignited by lightning or volcanic eruptions. The subsequent reductions in vegetation cover would have resulted in increased rates of erosion of a kind which might well be mistaken for products of human activity were archaeologists in any doubt about the time that humans first arrived in New Zealand. It is especially worth pondering whether minor climate changes of the kind which brought about these episodes of forest burning and rapid hillslope erosion in New Zealand (and evidently in New Caledonia – see Chapter 11) *before* humans arrived first might not also have occurred here and elsewhere *after* they arrived. For if such was the case, then it appears increasingly plausible that what many authorities have presumed to be a sign of human impact may instead be a decidedly non-human impact (Nunn, 1997a).

Owing to tectonic movement, particularly coseismic uplift, it has proved difficult to isolate the course of late Holocene sea-level changes around New Zealand. The most compelling attempt was that by Jeremy Gibb (1986) in the 'stable' estuaries of Blueskin and Weiti bays for which he concluded that there had been no significantly higher sea level since ~6 ka BP.

14.9 MIDDLE HOLOCENE ENVIRONMENTS OF THE PACIFIC ISLANDS

El Junco lake in the Galápagos islands grew shallower during the period 6200–3000 BP compared to the times before and after (see Figure 13.4). This is consistent with higher temperatures and/or reduced precipitation (Colinvaux, 1972). In contrast, lakes on Mangaia in the Cook Islands reached their maximum depth 6500–4500 BP as a result of increased precipitation (Ellison, 1994).

A catchment disturbance on Mangaia about 3270 BP may have been associated with a tropical cyclone and/or El Niño event. The latter suggestion is intriguing given the same explanation for a major episode of flooding about the same time in northern coastal Peru (Wells, 1990; see above). It is conceivable that the ~3270 BP disturbance on Mangaia was associated with humans although artefactual evidence suggests they did not arrive here until some 2000 years later (see Table 12.2).

Similar explanations have been flirted with to explain middle Holocene environmental changes elsewhere in the Pacific islands. Analyses of cores from Saint Louis Lac on New Caledonia show a high influx of charcoal 4650–4350 BP but Janelle Stevenson and John Dodson (1995) did not attribute this to human vegetation burning since it was not followed by persistent vegetation change. They concluded that it is more likely to have been the product of a single, naturally caused

forest fire which coincided with an exceptionally dry period. This is consistent with the archaeological evidence on New Caledonia. In contrast, the suggestion by Wendy Southern (1986) that humans may have been responsible for fires in lowland Vitilevu island in Fiji as early as 4300 BP conflicts with archaeological evidence pointing to initial human arrival only about a thousand years later.

In fairness, it should be made clear that the early history of humans in the Pacific islands is far less well known than for many parts of the Pacific Rim. Dates for the earliest known occupation (as in Table 12.2) must be regarded as tentative until all islands in a particular group have been adequately surveyed. The chances of finding the earliest settlement site in a particular archipelago after only a few such surveys are slim. Then there are the possibilities that the earliest inhabitants left no unambiguous record of their presence (particularly artefactual) or, if they did, that this record has been effectively obliterated either by rebuilding on the same site or by natural processes.

There is little doubt that all major Pacific island groups were affected by a higher-than-present sea level during the middle Holocene. Representative data are given in Table 14.2 and a typical coastal section shown in Figure 14.3. The picture has not always been this clear, and earlier views had such ramifications for an understanding of other environmental changes, particularly on Pacific island coasts, that these are discussed further in the following section.

14.10 MIDDLE HOLOCENE SEA-LEVEL MAXIMUM: RESOLVING POLARIZED VIEWS

From the 1960s, when the first regional compilations of Holocene sea-level change were produced, until recently, ideas about the nature of sea-level changes in the Pacific Ocean during the middle Holocene in particular became increasingly polarized. Oceanic islands played a critically important role in the debate following the recognition by Arthur Bloom of their potential as dipsticks, recording changes in their own level or

Table 14.2   Representative sea-level maxima during the middle Holocene from Pacific Islands

Island (group)	Magnitude (m)	Time (ka BP)	Source of information
Fiji	1.00–2.00	3–2	Nunn (1990a)
French Polynesia	1.00	2–1.5	Pirazzoli and Montaggioni (1988)
Kosrae, Micronesia	1.00	3.7	Kawana et al. (1995)
Mangaia, Cook Is.	1.70	4–3.4	Yonekura et al. (1988)
New Caledonia	2.06	5.5	Cabioch et al. (1989)
Oahu, Hawaii	2.00	4.7–2.1	Jones (1997)

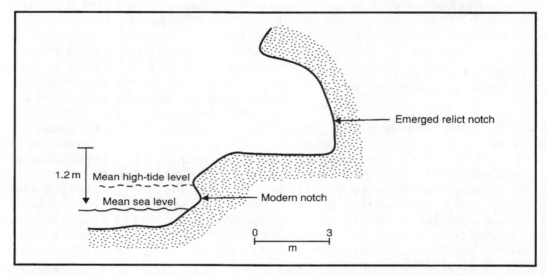

Figure 14.3   Emerged shoreline from the central west coast of Yaukuvelailai island, southern Fiji. This view is typical of shorelines which were excavated when the sea level in this part of the Pacific was ~1.5 m higher than today some 3000 years ago
After Nunn (1992)

that of the sea around them: 'a small island, rising steeply from a deep ocean basin, can not be differentially warped by the isostatic response of the adjacent basin floor to fluctuations of sea level' (Bloom, 1970: 146).

It was this principle, which has stood the test of time, that led Bloom to investigate Holocene sea-level history around certain atolls of the northwest Pacific, from which he developed his view that sea level had risen throughout the entire Holocene – an idea which, in the Pacific, has not stood the test of time. Bloom's 'Micronesia Curve' (Figure 14.4B) conflicted with the view of Fairbridge (1961) that sea level reached a maximum during

the middle Holocene and has since declined (Figure 14.4A).

The resolution of this debate has been a long time coming and has involved meticulous field-work in many Pacific island groups. From all, without exception, evidence has now been found for a sea-level maximum during the middle Holocene (Nunn, 1995, 1998c). It is helpful to understand how a contrary view could have prevailed for so long.

The principal evidence that Bloom and others used in support of their Micronesia Curve in the Pacific Islands was an absence of emerged coral reef on those islands. Certainly, the most plausible

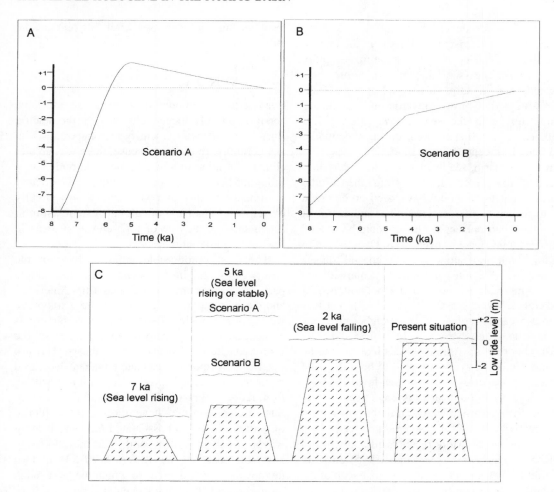

Figure 14.4   Illustration of how the Holocene sea-level record could be misinterpreted by depending on the evidence of emerged reef. (A) Scenario A shows the now generally accepted course of Holocene sea-level change in the Pacific Islands. This scenario involves a higher-than-present sea level peaking during the middle Holocene. This graph is based on results of retrodictive modelling and has been confirmed in essence by many empirical studies. (B) Scenario B shows the 'Micronesia Curve' which was once believed to apply to most of the Pacific, but has since been shown to be erroneous for this region. (C) This shows a 'catch-up' reef at various times during the Holocene. At 5 ka, the reef is still catching up and its level bears no relation to the sea surface; either scenario A or scenario B could apply. The late Holocene fall of sea level allowed such reefs to catch up. An absence of emerged reef in such a situation should not be readily equated with scenario B, especially when other evidence suggests that scenario A applied

After Nunn (1994a)

explanation of a modern reef flat lacking any emerged reef is that this flat never grew higher (at least within the Holocene). Even accepting this argument to be true – and it may not be had all the emerged reef been eroded away – there is still a

conceptual gulf to be bridged before this can be accepted as evidence that sea level (rather than reef) was not once higher than it is today.

Reefs are generally good indicators of palaeo-sea-level position since they commonly grow to close to low-tide level. Yet, during the early Holocene sea-level rise, not all reefs in the Pacific were able to keep up with the sometimes rapidly rising sea surface (see Chapter 13). Some reefs drowned, and remain submerged today. Other reefs struggled to grow upwards and eventually 'caught up' with the sea surface after it had stabilized, and perhaps fallen, during the middle (and late) Holocene (Figure 14.4C).

In this scenario, sea level would have certainly exceeded its present level during the middle Holocene but there would have been no emerged reef to testify to this. Instead other indicators of this higher-than-present sea level would be needed to discern this course of events. On Oahu island in Hawaii, such indicators include coastal dunes, notches in aeolianite and basalt, emerged sea stacks, emerged sea caves and emergent beach-rock (Jones, 1997; see also Figure 14.3). Yet it is the mid-ocean intraplate atolls which because they are subsiding only very slowly – indeed barely measurably over short time periods like the middle and late Holocene – are believed to hold the least ambiguous record of Holocene sea-level changes. This was why Bloom and his team (see above) directed their efforts towards atolls but their error was equating a lack of visible emerged reef with a lack of emergence. More recent work, anticipated by Rhodes Fairbridge (see Nunn, 1995) and Jim Schofield (1980), has shown that most such atolls are actually underlain by *in situ* reef conglomerate emerged significantly above modern reefs as a

result of late Holocene sea-level fall (Dickinson, 1999).

14.11 SYNTHESIS

Even with the amount of information available about middle Holocene climates in the Pacific Basin, it is difficult to identify regional trends with any certainty. In a general sense, there was greater climate variation in low and middle latitude areas. In higher latitudes, modern climates had evidently become established and were altered only slightly by climate changes of global extent; similar statements appear true of high-elevation areas in low and middle latitudes.

There is a noticeable contrast between the aridity of the middle Holocene in low and middle parts of the eastern Pacific Rim (the Americas) and the contemporary wetness of many parts of the western Pacific Rim. This is explicable largely as a function of the expansion in influence of the southeast (Easter Island) and northeast (Hawaii) Pacific anticyclones associated with a decreased intensity of ocean (gyral) circulation resulting from reduced Equator–Pole temperature gradients. Such a condition may have allowed the influence of warm currents and associated moisture-bearing winds to extend farther south than they do today along the coasts of the western Pacific Rim. Two unknowns are the degrees to which the East Asian monsoon and the ENSO phenomenon may have been affected by these changes.

CHAPTER 15

The Late Holocene (3 ka – present) in the Pacific Basin

... that Earth now
Seemed like to Heaven, a seat where gods might dwell
John Milton, *Paradise Lost,* VII

15.1 INTRODUCTION

Commentaries on reconstructions of past environments often focus on the degree of difference they exhibit compared to modern ones: 'rainforests now covered with ice', 'deserts where agriculture once flourished', 'islands whose sinking is now recorded only by reefs'. Museums worldwide have won praise for imaginative three-dimensional restorations of ancient environments, but the gloss wears a bit thin, the rhetoric becomes somewhat muted, as we approach the present. Aside from the effects of humans on the environments they inhabited during the late Holocene, comparatively little changed during this interval.

The fascination with the late Holocene is thus primarily accounted for by the rapidly changing nature of the interactions between humans and their environments. Many people in the modern Pacific Basin live in almost wholly artificial environments in which talk of unregulated environmental change may appear as alien as predictions concerning the future control of nature by humans might have appeared to the inhabitants of the same area at the start of the late Holocene some 3000 years ago.

Yet small subtle environmental changes occurred throughout the late Holocene independently of humans and, because we are dealing here with such a short and recent slice of time, we can identify these changes more clearly than is possible for earlier times. More importantly, it is possible to determine in many cases the precise effects which environmental changes during the late Holocene had on humans. And, as the present is approached, it also becomes possible to identify and evaluate the efforts by humans to counter (the effects of) particular environmental changes.

Although questions concerning the human versus non-human causes of environmental change have been raised for earlier times (Chapters 12–14), it is to the late Holocene that they are most applicable. With this in mind, the first section of this chapter reviews the status of humans in the Pacific Basin around the start of the late Holocene. The sections which follow consider the nature of late Holocene environmental change, particularly its effects on humans and/or the responses it evoked from them, for each of the subdivisions of the Pacific Basin used as a basis for discussion throughout this book. Within each of these sections, there are subsections dealing with specific issues including for each subdivision an account of contemporary environmental issues. Obviously these are topics to which whole books could easily be (and have been) devoted yet here, in keeping with the overall theme of maintaining human impact within a realistic context in this book, these discussions are brief and not intended to be all-encompassing. In particular, it is not felt necessary to make reference in every instance to environmental problems, such as population growth and 'global' warming, whose effects are not specific to the Pacific Basin. Following the regional sections, some issues of key importance to an understanding of late Holocene environmental change in the Pacific Basin are aired. These are followed by a discussion of the environmental impacts of volcanism and tectonics during the entire Holocene, some of which was deferred from Chapters 13 and 14. Suggested directions for future research are given in Chapter 16.

For the first time in this book, ages expressed in calendar years are given as BC and AD in this chapter. Radiocarbon ages continue to be denoted by BP.

15.2 HUMANS IN THE PACIFIC AT THE START OF THE LATE HOLOCENE

With the exception of Antarctica, humans had reached every part of the Pacific Rim by the beginning of the late Holocene (3 ka). Within the vast ocean, the process of island colonization was underway although some island groups were settled only within the last thousand years (see Table 12.2).

Late Holocene hunter-gatherers continued to have comparatively minor impacts on the environments which they occupied. Yet many of the earliest late Holocene humans in the Pacific Basin were agriculturalists, some benefiting from thousands of years of experience in modifying environments to better satisfy their needs. Owing to generally low population densities, the impacts of most of these agriculturalists at this time were neither widespread nor especially profound compared to what followed.

Errors in identifying human impact during the earliest part of the late Holocene may have arisen from the practice of extrapolating more recent, better observed behavioural traits back through time. For example, a frequent reaction of foreign minds confronted by subsistence agricultural practices is to assume, because of their apparent lack of sophistication, that these have existed unchanged since their inception (Lourandos, 1997). A good example comes from Australia, a continent which people may have inhabited for more than 100 000 years (see Chapter 12). The

first European observers of this continent's native people reported their apparent predilection for burning vegetation, assuming this to have been a tendency unchanged for hundreds of generations. This may be mere wishful thinking.

Barely a mention is given of non-human factors in many commentaries on the cause(s) of environmental change in areas occupied by humans during the late Holocene. Humans are widely and popularly regarded as having been the main agents of change, a maxim so evident to some authors that they have no need to justify it. It is worthwhile exploring the background to this view briefly, especially as this book takes a contrary stance.

15.2.1 Late Holocene Human History: Philosophical Considerations

Tradition is venerated beyond its merit in some academic fields. Principles enunciated a hundred years or more ago may be commonly considered sacrosanct and rarely questioned for no reason other than their antiquity. This may be justified in the experimental sciences but is less so for empirical ones because new observations may contradict conclusions reached from earlier observations.

This is precisely what has happened with recent environmental change, particularly studies of its principal causes in regions (such as most of the Pacific Basin) which were beyond the direct influence of most centres of scientific endeavour during its formative years in the 19th century. Much early thinking about environmental change operated within a framework of solely human causation; a hundred years ago, few scientists considered non-human causes worthy of consideration. The antecedents of this lay in the largely anthropocentric world view of the Christian Bible, its injunctions to 'subdue' the Earth and to 'have dominion' over every living thing, and the tendency of most pre-20th century European scholars to reconcile their deductions with its teachings.

Throughout the 19th century, the ideas that humans had a god-given right to mould the Earth to their own ends and that they had done so to the almost total exclusion of other causes became widespread in parts of the Pacific Basin. It was given a further bolster – particularly in North America – by the works of George Perkins Marsh who recognized the dangers associated with the long-term human exploitation of the environment and preached the need for responsibility[1] (Lowenthal, 1958). Outside the academic world, where contrary ideas have occasionally reared their heads, the view that humans have been the principal cause of environmental change and have been so by divine right has persisted. Today, although many environmental changes, particularly those that have occurred within the past few hundred years, are recognized as having been undesirable, the view that these changes were overwhelmingly of human causation persists. In the past few decades, their undesirability has been highlighted by exponents of eco-theology who find parallels for the 'fall of man' in recent environmental degradation (Nelson, 1993).

Most educated people today bewail the pace of 20th-century environmental change, the most extreme examples of which have manifestly been caused largely by human mismanagement. To many such people, a logical corollary to this view is the belief that modern humans have been responsible for most environmental changes during the time they have inhabited the Earth. Yet even for the entire late Holocene, there are considerable dangers in regarding humans as having been the main or only cause of change in the environments they occupied. Such dangers include the resultant tendency to underrate the potential of future non-human changes, and to believe that the eradication of human impact would lead to the appearance of an unchanging

environment in which humans could live in harmony with nature – a return to Eden – even at their present levels of sophistication.

The potential magnitude of this problem can be measured by the fact that in 1994, the President of the Geological Society of America, William R. Dickinson, devoted his Presidential Address to the subject of the Holocene. He made a classic statement of the dangers associated with an overwhelmingly anthropocentric interpretation of environmental change (Dickinson, 1995: 1):

> In a time of well-founded environmental concerns, the widespread impression that civilization is the only disjunct influence on an otherwise fixed tapestry of nature is a dangerous misperception that can lead to much folly. As we [professional geoscientists] use our collective knowledge of geoscience to foster rational decisions about environmental issues, we have no more important task than to dispel the illusion of global constancy held by the mass of society.

Dickinson's words were timely and underpin the remainder of the discussion in this chapter, in which human impact is recognized as only one of many causes of late Holocene environmental change.

15.2.2 Humans in the Americas at the Start of the Late Holocene

It is plausible to suppose that most parts of the Americas had been visited by humans by the start of the late Holocene. For most humans in Pacific North America, a dependence on fishing, hunting and gathering remained an important part of their lifestyle at least until European contact. Certainly there were some communities which developed agriculture of comparable sophistication to that being practised in Central America, but it was more of a struggle in the generally less hospitable environments of the modern western United States. Some exceptions occurred along the well watered parts of the Pacific coast: for example, the Chumash Indians of southern California whose sedentary, securely agriculture-based lifestyles led to political differentiation in the early part of the late Holocene (King, 1990).

---

[1] Marsh must be viewed within the milieu of 19th century North America where the view was commonly held, notably by some American presidents, that almost every change which humans made to the environment was by definition beneficial. Rather Marsh argued for the sensible management of the environment, a view which gave birth to the conservation movement in North America

The same general picture does not apply to Central America, one of the world's three 'cradles of civilization' where a variety of plants had been domesticated earlier in the Holocene. As in the valleys of the Huanghe and Yangtze in China (and in Mesopotamia), this situation led to the comparatively rapid development of farming communities, and later, when enough food was being produced to feed non-food producers, larger settlements. This tendency was helped in parts of Central America by the richness of many soils, and the comparative ease of irrigation.

Most agricultural innovations developed in Central America spread along the west coast of South America more readily than into North America, the harsh arid environments of central and northwest Mexico discouraging human movement and cultural interchange. Along the Pacific coast of South America, stretching into the high Andes in many places, domestication of plants and animals was well underway by the start of the late Holocene.

Of particular interest to the whole question of early human–environment relations around the start of the late Holocene is the discovery that soil erosion associated with contemporary human activities in the highlands of central Mexico was as great as that following Spanish arrival in AD 1521 (O'Hara et al., 1993). This demonstrates that early humans were capable of causing erosion comparable in magnitude to that following the Spanish introduction of plough agriculture, but the questions as to how widespread such practices were and what contributions they made to environmental change remain unanswered.

Farther south, Piperno (1994: 323) used a study of sedimentation rates (as a proxy for soil erosion) in the Darien rainforest of Panama to suggest that maize agriculture around the start of the late Holocene caused 'extensive local disturbance'. Yet any suggestion that such changes were associated exclusively with centres of cultural development must be questioned since palaeo-botanical signs of 'significant environmental modification associated with prehistoric agriculture are now emerging from areas of the lowland tropics that did not witness the development of high

civilizations, or that were on the periphery of complex societies' (Piperno, 1994: 324).

Although agriculturalists lived in most parts of Pacific South America by the start of the late Holocene, many inhabitants of this region depended on hunting and gathering, a reflection in the highlands in particular of the need for technological innovation before these environments could be successfully farmed full-time. The emergence of complex societies ('civilizations') in this region during the late Holocene would not have been possible in the absence of a secure agricultural base arising from such innovations.

In many parts of South and Central America, the rise of metallurgy and metalworking – chiefly for the production of ornaments – was a measure of increasing social complexity. Copper, gold and silver were the main metals worked in prehispanic times, most ores being obtained from superficial, particularly placer, deposits. The required technology is thought to have developed first in the central Andes (Chimú–Mochica area in Figure 15.1) about 1500 BC whence it spread both south and north reaching Mexico by AD 650 (West, 1994). The environmental impacts of early mining were slight yet foreshadowed in kind those which followed. Today large areas which have been strip-mined or excavated are fringed by spoil heaps and a commonly ramshackle mining-associated infrastructure, all of which pose a variety of threats to the areas' inhabitants.

15.2.3   Humans in East Asia and Australasia at the Start of the Late Holocene

Agricultural developments along the valleys of the Huanghe and Yangtze rivers in eastern China during the later Holocene were centred on millet, then later rice, soybeans and tea, along with dogs and pigs. The earliest dynasty recalled by Chinese tradition is the Xia (~4.2–3.7 ka) but little is known about this. Its successor – the Shang dynasty (~3.7–3.1 ka) – is better known and apparently marked an abrupt advance in many aspects of cultural development.

Anyang in the lower Huanghe valley was the capital for some 300 years during the Shang

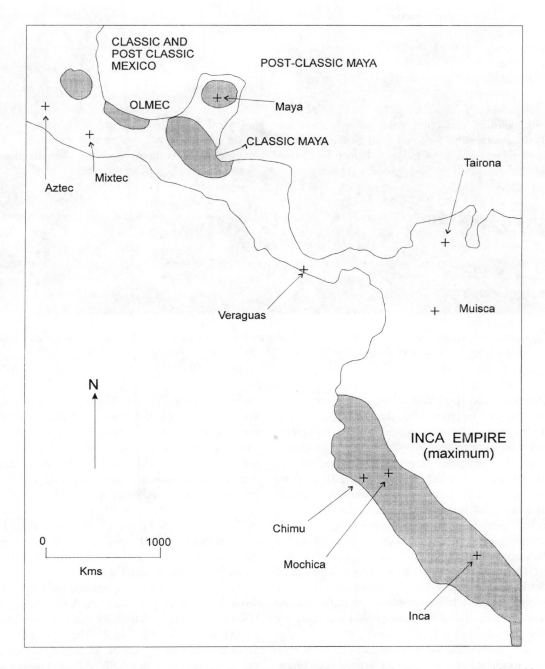

Figure 15.1   Major prehispanic centres of cultural development and innovation (shaded areas), and gold-working centres and peoples (crosses) in northern Pacific South America and southern Central America. Note that the Inca Empire extended farther south than shown
Compiled from Lambert (1987) and other sources

Plate 15.1   Excavations at Anyang 1928–1937. Centrally located in the lower Huanghe valley, Anyang was the last capital city of the Shang dynasty in China, overthrown between 1122 BC and 1027 BC.
Institute of History and Philology, Academia Sinica

dynasty, and included large wattle-and-daub houses and temples with thatched roofs (Plate 15.1). Such cities were primarily religious centres but also had sectors for other non-food producers like potters and metalworkers. A warmer, moister climate during the life of Anyang is testified to by the use of water buffaloes as domestic animals although cattle were also kept for sacrificial offerings and scapulimancy, evidence of which are the numerous inscribed oracle bones found at this site.

The Shang dynasty was overthrown around 3.1 ka and replaced by the Zhou dynasty. The subsequent growth in the size of cities, considered a crude indicator of increasing cultural complexity and spread, is shown in Figure 15.2.

The first furnaces hot enough to melt iron were developed some 2.5 ka in China and would have led to an upsurge in the demand for iron ore and fuelwood which, along with an increasing demand for domestic firewood and potter's clay, led to human impacts on a much greater area than during the middle Holocene. Yet it was agriculture which still placed the greatest demands on the environment. Around 2 ka, China was home to some 60 million people, 90% of whom were agriculturalists growing mostly millet in the north and rice in the south. Canals supplied water for irrigation and carried away excess when flooding threatened.

The situation to the north in the Russian far east was different; signs of domesticated animals some 4.5–4 ka are known but arable farming did not commence until around 1.2 ka.

The practice of shifting agriculture had spread throughout the New Guinea highlands during the Holocene and was well established on most parts of the island by 3 ka. In Pacific Australia, the situation was quite different, the human occupants of the region being nomadic and, in consequence of low population densities, having little permanent effect on most of its environments.

Sea-level rise during the earlier Holocene probably displaced large numbers of coastal and

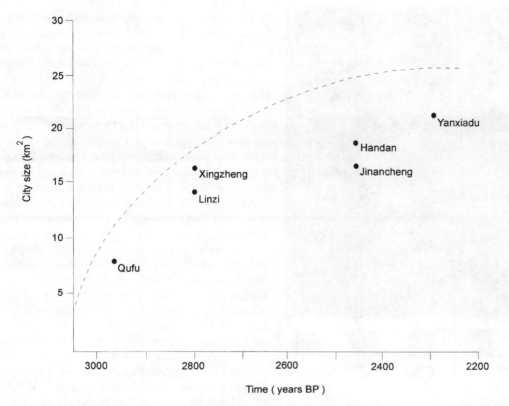

Figure 15.2   Changes in the size of cities in eastern China during the early part of the late Holocene
After Chen (1994) with permission of Chen Shen and Antiquity Publications Ltd

lowland people living in the Huanghe and Yangtze valleys and their surrounds including the land bridges connecting the mainland with what are now offshore islands. This event may have provided an impetus for migration out of this region which led eventually to the initial colonization of many Pacific islands by Austronesian language speakers[2] around the beginning of the late Holocene (see below).

15.2.4   Humans in the Pacific Islands at the Start of the Late Holocene

Linguistic evidence has been used to trace the migration of people during the middle Holocene out from what are now southern China and Taiwan through the archipelagoes of southeast Asia, then either west as far as Madagascar, or east through the Pacific islands[3]. Almost all Pacific people today can trace their ancestry to a founder group – identifiable earliest on Halmahera island – which has been named the Lapita people, after the place

---

[2] The earlier occupants of Papua New Guinea and Solomon Islands (like the first Australians) were Papuan language speakers. There is ample evidence that they coexisted in places with later Austronesian-speaking arrivals (Kirch, 1997)

[3] The 'either' and 'or' are important in this sentence, particularly to explain the popularly held misconception in Fiji that the indigenous people arrived from Africa. In fact, as the linguistic trail shows clearly, both these Fijians and certain African groups can be traced back to a common root which split in the southern East Asia region during the early Holocene

Plate 15.2 Pottery sherd decorated in the style of the Lapita Cultural Complex, collected from Mago island, eastern Fiji. The sherd is from a shallow carinated bowl and was found in deposits dating from ~2950 BP
Geoff Clark

in New Caledonia where their distinctive pottery was recognized first (Kirch, 1997; Plate 15.2).

The Lapita people were among the most expert sea travellers of their time, able to sail long distances across the ocean on large rafts or canoes. Although dependent largely on ocean resources for subsistence, the Lapita people were also horticulturalists and carried on their ocean voyages plants and animals intended to help support their accustomed lifestyle on new islands. On most islands where they settled, they planted various types of aroid (particularly *taro* – *Colocasia esculenta*), yams (*Dioscorea* sp.), bananas and breadfruit. It is unclear why rice was not introduced to the Pacific islands by the Lapita people for it was the principal staple food

of their ancestors in East Asia and would have been suitable for many Pacific islands using adaptive strategies similar to those used today in East Asia (see Plate 14.1) and elsewhere. Weeds were inadvertently introduced to many Pacific islands at this time through seeds in the mud packed around corms of *taro* to keep them viable after the long sea crossings. Other seeds carried in the guts of people and their accompanying animals were also dispersed across island groups in this way. The resulting environmental changes have been more rapid and more severe than might be expected owing to the uncommonly vulnerable nature of oceanic island ecosystems (Nunn, 1994a).

The Lapita people also carried pigs, chickens and rats which ran wild when an uninhabited island was reached. Most feral pig populations on Pacific islands originated in this way. When assigning causes of contemporary change, the potential of rats and feral pigs to transform environments independently of humans should not be underestimated[4] (Nunn, 1997c).

There is some disagreement about the extent of environmental change associated with the agricultural demands of Lapita people on Pacific islands. Patrick Kirch (1997: 203), for example, argues that they 'transformed Oceanic rainforests into a landscape of high productivity', but the present writer has emphasized the ambiguity of the supposed evidence for this, particularly in the eastern Lapita settlements (Nunn, 1994b), arguing that the impact of the Lapita people on these environments was far less. The debate is hampered by a lack of objective evidence, but studies are underway to help resolve it.

The Lapita colonization of the Pacific islands had begun by the start of the late Holocene; earliest known settlement dates were given in Table 12.2. The rapidity with which Lapita people colonized most of New Caledonia, Vanuatu, Fiji, Tonga and Samoa has been likened to an express

---

[4] This situation has more recent analogues involving goats, pigs and rabbits introduced deliberately as sources of food on hitherto (effectively) uninhabited islands (Nunn, 1994a)

train[5] (Diamond, 1988) and may have been in response to changing environmental conditions (such as sea-level rise – see above) and/or intertribal conflict in the lands they had occupied to the west.

The former idea is worth dwelling on briefly. For South America and elsewhere, it has been suggested that extreme climatic events which deprive people, albeit temporarily, of resources on which they depend provide stimuli for lifestyle changes. Imagine that a severe El Niño event affected the western Pacific around the start of the late Holocene. The combination of increased ocean-surface temperatures and lowered sea level may have led to widespread reef death, depriving coastal people of an important food source. On land, severe drought may have devastated crops prompting their cultivators to set sail in the hope of finding an island not so affected. At present, appropriate archaeological and palaeoclimate data from the western Pacific are not sufficient to test this idea.

15.3 OVERVIEW OF THE LATE HOLOCENE IN THE PACIFIC BASIN

In most parts of the world, the late Holocene marks a time during which climate deteriorated relative to that of the climatic optimum of earlier Holocene times. In some places, a marked cooling named the Neoglaciation, and various other events referred to by name, have been recognized. Owing to their uncertain duration, extent and character, such labels are avoided in this book except in the account of the last 1200 years when Pacific-wide events can be recognized.

The nature of climate change varied from low to high latitudes during the late Holocene in the Pacific Basin. In most low latitude environments, aridity increased. In most middle and high latitude areas, conditions became wetter. The global rise in methane during the late Holocene is attributed to the massive increase in wetlands in these regions (Blunier et al., 1995).

A recurring theme in the discussions below is the potential for mistaking a palaeoclimate change for a human impact or *vice versa*. As seen above, this issue has far-reaching implications and is a long way from being resolved in many places. While most commentators accept that climate change occurred alongside and independently of human impact during the late Holocene, others have gone a step further by proposing that a significant part (1°C) of the global cooling experienced during the late Holocene was caused by human impacts on terrestrial environments, particularly desertification, salinization and deforestation (Sagan et al., 1979).

Within the last 1200 years or so, three periods of climate–environment change over which humans are thought to have had no influence are recognized in most parts of the Pacific Basin. These are the Little Climatic Optimum (~1200–650 BP), the Little Ice Age (~650–150 BP), and the period of recent warming (~150 BP to present). While the Little Climatic Optimum (or Medieval Warm Period) may not have been global in extent, its legacy in most parts of the Pacific Basin is clear (Hughes and Diaz, 1994) which justifies its recognition as a region-wide event. Evidence for a global Little Ice Age is better (Grove, 1988). Lasting perhaps only a hundred years, the time of transition between the Little Climatic Optimum and the Little Ice Age (around 650 BP or ~AD 1300) is suggested as having been a time of unusually rapid change which left its mark on many Pacific Basin environments (see section 15.13.2).

15.4 LATE HOLOCENE ENVIRONMENTS OF ANTARCTICA

15.4.1 Early and Middle Late Holocene

Since the Antarctic continent is so cold year-round, slight changes in temperature might not register in environmental records as clearly as they do in places with larger annual temperature ranges. On the other hand, changes in precipitation might do so for it is only by increasing precipitation that ice mass can grow and *vice versa*.

---

[5] A singularly unhelpful metaphor for most Pacific Islanders, uncharacteristically insensitive of Jared Diamond's writing

An abrupt abandonment of penguin rookeries along the Victoria Land coast during the early part of the late Holocene was attributed to the rapid extension of sea ice associated with cooling (Baroni and Orombelli, 1994). In contrast, on James Ross Island, just off the Antarctic Peninsula, the start of the late Holocene is marked by the disappearance of glaciers and the shrinkage of once-large meltwater lakes to small brackish water bodies containing the diagnostic crustacean *Branchinecta gaini*. The arid, cold conditions which brought about this change lasted from 3000 BP until around 1200 BP when glaciers returned and lakes expanded once more. This has been ascribed to a period of increased humidity and warmth which was, nonetheless, neither as wet nor hot as conditions here during the earlier Holocene temperature optimum (Björck et al., 1996).

15.4.2   The Last Millennium

A renewed phase of penguin colonization of Victoria Land coast sites occurred 1250–650 BP registering the effect of the Little Climatic Optimum in Antarctica (Baroni and Orombelli, 1994). The abrupt cooling and increased precipitation at the end of this period signal the start of the Little Ice Age. Although peripheral to the study area, the high-precision ice-core record from the Law Dome shows a warm period ~1650–950 BP followed by a cooling marking the Little Ice Age. After a slight warming 550–350 BP, temperatures reached a minimum around 150 BP since which times they have been rising (Morgan, 1985).

Oxygen-isotope measurements on ice cored at Byrd Station suggest that a cooler period occurred 900–500 BP and that warming began ~400 BP reaching a maximum about 150 BP (~AD 1800) since which time temperatures have been falling. An ice-core record from the Antarctic Peninsula shows a decrease of ~2°C in temperatures since a warm maximum in AD 1850 (Aristarain et al., 1990). These records are somewhat anomalous both regionally and globally, and may have been influenced by the distinctive isotopic composition of the snow deposited (Langway et al., 1994). Yet assuming them to be legitimate palaeoclimate indicators, two opposing trends can be recognized in Antarctica for the last 500 years (Mosley-Thompson, 1992). The first, which includes records from the Antarctic Peninsula and Byrd, is illustrated by the Siple record in Figure 15.3. Conditions at Siple were warmer during the Little Ice Age and have been cooling since its termination. This is the opposite situation to that at the South Pole and at Quelccaya in Peru (Figure 15.3), discussed in more detail below. The opposition between ice-core records from Antarctica may represent intraregional climate variations, though the uncertainty underscores the problem of obtaining regionally valid information about climate change within the Antarctic region from just a few sites.

The effects of recent 'global' warming have been manifested through much of Pacific Antarctica by an increased rate of ice calving. An increase in exposure of ice-free terrain over the past 150 years in parts of the Transantarctic Mountains (Mayewski et al., 1995) may be a response to the same phenomenon.

15.4.3   Contemporary Environmental Issues

The importance of knowing something about the nature and effects of late Holocene climate change in the Antarctic segment of the Pacific Rim is that this was the only part to have been largely unaffected by humans during this time. Yet increasingly within the last two hundred years or so, humans have left their mark on this region be it through the mass slaughter of many of its largely 'naive' fauna or its by-products. The opening of an 'ozone hole' above the continent, attributed to the release of chlorofluorocarbons into the atmosphere since the 1940s, is cause for concern and is being monitored closely.

A combination of the multiple claims to portions of Antarctica, its inhospitable yet indisputably fragile environment, and its mineral potential have led to an international agreement not to exploit the latter until at least AD 2041. Moves to declare the continent a World Park in

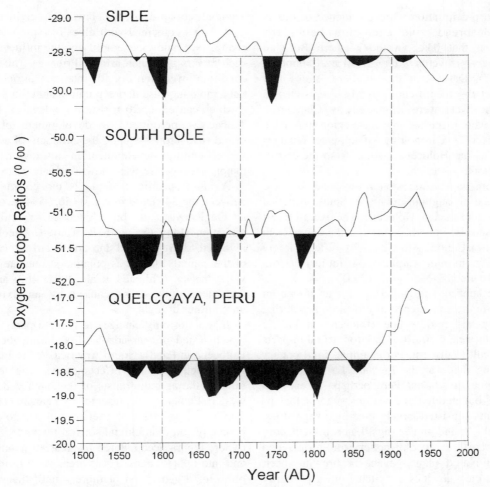

Figure 15.3   Variations in palaeotemperatures over the past 500 years derived from ice cores in Antarctica and
South America: Siple station, Antarctica; South Pole; Quelccaya, Andean Peru
After Mosley-Thompson (1992) with permission of Routledge Publishers

perpetuity are underway; a Whale Sanctuary has been declared from Antarctic shores to 40°S in most of the Pacific Ocean.

15.5   LATE HOLOCENE ENVIRONMENTS OF SOUTH AMERICA

15.5.1   Early and Middle Late Holocene

The climate of southernmost South America became increasingly cool and wet during the late Holocene. The resulting expansion of coastal wetlands and increasing dominance of upland forest by *Nothofagus* was also a response to increased cloud cover and increased storminess, which removed less hardy species (Heusser, 1989). Farther north, several episodes of cooler, probably wetter conditions separated by warmer, probably drier intervals occurred. Throughout most of the Andes, glaciers expanded 4700–4200 BP, 2700–2000 BP and during the Little Ice Age (Mercer, 1982; Coltrinari, 1993). These episodes are interpreted primarily as a response to cooling although most may also have been associated with increased precipitation. Direct evidence of this is not widespread, one exception being the conclusion that lakes in the northern Bolivian Andes became wetter after ~2300 BP (Abbott et al., 1997). Data from the southern Andes that contradict the identification of these episodes as cool (and wet) were discussed by Clapperton (1993a) who suggested that this region may have experienced a chronology of climate changes during the late Holocene distinct from the rest of Pacific South America.

Modern vegetation conditions had become established throughout montane South America by the start of the late Holocene. The overall forest decline and rise in herbaceous taxa thereafter, both here and in Central America, may reflect largely increasing human impact, particularly from livestock grazing (Markgraf, 1993).

Since hunting and gathering was an important element in the subsistence of most humans during the early and middle late Holocene in Pacific South America, their direct environmental impacts are difficult to discern. For example, fired pottery was being made at the lowland La Florida site (near Lima) in coastal Peru perhaps as early as 1700 BC but midden excavations suggest that its inhabitants subsisted largely on fruits, vegetables, fish, shellfish and guinea pigs. Later in the history of these areas, the use of pottery for storing food from cultivated plants, even for brewing beer, becomes clear as does a switch from a primarily marine to a farming economy based largely on maize.

Ceramics appear to have been used only later among most Andean montane communities where hunting of deer and wild camelids remained an important part of subsistence systems in spite of the spread of maize and other cultigens and, most importantly, that of appropriate techniques for intensifying the growth of such crops in such places (Morris and von Hagen, 1993). Agricultural intensification led to the growth of centralized societies, one of the most elaborate of which was Tiwanaku on the Bolivian–Peruvian *altiplano* around the southeast shores of Lake Titicaca (Binford et al., 1997). Yet both the emergence of agriculture in this area (~1500 BC) and the

eventual collapse of the Tiwanaku civilization (~AD 1100) coincided with times of rapid climate change (see below), exemplifying the influence of such factors in human affairs. Prior to ~1500 BC, conditions were too dry for intensive agriculture but a rapid increase in precipitation about this time – which caused a >20 m rise in the level of Lake Titicaca – permitted the development of the raised-field crop systems, the products of which fuelled cultural development in this region. The major urban centre was shown in Plate 2.2.

A more specific example is the coincidence between the 30 year drought AD 562–594 recorded at Quelccaya and the Moche IV–V cultural transition along the north Peru coast. The latter involved 'drastic inland and northward relocation of the capital, as well as population nucleation at valley necks' (Shimada et al., 1991: 262) and is consistent with stress caused by an extreme unremitting drought.

It was not only changes in climate parameters which tested human adaptability during the late Holocene in Pacific South America. A fascinating reconstruction of the effects of periodic tephra falls on the agriculturalists of the Jama valley in western Ecuador was presented by Pearsall (1996). The long-term effect of a tephra fall was to drive people off the flat alluvial soils of the valley floor (where tephra accumulated in greatest quantities) and into upland areas. These areas were farmed – often for a generation or more – until the valley floors once again became suitable for growing crops.

Aside from visible alterations to the land surface (see Plate 15.4) and the overall pattern of vegetation change, the precise effects of humans on the pre-modern late Holocene environments of Pacific South America are not easy to discern. For example, variations in dust concentrations in Quelccaya ice cores (see next section) may reflect changes in the practices of agriculturalists in this part of the Andes; two dust peaks AD 600 and AD 920 are prominent.

The effects of a late Holocene fall of sea level along the Pacific coast of South America are obscured by contemporaneous tectonic activity in places, although are clearer where uplift rates fell during the late Holocene (as in north-central Chile; Ota and Paskoff, 1993) and/or around river mouths. The later Holocene outgrowth of the Santa Delta in Peru is a good example of the effects of sea-level change on coastline morphology and human settlement (Figure 15.4). At the time of the Holocene sea-level maximum (Las Salinas Period, 5000–1800 BC), a deep bay with islands existed; human settlement was confined to its bedrock margins. In the early Suchimancillo Period (0–AD 200), the bayhead shoreline had prograded but settlement was still restricted to higher, less waterlogged areas. Under the influence of both a falling sea level and perhaps increased catchment precipitation (to explain accelerated delta growth), the shoreline during the Guadalupito Period (AD 400–650) was farther west than before and presumably sufficiently well drained for settlements to be established in three lowland areas. During the Little Climatic Optimum (Early Tambo Real Period, AD 1150–1350), the shoreline had prograded still farther west and settlements had become clustered along the southwest-facing part of the shoreline, probably because of an abundance of marine resources nearby.

15.5.2 The Last Millennium

For northern Patagonia, an extraordinarily precise picture of climate change over the past millennium (Table 15.1) was obtained by Villalba (1990) from dendrochronological analysis of alerce (*Fitzroya cupressoides*), a tree which can live for 1800 years. The absence of a protracted Little Ice Age in these data is contradicted by other evidence from this region, discussed below, suggesting that they reflect only local changes akin to the Siple record shown in Figure 15.3. In the same area, dendrochronological analyses have also been used to indicate shifts in the west–east storm belt along the South American coast. Santiago lies close to the northern limit of this belt today, and its precipitation record has proved a sensitive recorder of such shifts (Boninsegna, 1992). Droughts AD 1270–1450 and AD 1600–1650 imply

a southward shift perhaps associated with a strengthening of the southeast Pacific (Easter Island) anticyclone.

The Little Ice Age is well marked by glacier expansion throughout Pacific South America (Coltrinari, 1993). For example, Mercer and Palacios (1977) dated moraines in Andean Peru to AD 1350–1650 marking Little Ice Age advances of glaciers. Documents and sketches made since the Spanish conquest of the northern Andes from about AD 1531 allowed Hastenrath (1981) to determine that, during much of the past 400 years, valley glaciers had extended farther downslope than they do today, a view borne out by the accounts of many travellers in the area during the early 20th century.

Some of the most profound Holocene environmental changes in the usually quite dry parts of modern Pacific South America came about as a result of massive short-lived precipitation increases associated with El Niño. Dates of several events during the late Holocene were acquired by Lisa Wells (1990) in her study of Rio Casma floods in northern Peru; many produced major societal changes. Along the Santa coast in the same dry area, beach ridges formed during late Holocene El Niño events as a result of the massively increased river loads, which characterized these events, being dispersed along the shoreline (Sandweiss, 1986; see Figure 15.4). An extended treatment of El Niño and its effects on Pacific Basin environments is given in a separate section towards the end of this chapter.

Some of the most illuminating and influential studies of last-millennium climate change in the Pacific Basin have come from cores through the Quelccaya Ice Cap, northwest of Lake Titicaca in the Peruvian Andes (Thompson and Mosley-Thompson, 1992), and Huascarán col to the northwest (Thompson et al., 1995). The 1500 year Quelccaya record comes from two cores dated by tephra and by counting of annual ice layers (Plate 15.3) and analysed in several ways. The Little Ice Age has one of the clearest signatures in all these cores and was followed by a strong warming in the last 200 years shown in Figure 15.3.

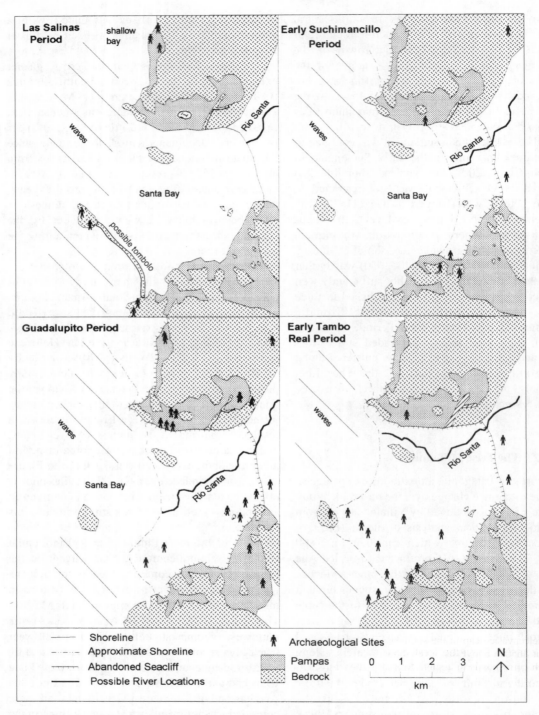

Figure 15.4   Late Holocene growth of the Santa Delta, Peru, showing changes in shoreline position and settlement pattern. Note the gradual westward movement of sites associated with shoreline progradation
After Wells (1992)

Table 15.1 Climate episodes recognized in northern Patagonia from dendrochronological analysis of long-lived alerce (*Fitzroya cupressoides*) trees After Villalba (1990)

Dates	Climate conditions
AD 900–1070	Cold, moist
AD 1080–1250	Warm, dry
AD 1270–1670	Cold, moist (minimum, AD 1340) (minimum, AD 1650)
AD 1720–1790	Warm
AD 1800–1820 (?)	Cold
AD 1850–1890	Warm

Plate 15.3 The 50 m ice cliff along the margin of the Quelccaya ice cap in the southeast Peruvian Andes about 5650 m above sea level showing layers, each of which corresponds to a single wet–dry annual cycle of ice accumulation. This photograph was taken in 1977; by 1995, this ice cliff had melted as part of a rapid and accelerating recession of the ice cap
Lonnie G. Thompson, Byrd Polar Research Center, Ohio State University

Maize agriculture had become widespread throughout the northern Andes by 2000 BP with warmer, wetter conditions in the last millennium encouraging its spread. Farther south, in the Bolivian Andes, this was inhibited by the onset of drier conditions during the Little Climatic Optimum, a situation which may also have affected the southern Andes (Graf, 1981).

The spectacular collapse of the Tiwanaku civilization (see above) was coincident with the beginning of an interval of severe aridity, indicated ~AD 1100 in ice-core records from nearby Quelccaya and, later, by a fall in the level of adjacent Lake Titicaca (Binford et al., 1997). These changes prompted the abandonment of the raised-field systems and urban centres, and the dispersal of the Tiwanaku people.

The emphasis in other areas during the Little Climatic Optimum was on adaptation to the new environmental conditions, especially freshwater shortages, which – in perhaps its most famous manifestation – produced the Chimu–Mocha Intervalley (La Cumbre) Canal about AD 1200 (Ortloff et al., 1982; see also Chapter 2). This canal was rendered obsolete by tectonic movements (uplift and warping), exemplifying the potential role of these in short-term landscape change in this region (Browman, 1983).

The Inca (Inka) Empire covered much of Pacific South America (see Figure 15.1) by the time of Spanish arrival in AD 1531. The rise of the Incas had been both recent – beginning only around AD 1438 – and rapid, notable for the spread of many innovative land management practices (Plate 15.4) which permitted the operation of an efficient centralized system. The Spanish interest was motivated largely by a desire for precious metals to which end they simply took over many aspects of Inca communication and land management. In later posthispanic times, as the numbers of Spanish settlers grew, wheat and barley were introduced to highland areas along with grapevines, olives and figs in lowland oases. Generally, the introduction of European range animals had less of an impact here than in Central America because the alpaca and llama were already domesticated.

Plate 15.4   The Inca people of the Andes were among many early human inhabitants of the Pacific Basin who rendered steep slopes in areas of low rainfall suitable for intensive cultivation by terracing. This photo shows one of the most unusual examples of Inca terrace building at Moray, northwest of Cuzco in the Peruvian Andes Shippee-Johnson, courtesy of Department of Library Services, American Museum of Natural History. Negative no. 334812

15.5.3   Contemporary Environmental Issues

Owing to high population densities, many Andean environments are under more stress today than montane environments anywhere else in the Pacific Basin. In parts, irreversible soil degradation has taken place; elsewhere any further agricultural development will jeopardize future food production. One of the few studies to base such statements on empirical data was carried out in the Rio Ambato catchment in the Ecuadorian Sierra, and found that its 'productive agricultural future ... may range from 10–75 years under current land-use practices' (Harden, 1988: 339). One solution is 'a return to the farming practices of the prehispanic era that have been largely forgotten' (Lauer, 1993: 165) although the practicality of this suggestion in an irreversibly changed world might – as for other parts of the Pacific Basin – be doubted.

Some nationally sponsored efforts to irrigate dry areas of lowland Pacific South America have been criticized for failing to account for long-distance aeolian deposition from coastlines rendered chronically unstable because of the effects of ENSO and tectonically induced change

(Moseley et al., 1992). Yet such developments are urgently needed to sustain a growing lowland population for many of whom traditional sources of subsistence have gone. Although land degradation figures high among the causes of these, the collapse of the massive Peruvian anchoveta fishery has also been felt acutely. This fishery was harvested heavily during the 1960s, 1970 yields exceeding 12 million tonnes. Owing to such unsustainable levels of exploitation, this fishery never recovered from the effects of the 1972 El Niño; even in 1985, the catch was less than 1 million tonnes (Hilborn, 1990).

15.6 LATE HOLOCENE ENVIRONMENTS OF CENTRAL AMERICA

15.6.1 Early and Middle Late Holocene

The difficulty of obtaining unimpeachable evidence of climate change from Central America noted for earlier times becomes even more acute in the late Holocene because of the overall increasing population and their increasingly profound and extensive environmental impact. For this reason, some palaeoclimatologists have focused on the glacial record from the high volcanic mountains of central Mexico (Bradbury, 1982) and, although this remains poorly known, the Malinche IV glaciation 3000–2000 BP indicates late Holocene cooling and/or increased wetness.

It was towards the end of the first millennium BC that maize productivity in Panama increased and people moved from the hills to the alluviated valley floors. It has been argued here that the former had been 'overburned' implying that the associated soil erosion and fertility decline were factors in the move to the valley floors[6] (Cooke et al., 1996). Other researchers have attempted to tie

---

[6] Similar arguments marshalled to explain changes in late Holocene settlement pattern on some Pacific islands have been criticized for their failure to account for the effects of contemporary sea-level fall in making valley floors and coastal plains more attractive to human settlers (Nunn, 1994a)

human to non-human changes. For example, late Holocene fires in Costa Rica ~2430 BP and 1180–1110 BP were thought by Horn and Sanford (1992) to have been among those started by people, yet to have spread more widely than most because they were lit during exceptionally dry periods, possibly associated with El Niño events.

The Maya civilization was one of the earliest in Central America, the first signs dating from perhaps 2000 BC, the Classic period lasting from AD 250–800 (Simmons, 1989). Population reconstructions from 300 BC are given in Table 15.2. The diversity of the agricultural foundations of the Classic Maya phase rendered them more resilient to climate change than those of other comparable civilizations; Maya agriculture was both upland and lowland, involving both shifting cultivation, terraced and raised-field farming, supplemented by arboriculture which produced a largely cultural landscape in many parts of modern Honduras, Belize and Guatemala (see Figure 15.1). The collapse of the Classic Maya phase AD 800–900 was associated with a sudden population decline (see Table 15.2) and the abandonment of cities and farmlands which became overrun with tropical forest and have since been only sparsely peopled. The reasons for this collapse have been hotly debated yet remain inconclusive, explanations involving population overshoot or environmental collapse being problematic (Turner, 1990).

The influence which climate change had on human lifestyles during the late Holocene in Mexico is exemplified by the Puebla–Tlaxcala area (Figure 15.5). Population increase 2750–2250 BP was manifested by increases in the number of settlements and by the spread of irrigated agriculture supplementing rainfed maize and beans. The following period (2250–1850 BP) was characterized by a well managed environment with few signs of damage. The time of least soil erosion was during the Tenanyecac period when adequate moisture allowed conservatory practices to operate with optimum efficiency. Falling moisture levels combined subsequently with population growth to cause a breakdown in environmental management. A further increase

Table 15.2   Population reconstructions for the central Maya lowlands
After Turner (1990)

Year	Total population (thousands)	Population density (persons/km^2)	Growth/decline (%/year)
300 BC	242	10.6	–
AD 300	1020	44.9	+2.4
AD 600	1077	47.4	+0.018
AD 800	2663/3435	117.2/151.2	+0.45/+0.58
AD 1000	536	23.6	−0.89/−0.93
AD 1200	285	12.5	−0.32
AD 1500	104	4.6	−0.34
AD 1700	62	2.7	−0.26
AD 1850	8	0.4	−1.4
AD 1900	6.8	0.3	−0.33
AD 1950/60	14.8	0.7	+1.4
AD 1970	48.7	2.1	+7.9

in aridity led ultimately to population dispersal in Tlaxcala times (Lauer, 1993).

15.6.2   The Last Millennium

A severe drought affected much of Mexico around 1000 BP resulting in depopulation of its drier, northern part. A return to moister conditions was marked by the intensification of agriculture in many areas around 950 BP. In the Puebla–Tlaxcala area of Mexico (see above), this process was driven by population growth and associated with warming (Little Climatic Optimum) yet also led to increased environmental degradation. These processes climaxed around 650 BP (~AD 1300) and the numbers of settlements declined thereafter along with the area of cultivated land because of the climate deterioration associated with the start of the Little Ice Age (Lauer, 1993).

The Malinche V glaciation on various mountains fringing Mexico's central valley coincides with the Little Ice Age (Heine, 1978). Elsewhere in this region, increasing rainfall and/or lower temperatures led to a rise in lake levels ~AD 1325–1521, a time which was followed by a series of prolonged droughts ~AD 1521–1750. After this time, precipitation increased once more until the time of the comparatively well documented drought lasting several decades around AD 1800. Humans are implicated in subsequent lake-level

fluctuations in this area (Metcalfe, 1987; O'Hara, 1993).

Just prior to Spanish arrival in AD 1521, there were two types of human occupation in Central America. The first was simple farming, dominated by slash-and-burn agriculture and the growing of maize, beans and squash in various parts of this region, often where older civilizations once existed and had left a legacy of depleted soils and dispersed people. The second was that of the various civilizations, notably the Postclassic Maya in tropical Central America, and the Aztecs and the Tarascan people in Mexico. Like the Inca of Pacific South America (see above), the Aztec domain included several large cities[7] into which tribute in the form of agricultural produce flowed regularly[8]. All these civilizations were based on similar systems of agricultural intensification centred on systems of raised fields, particularly around the fringes of shallow freshwater lakes.

---

[7] The largest Aztec city, Teotihuacan, developed because volcanic eruptions had displaced scattered groups of people causing them to become concentrated in this area, where there was abundant freshwater, obsidian deposits, an extensive lake system, and easy access to the Atlantic coast (Manzanilla, 1997)

[8] Yet in the century or so before Spanish arrival, the Aztec belief that the end of the world was imminent was strengthening as a result of an increasing number of extreme events (floods and droughts) which strained even their unusually competent management of the environment

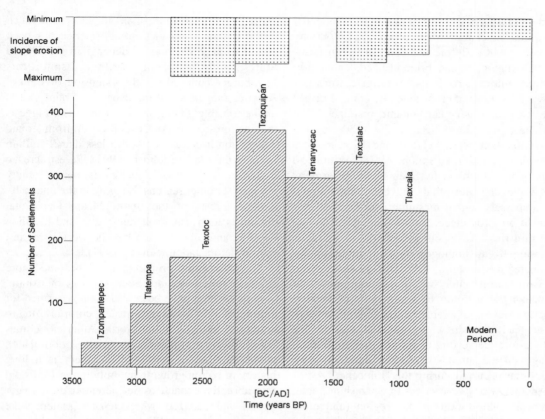

Figure 15.5 Relationship between culture phases, number of settlements, and slope erosion in the Tlaxcala region of Mexico. Note how erosion increased with increasing population density in the early phases, followed by the Tenanyecac Phase in which human management of the landscape, aided by a suitable climate, was so effective that minimal slope erosion occurred. Subsequently, increasing population density combined with the onset of warmer drier conditions led to a resurgence of slope erosion, which has been a feature of this environment ever since.
After Lauer (1993) Reprinted from *Mountain Research and Development*, 13(2), p. 162, by permission
© The International Mountain Society and United Nations

Spanish settlement in Central America[9] was initially driven by a desire for precious metals but, unlike the situation in much of Pacific South America (see above), this was tempered here by a realization of the potential of much of the region for both arable (particularly wheat) farming and ranching. The replacement of digging sticks by animal-drawn ploughs led to an abrupt increase in erosion in many places as more ground was rendered more vulnerable to erosion during heavy rain. Introduced animals – cattle, sheep, horses, pigs and mules – reproduced rapidly and many became feral. The effects of this included the trampling of crops by herds of cattle, a problem which forced many villages to move to more marginal lands where cattle were less prone to roam. Degradation accelerated throughout much of the Central Valley of Mexico as livestock numbers reached unsustainable levels in the decades after initial Spanish settlement.

An excellent example of the environmental consequences of overstocking and indiscriminate

_____

[9] The Spanish Conquest of Mexico precipitated the 'greatest environmental catastrophe in human history' (Simon, 1997: 4)

grazing of sheep was detailed for the Valle del Mezquital in central Mexico AD 1521–1600 by Melville (1990). Before the Spanish began farming this region, it had been utilized, apparently without major problems, by agriculturalists growing a variety of crops aided by comparatively sophisticated water-management practices. The first sheep were introduced in the 1530s, by the late 1550s there were 421 200, and 15 years later more than 2 million (Simon, 1997). Within 50 years of its settlement by Spanish sheep farmers, 'the Valle del Mezquital had been transformed into a sparsely settled mesquite desert considered to be fit only for sheep, and the decimated Indian communities were congregated in villages separated by an homogenous vista of mesquite-dominated desert scrub' (Melville, 1990: 73). The environmental and social consequences of European colonization of many parts of Central America were comparable to situations elsewhere in the Pacific Basin where colonialist arrogance and ignorance brought about environmental changes of magnitudes exceeding those of many others that occurred during the Holocene.

Another widespread effect of Spanish and later European colonization in this region (and other parts) of the Pacific Basin was that the numbers of indigenous people declined, most numerously through the introduction of unfamiliar diseases[10], or were driven from the lands they had long occupied. This led to many settlements and agricultural lands being abandoned. In places such as the Darien rainforest in Panama (Piperno, 1994), this led to regeneration of the forest cover but more commonly produced landscapes dominated by largely exotic weeds where, in the absence of human management, erosion often increased.

15.6.3 Contemporary Environmental Issues

Central America has been the third largest supplier of beef (after Australia and New Zealand) to the

United States for more than 40 years and much recent deforestation in Central America has been linked to an escalation of demand in the United States. Although some writers dispute this 'hamburger connection', the extension of Central America rangelands at the expense of rainforest is unquestionable. For example, one estimate suggests that the forested area of Costa Rica fell from around 3.4 million hectares in 1940 to less than 1 million hectares in 1987 (Edelman, 1995). Forest acreage elsewhere in Central America is being rapidly reduced for other reasons. Logging in the unusually diverse forests of the Sierra Madre Occidental continues at unsustainable rates; at around 1 million hectares annually, Mexico has the world's highest rate of deforestation (Simon, 1997).

Environmental problems have become more acute in recent decades even in areas of Central America which have been traditionally farmed for centuries. A good example, comparable to situations elsewhere in North America, comes from southern Honduras (Stonich, 1995). Here, along the Pacific coastal plain, there is a long history of cotton growing but only in the 1950s did this abruptly expand as its potential as a major export was realized. Many tenant farmers were evicted, many smallholders were bought out, wholly or partly, to make way for massive, highly mechanized plantations. This led to a massive displacement of largely landless, jobless people. Those who remained in the area as subsistence farmers commonly had less land to farm, and had consequently to intensify its usage or begin to work marginal lands in order to produce sufficient food. Today, much less cotton is grown; many former cottonfields have poorer soils than they once had, the legacy of decades of overworking and excessive use of fertilizers and pesticides.

Elsewhere in Central America, there are good examples of the widespread effects of the 'Green Revolution'. One of the most extreme situations is found in the mountains of Oaxaca in southern Mexico, claimed as 'the most eroded landscape on the planet' (Simon, 1997: 35). A long history of erosion and, more recently, changes in agricultural techniques designed to maximize yields set the stage for the contemporary crisis in Oaxaca. The

---

[10] The introduction of smallpox to Mexico in 1520 has been traced to a single infected person who landed at Veracruz with Cortez. Perhaps more than 100 000 Aztecs died from the disease within a few months (Simon, 1997)

advent of chemical fertilizers in the mid-1970s, intended to boost corn yield, led within a few years to farmers depending on these for crops of comparable yield to those of the early 1970s. Increasing debts were incurred when it became necessary to purchase fertilizers, a situation which has led to many farms being abandoned and the people seeking wage employment elsewhere. Abandoned farms are often the sites of the most rapid degradation.

As elsewhere in the Pacific Basin, 20th century economic growth in parts of Central America has come with a huge environmental price tag which politicians have been reluctant to scrutinize too closely. The environmental cost of Mexico's economic boom between the 1940s and 1970s is difficult to overstate (Simon, 1997).

15.7 LATE HOLOCENE ENVIRONMENTS OF NORTH AMERICA

15.7.1 Early and Middle Late Holocene in the United States

The overall late Holocene trend of climate in the Pacific segment of the United States was one of cooling beginning at least ~5000–3000 BP in most parts; temperatures in the White Mountains of California, for example, have fallen 2–3°C in the last 6800 radiocarbon years (see Figure 13.2). Palaeoprecipitation is less easy to generalize about, although a major drought affected most of this area around 2000 BP (Thompson et al., 1993). Some of the most detailed work has been in the Sierra Nevada and nearby White Mountains, where glacial advances 2800 BP, 2200 BP and 1850 BP (Scuderi, 1987) coincided with cool, wet periods both here (LaMarche, 1973) and along the California coast (Davis, 1992).

Changes in effective precipitation are regarded as the main control of lake-level fluctuations at Mono Lake, California, for which a late Holocene record was obtained by Stine (1990). The detail of this record surpasses that for most other lakes in the region and shows a number of variations linked to solar activity (Figure 15.6). At San Joaquin Marsh, also in California, pollen indicative of alternating saltwater and freshwater conditions was interpreted by Davis (1992) as signalling times of warm, dry and cool, wet periods respectively. Two of the latter, 2800 BP and 2300 BP, match times of ice advances in the Sierra Nevada (see above).

Although human influences on the environment were not as great in this part of the Pacific Rim compared to certain others during the early part of the late Holocene, humans may nonetheless have brought about significant change, although, as elsewhere, this is not always easy to identify unequivocally. A good example is provided by studies of Santa Rosa island off the California coast, where charcoal dating from 3250–1500 BP could be the result of either aboriginal burning, a drier climate or both (Cole and Liu, 1994). Humans undoubtedly burned large areas of vegetation in the western United States during the Holocene but there are various views about whether humans are implicated in the origin or spread of grassland–savannas (Whitney, 1994). The situation in this region is likely to have been similar to parts of Australia, discussed in more detail below.

Climate change during the late Holocene caused variations in the position and configuration of the North Pacific gyre manifested by changes in the strength of the cool south-flowing California Current. When the latter was strong, the vegetation of the coastal area was much as today. When the California Current was weaker, a north-flowing tongue of the warm North Equatorial Current reached southern California and temperatures rose eliciting a response in the vegetation (Fairbridge, 1992).

15.7.2 Early and Middle Late Holocene in Canada

Owing to the mountainous character of most of the Pacific margin of Canada and the dearth of palaeoclimate studies carried out there, both the nature and degree of information available for the late Holocene are less than for the corresponding part of the United States. Most palaeoclimatic inferences come from studies of glacier fluctuations, commonly dated by the ages of trees

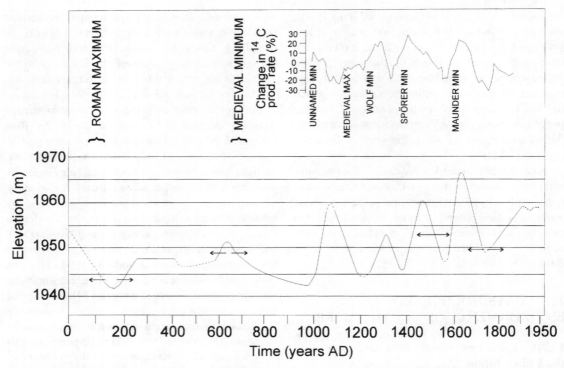

Figure 15.6   Changing levels of Mono Lake, California (bottom) are correlated with changing sunspot activity.
Specifically, high lake levels generally occurred at times of low sunspot activity represented in the upper graph by
changes in the production rate of ^{14}C
Reprinted from *Palaeogeography, Palaeoclimatology, Palaeoecology*, 78, Stine, S. 'Late Holocene fluctuations of
Mono Lake, eastern California', 333–381. Copyright 1990, with permission from Elsevier Science

overridden by glaciers (see Plate 13.3). The study
by Luckman et al. (1993) suggests that glaciers
extended downslope throughout the Pacific Rim in
Canada during the early part of the late Holocene
3100–2500 BP largely in response to falling
temperatures.

The overall cooling experienced in Pacific
Canada during the late Holocene has been tracked
by the southward movement of boreal spruce
(*Picea glauca*) forest, just as early Holocene
warming was tracked by its northward movement
(Ritchie, 1984).

15.7.3   The Last Millennium in the
United States

The Little Climatic Optimum shows up in the rise
of sea-surface temperature in southern California

~AD 1150–1300 which was associated with severe
droughts (Arnold and Tissot, 1993). As in the
Pacific Islands (see below), there is abundant
evidence here that these climate changes produced
environmental stress and increased (or initiated)
conflict between its human inhabitants, and that
environmental change became an important factor
in the development of social complexity in this
region (Colten, 1994).

Cool and wet conditions at San Joaquin Marsh
since ~AD 1390 coincide with the Little Ice Age
along the California coast (Davis, 1992). Glacial
advances marking wet phases of the otherwise
generally dry Little Ice Age in the Sierra Nevada
and White Mountains occurred AD 1480, AD 1620,
AD 1710 and AD 1820 (Scuderi, 1987). A study
using lichenometry on moraines around Mount
Rainier concluded that there had been 15 periods

of glacier recession, associated with prolonged periods of unusual aridity, since AD 1323 (Burbank, 1981).

Palaeotemperature series for the western United States developed from tree-ring data show no clear overall cooling during the Little Ice Age; despite prolonged periods of cooling, the coldest time in this region during the last ~500 years was in the late 19th and early 20th centuries (Fritts and Lough, 1985).

The shortcomings of European colonization of the Pacific Rim in the United States were expressed lucidly by Bunting (1995: 413), writing about western Oregon:

> The region's history from settlement to industrial society was largely shaped by the tensions and interaction between an economy of nature and a capitalist market economy. Northwest settlers understood imperfectly both economies, but thought they could control them. They could not, and their failure to comprehend either economy seriously flawed their own goals of independence and security.

Illustrations of the nature of the logging about a century ago in the northwest of the United States (and Pacific Canada) are shown in Plate 15.5. While most logging at this time was carried on with scant regard for its environmental impact – analogous to the modern practice in other parts of the Pacific Basin – it is now subject to strict legislation drawn up with a view to long-term maintenance and renewal of forests along the Pacific Rim in North America.

15.7.4 The Last Millennium in Canada

Wider tree-ring widths AD 900–1300 suggest that conditions during the Little Climatic Optimum were more favourable for tree growth in the southern Canadian Rockies than subsequently, but data are few so this conclusion should be regarded as preliminary (Luckman, 1994).

Early and middle late Holocene advances of glaciers throughout the Canadian Cordillera (see above) were exceeded by those of the glaciers which advanced during the Little Ice Age. This situation is different to that in other parts of the Pacific Basin, and was ascribed by Brian Luckman (1993) to the effect of short-term climate fluctuations being imposed on a long-term decrease in summer solar radiation.

Little Ice Age records are the most abundant palaeoclimate records from Pacific Canada and most end with the marked warming characterizing the last hundred years or so. Historical recession of glaciers has been occurring throughout montane Pacific North America (although see Plate 16.1) – indeed, it is because this recession has exposed the record of earlier glacier fluctuations that researchers have been able to unravel much of the earlier environmental history of this region. A good example of the implications of glacier recession for environmental change comes from the Saint Elias Mountains which straddle the Canada–Alaska border. The Grand Pacific Glacier receded 24 km between 1879 and 1912; with the removal of buttressing ice, valley-side slopes have collapsed in this area, which is now being uplifted at rates as high as 3 cm/year owing to isostatic rebound (Clague and Evans, 1993).

15.7.5 Contemporary Environmental Issues Along the Pacific Rim in North America

Despite its recent history of rapid environmental change associated with unsustainable human demands, the greater part of Pacific North America is now being managed sustainably. Resources once squandered are now being replenished; strictly enforced legislation now governs most extractive industries with the intention of minimizing environmental impacts. In most of this region, the potential impact of non-human phenomena like volcanism, earthquakes (see below) and extreme weather conditions associated with ENSO are of greater concern to environmental managers.

There are of course many environmental problems in this region including air and water pollution. Of those which are most associated with environmental change, one of the best examples is the central valley of California, site of a latter-day 'hydraulic society' in which agriculture has become dependent largely on irrigation. The

Plate 15.5   Ox team at log dump, Grays Harbor, Washington in the 1890s (upper), Mumby Company logging camp at Bordeaux, Washington, northwest United States in 1904 (lower). Although most North American forests are now sustainably managed, much logging elsewhere in the Pacific Basin is effectively unregulated and commonly a cause of rapid environmental change similar to the situation in much of Pacific North America at the time these photographs were taken

Special Collections Division, University of Washington Libraries. Negative nos UW 187 (upper), UW 1765 (lower)

problem of land subsidence associated with overpumping of groundwater is acute (see Table 15.7) and the process of increasing mechanization has brought a series of environmental and social problems (Worster, 1992).

15.8 LATE HOLOCENE ENVIRONMENTS OF BERINGIA

15.8.1 Early and Middle Late Holocene

Given that modern vegetation was established largely by the start of the middle Holocene (see Chapter 13), dramatic vegetation shifts do not show up in many Beringian pollen records during the late Holocene. The changes that have been detected suggest that much of the region's climate became both wetter and cooler throughout this period (Heusser et al., 1985).

There are other indications of late Holocene climate change in Beringia. For example, glacier expansion in the Fairweather Mountains of southeast Alaska occurred 3500–2500 BP in response to higher available moisture levels (Mann and Hamilton, 1995). Changes wrought to beach ridges in northwest Alaska provide another proxy of climate change. Stormy conditions producing erosion of beach ridges characterized the early part of the late Holocene 3300–1700 BP. Then followed a period of less stormy conditions 1700–1200 BP marked by beach-ridge progradation linked to a southerly shift in North Pacific storm tracks (Mason and Jordan, 1993).

The expansion of crowberry (*Empetrum*) in the Aleutian Islands during the late Holocene suggests progressively moister conditions (Heusser, 1978). This may have led to increased organic productivity, particularly on strandflats, which may explain in part the more extensive use of Aleutian island coasts by humans during the late Holocene (Black, 1980).

15.8.2 The Last Millennium

The recent recession of glaciers in the Kenai Mountains of southern Alaska has uncovered deposits which have been used to trace the history of earlier ice fluctuations. A period of glacier expansion around AD 600 was followed by the better-documented Little Ice Age AD 1300–1850. By comparing the behaviour of maritime glaciers, which are more sensitive to precipitation variations here, with that of continental glaciers, which fluctuate more in response to temperature change, it has been possible to determine which advances during the Little Ice Age were primarily the result of precipitation change and which were caused largely by temperature change (Table 15.3).

Similar conclusions about the Little Ice Age have been reached from tree-ring data in this region. In central Alaska, unusual summer cold is indicated AD 1620–1710 (Jacoby et al., 1985), a time almost exactly coincident with the Maunder Minimum period of reduced sunspot activity (Eddy, 1977), discussed in more detail below.

Glacier expansion AD 1800–1900 in the Fairweather Mountains (Mann and Hamilton, 1995) probably represents the time at the end of the Little Ice Age when moisture availability had increased sufficiently to allow ice fields to expand. Since this time, warming has produced glacier recession in many places although, as shown in Plate 16.1, the signals may be mixed.

Radiocarbon evidence for significant shifts in human settlement and culture in southern Alaska were identified by Mills (1994). The late Holocene shifts 950–850 BP and 720–670 BP may signal changes in climate, as was the case elsewhere in the Pacific Basin. Owing to low population

Table 15.3   Glacier advances in the Kenai Mountains, southern Alaska, during the Little Ice Age sorted by primary cause
After data in Wiles and Calkin (1994)

Age of glacier advance	Principal cause
AD 1420–1460	Increased winter precipitation
AD 1440–1460	Lower summer temperatures
AD 1640–1670	Increased winter precipitation
AD 1650–1710	Lower summer temperatures
~AD 1750	Increased winter precipitation
AD 1830–1860	Lower summer temperatures
AD 1880–1910	Increased winter precipitation

densities and the persistence of hunter-gatherer lifestyles, discernible pre-European human impacts on Beringian environments were few and localized. Some impacts were caused by visitors, a good example being the Great Northern Steller's Sea Cow (see Plate 7.3), eaten to extinction by Russian fur hunters in the vicinity of Bering Island AD 1741–1768.

15.8.3 Contemporary Environmental Issues

Rapid economic and social development in parts of Beringia during the past 25 years or so raised issues and brought problems which were encountered decades earlier in other parts of the Pacific Rim, and many of the lessons learned from these experiences have been successfully brought to bear on the Beringian environment. Much of the development in Beringia has been associated with the petroleum industry, and key environmental concerns involve pollution of land and sea. One of the most regrettable incidents was the oil spill from the *Exxon Valdez* on 24 March 1989 which left a layer of oil up to 20 cm thick along 2000 km of the Prince William Sound coastline.

15.9 LATE HOLOCENE ENVIRONMENTS OF EAST ASIA

15.9.1 The Early and Middle Late Holocene in China

Warm dry conditions in most of Pacific China during the period 5000–2500 BP were succeeded by cool dry conditions (Winkler and Wang, 1993). A range of written evidence, extraordinary for its antiquity, allows insights into the climate changes of the late Holocene in East Asia unmatched elsewhere in the world. Much of this evidence is inferential, from reports and interpretations of events ('five thousand soldiers were able to cross the frozen lake', 'what had the Emperor done to make the winter so harsh') or from phenological data (dates for the flowering of shrubs or the arrival of migrant birds). Such sources have allowed a number of stages within the late Holocene history of China to be identified (Zhang, 1988).

The 'first warm stage' ~3000–1000 BC in China was warmer than at present and rainfall was more abundant, conditions which encouraged agricultural experimentation. This stage includes much of the Holocene Climatic Optimum (see Chapter 14). Among the many indications of warmer environments were the abundant bamboo thickets along the sides of the Huanghe valley where, since about AD 1300, it has been neither warm nor moist enough for bamboo to grow. Lakes and marshes in the Yangtze valley were far more extensive at this time (Yates, 1990).

The first cool stage ~1000–850 BC was cooler and drier than today; in 903 BC and 897 BC, unusually, the Hanshui River froze.

In the second warm stage ~770 BC–AD 0, subtropical plants such as bamboo also grew farther north than today. Conditions were probably also wetter. It was during this stage in northern China that the plough was first developed, an innovation which must have aided the spread of dryland agriculture particularly at a time when the growing season was 30–40 days longer here than today (Yates, 1990).

During the second cool stage ~AD 0–600, ice could be stored in Nanjing without recourse to special storage houses of the kind which had functioned in earlier times. Several thousand troops and their supplies crossed the Bohai Gulf on ice. Temperatures are estimated to have been 2–4°C lower than today, and records of floods indicate that the climate was generally drier (Lamb, 1977).

The Han Dynasty (206 BC–AD 220) spanned these last two stages. Cultural activities were centred around Jiangxi on the south bank of the Yangtze River. Cut wood was the main export from this to other areas of the region at the time (Li, 1988), a process which began the remarkable late Holocene deforestation of China, discussed in a separate section below.

The third warm stage AD 600–1000 was marked by a longer growing season than at present; unlike today, oranges could be grown in Xi'an. This coincides with the second phase of lake expansion (see below).

Despite the alternation between warm and cool phases during the period ~3000 BC–AD 1000, there is some evidence that these were superimposed on an overall cooling trend in China similar to that believed to have characterized most other parts of the Pacific Basin during the late Holocene. The changing northern limit of elephants in China provides an unusual record of this cooling trend. During the first warm stage, elephants lived north of the Huanghe; in the second warm stage, they could be found only as far north as the Huaihe, between the Huanghe and Yangtze. By the third warm stage, elephants were found only south of the Yangtze (Zhang, 1988).

Late Holocene lake-level fluctuations in China were synthesized by Fang (1993) who identified periods of expansion 500 BC–AD 0 and AD 650–950 which fall within the second and third warm stages respectively.

The importance of rice to East Asian economies during the late Holocene is clear from contemporary histories although its environmental impacts have not been clearly spelled out. Rice clearly fuelled the major economic expansions of the period. For example, practices such as the planting out of seedlings (illustrated in Plate 14.1) began in the 8th century, its spread leading to increased rice yields and an acceleration in population growth in China between the 10th and 13th centuries (from ~53 to 100 million people; Gernet, 1996).

Tea appears to have entered the Chinese pharmacopoeia during the late middle Holocene[11], its popularity rising rapidly during the Han Dynasty when its cultivation was confined mostly to mountain plantations (Evans, 1992). By AD 300, demand for tea had grown so much that new hill-tea plantations were created along the Yangtze valley. Through most of the rest of China's history, tea was the upland counterpart to lowland rice farming in many areas.

Sea level may have been stable for most of the late Holocene in coastal China, although the possibility remains that evidence for its contemporary fall, as elsewhere in the Pacific Basin, remains undiscovered.

15.9.2 The Early and Middle Late Holocene on Offshore Islands of East Asia

The warm dry conditions of the middle Holocene in Japan were followed by warm wet conditions during at least the early part of the late Holocene. This is shown by the expansion of *Cryptomeria*, which requires heavy winter snowfall, throughout Japan as far north as 39°N (Tsukada, 1986; Winkler and Wang, 1993).

The presence of humans practising comparatively sophisticated agriculture often in association with marine resource exploitation (Plate 15.6) throughout much of this region obscures the record of natural change during the late Holocene. 'Millet-like' pollen first appears around 3 ka at Ukinuno Pond, considered a diagnostic site by Tsukada (1986), along with that of cantaloupe (*Cucumis melo*).

The evidence for late Holocene sea-level fall is generally less equivocal on offshore islands than on the East Asian mainland although the tectonic overprint is sometimes difficult to see past. For example, sea level fell about 2 m around Kume in the Ryukyu Islands 3.0–1.2 ka BP yet this figure is imprecise because of late Holocene tilting of the island, which shows that the Holocene maximum sea level reached ~4 m in the northwest, and ~1.6 m in the south of the island (Pirazzoli and Delibrias, 1983).

15.9.3 The Last Millennium in China

The third cool stage AD 1000–1200 identified by Zhang (1988) was recorded by the freezing of several lakes. Freezing of the Grand Canal near Suzhou often prevented ships from sailing.

The fourth warm stage AD 1200–1300 coincides with the Little Climatic Optimum in many other parts of the Pacific Basin. Documentary records show that the cultivation

---

[11] Owing to the availability of tea leaves, and the likelihood that the earliest *Homo sapiens* in China knew how to boil water, it has been proposed that brewed tea was drunk 40 000 years ago at Zhoukoutien (Evans, 1992)

A

of citrus and the perennial herb *Boehmeria nivea* reached their northernmost limits in recorded history at this time; temperatures were generally 0.9–1.0°C higher than at present (Zhang, 1994).

The fourth cool stage – coincident with the Little Ice Age – began around AD 1400 and was at its coldest AD 1650–1700 when rivers and lakes froze regularly.

Climate changes during the last 500 years or so have been the subject of intense study by Chinese scientists. Selected records are given in Figure 15.7; the degree of variation within these generalized trends is illustrated for both temperature and precipitation by the climate of Beijing. There is considerable disagreement even between adjacent areas for this period, which makes it difficult to distinguish regional from local elements. This point is exemplified by the attempted correlations between indicators of moisture availability in Jilin and Beijing since AD 1826 (Table 15.4).

B

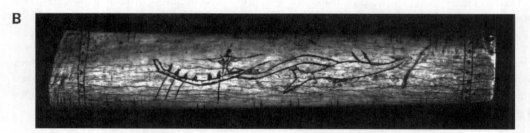

Plate 15.6   Depictions of whale hunting from the Okhotsk Culture (AD 600–900) of northeast Hokkaido, Japan. Okhotsk subsistence practices depended heavily on the marine realm and evidently included hunting large marine mammals yet the materials carved also give insights into the other animals hunted. The upper plate (A) is of a pair of antlers from a Yego deer (scale bar in centimetres); the lower (B) is from a shoulder-blade of a large bird made into a needle case (8.5 cm in length)
Yasuo Kitagamae

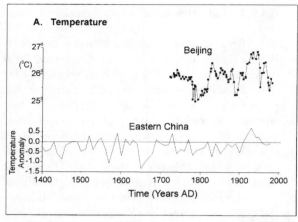

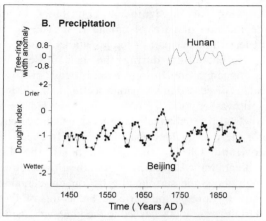

Figure 15.7 Climate change over the past ~500 years in eastern China. (A) Temperature changes for Beijing (after Zhang and Crowley, 1989) and eastern China (after Bradley and Jones, 1993). (B) Precipitation change as represented by tree-ring anomalies from Hunan (after Feng et al., 1993), and a wet–dry index for Beijing (after Zhang, 1988)

Although rice remains the dominant staple grown in Pacific China today, its dominance of river lowlands in traditional ways has lessened its contribution to environmental change during the last millennium. Although no data are known to the writer, other crops may have contributed more to such change. For example, in 18th century China, tea switched from being a largely domestic commodity to becoming China's chief export product. This would have been a reason for the spread of tea plantations in this region, perhaps – like Fiji sugar cane today (see Plate 15.14) – into marginal areas where regular planting could not be sustained. It is likely that many upland tea plantations – being located on wet steep slopes and visited regularly by weeders and pickers – became sites of severe erosion. Although its environmental legacy cannot be distinguished from those of other upland crops and particularly of deforestation, it is plausible to suppose that tea

Table 15.4 Latest Holocene humid–dry conditions in Jilin Province, northeast China, and Beijing From data in Feng et al. (1993)

Year (AD)	Jilin climate	Year (AD)	Beijing climate
1826–1835	Humid	1826–1870	Dry
1836–1843	Dry		
1844–1856	Humid		
1857–1865	Dry		
1866–1874	Humid		
1875–1883	Dry	1871–1894	Humid
1884–1897	Humid		
1898–1908	Dry		
1909–1918	Humid		
1919–1929	Dry	1895–1948	Dry
1930–1939	Humid		
1940–1950	Dry		
1951–1961	Humid	1949–1964	Humid
1962–1973	Dry	1965–1980	Dry

growing was a significant contributor to the degradation of parts of upland East Asia during at least the later part of the last millennium.

Deforestation during the last millennium continued a much longer-term trend in China and is discussed in a separate section below. Much soil washed off deforested slopes ended up in lowland river channels, a process which accounts for the rise in flood magnitude and frequency in lowland China during much of the last millennium. Many such rivers now flow at levels high above the surrounding floodplain, contained only by levees; the periodic breaching of these during large floods has led to the jumps in position of the lower channel of these rivers within recent recorded history (Figure 15.8). Today, as for most of the last millennium, the Chinese authorities control water flow within these rivers to reduce flood magnitude and to ensure water supply as needed.

15.9.4   Late Holocene Deforestation of China

Significant deforestation was occurring in parts of China during the Han Dynasty, more than 2000 years ago. The initial demand was principally for cut wood for mostly domestic purposes but increasingly during the late Holocene, to fuel furnaces for metalworking. Throughout the recent history of Pacific China, 'the link between an increase in population and forest clearance is unambiguous' (Menzies, 1994: 19). During the early part of the last millennium in particular, an increasing demand for agricultural land led to the clearance of forest from steepland areas with easily eroded soils. A connection between the resulting erosion, siltation of lowland river channels and increased flood risk was made by officials as early as the 13th century; the problem had become so acute by the 16th and 17th centuries in Fujian and Anhui provinces that state

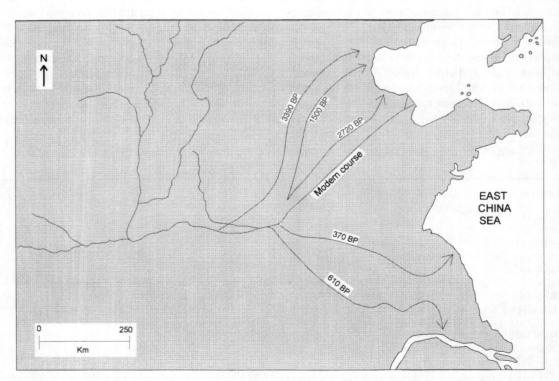

Figure 15.8   Changes in the course of the lower Huanghe, eastern China, during the late Holocene
Compiled from various sources

intervention was needed to resolve the resulting conflicts between highland and lowland communities. In a fascinating study of Hunan Province, Perdue (1987) showed that cycles of environmental degradation were linked directly to the changing degree of government intervention.

It has been difficult to get a true picture of the deforestation of China associated with 20th century industrialization (Smil, 1983). Particularly during Mao Zedong's 'Great Leap Forward', much accessible timber around villages and towns was used as fuel for backyard furnaces producing pig iron. Rapid population growth has caused demand for firewood to rise to unprecedented levels in recent decades.

Afforestation was for many years banned by law in China since it was considered to promote capitalism but this was reversed around 1980 with official encouragement for small-scale village-level tree planting. This has been backed by state initiatives to plant more trees, particularly to create a 'Great Green Wall' in the north of the country to prevent the southward spread of deserts. What afforestation has been carried out has not always been successful; Smil (1983) quotes an official statement that less than one-third of plantings survived, largely because of a lack of care in the method of planting and follow-up. Many trees being planted are introduced species; Australian *Eucalyptus* trees are the centrepieces of reafforestation schemes in south China (Simmons, 1989).

Not all the late Holocene deforestation of China can be attributed to humans. Fires are common in the drier parts of the country, and it can be assumed that these were more frequent and affected greater areas during drier, warmer times of the late Holocene.

15.9.5 The Last Millennium on Offshore Islands of East Asia

In what has been interpreted as an unambiguous indicator of human population growth and environmental impact, buckwheat (*Fagopyrum* spp.) pollen became common at Ukinuno Pond in Japan within the last millennium, coincident with a decline in numbers of forest species (Tsukada, 1986).

Finer details have been acquired in much the same way as for China (see above); a much-quoted example is the date on which cherry trees bloom (Table 15.5), an event still marked by a festival at Kyoto, Japan's ancient capital. Details of climate variations within the last millennium also come from records of floods and droughts (Lamb, 1977). The start of the Little Ice Age during the 15th century was, like the period AD 1750–1850, uncommonly cold and wet in northeast Japan but it is difficult to identify long-term trends in the available data.

An insight into prehispanic agricultural productivity of the Philippines was provided by Urich's (1995) study of Bohol island. So productive was the interior of this island that it was a major exporter of cotton and rice across the South China Sea. There are indications that, owing to intensive agriculture and high population densities, environmental degradation was spreading on Bohol by the time of Spanish settlement of the islands. This trend was accelerated by hostilities during the early Spanish period (AD 1565–1621) and the island has never regained its former condition.

Although rice remains the dominant staple in insular East Asia, and has produced characteristic landscapes, particularly in upland areas, it is less

Table 15.5 Mean blooming dates of cherry trees in Japan
After Arakawa (1957)

Year (AD)	Mean date (in April)
800–899	11.3
900–999	11.8
1000–1099	18.4
1100–1199	17.5
1200–1299	15.4
1300–1399	17.4
1400–1499	13.1
1500–1599	17.0
1600–1699	12.4
1700–1799	–
1800–1899	12.4

dominant than it was once, particularly as a result of European-facilitated introductions. For example, Ferdinand Magellan may have introduced maize (corn) from South America to the Philippines in AD 1521; if not, it arrived shortly afterwards. Today maize is well established as a second staple throughout the islands, particularly on dry soils.

The Pacific War left its mark on many parts of the Pacific Basin (see Plate 15.10). Perhaps nowhere was its environmental impact more abrupt or comprehensive than at Hiroshima and Nagasaki in Japan following the dropping of atomic bombs in August 1945.

15.9.6 Contemporary Environmental Issues in East Asia

Although most contemporary environmental issues in East Asia can be readily ascribed to human activity, there are others which cannot. Included among these are the effects of long-range dust deposition from inland Asia across much of the continental margin (Table 15.6).

The effects of the 'Green Revolution' have been most marked on rice farming in East Asia. Under the influence of new, faster-growing and/or higher-yielding cultivars, rice production has increased which has lessened the need for spatial extensions to rice-growing areas. Fertilizer use has brought mixed blessings. Cycles of dependency and debt comparable to those described for Central America (see above) have produced similar consequences in parts of China, South Korea and the Philippines but the extent of associated damage to the environment has been greatly outweighed by that caused by its commercial exploitation. Deforestation, described above for China, continues at unsustainable rates in many parts of this region and is contributing to massive upland soil loss and lowland river-channel and reservoir siltation.

Perhaps less than 10% of China's land area is now forested compared to 33% in the United States and 66% in Japan. Yet demand for forest products in Japan has never been greater, and is met largely (as elsewhere) by unsustainable and environmentally damaging logging in 'less developed' parts of the Pacific Basin. When considering the causes and potential solutions to such practices in these places, it is thus important to look beyond national boundaries and address the international pressures involved (Ofreneo, 1993).

Perhaps more than any other constituent region of the Pacific Basin, the environmental cost of

Table 15.6   Dust deposition in China, Japan and the North Pacific Ocean
After Inoue and Naruse (1991)

Place	Transport paths	Rate of dust deposition (mm/1000 years)
China		
Lanzhou	Jet stream	260
Luochuan	Jet stream	70
Beijing	Jet stream	100
Japan		
Modern	Jet stream	3.6–7.1
Last Glacial Maximum	Jet stream	13.5–22.9
North Pacific Ocean		
50°N	Polar easterlies, jet stream	0.8
6–50°N	Jet stream, tradewinds	0.4–2.0
11°N	Tradewinds, jet stream	0.1–0.7

rapid industrialization during the 20th century has been highest in East Asia. Even today, when the effects of particular practices are well known, their continuation in parts of this region points to the overwhelming primacy of economic over environmental considerations. One of the best examples may be Taiwan where economic 'success' has been accompanied by an 'environmental nightmare' (Chi, 1994). Other countries seem destined to follow the same path. Pollution of land, air and sea is widespread. One of the best-known examples of the latter was associated with mercury pollution of ocean waters and the fish from them which people ate at Minamata in Japan in the 1953–75 period (Smith and Smith, 1975).

Mining practices have also been of concern, particularly in the Philippines where, until very recently, the environmental consequences of mining were largely disregarded (Howard, 1993). Yet the escalating environmental crisis in parts of the Philippines is attributable primarily to forest clearance. Although an excellent example is that of mangrove forest clearance for domestic fuelwood and to develop aquaculture, the logging of upland forests has proved a more profound cause of environmental change. In 1989, the World Bank estimated that 90% of the most valuable old-growth dipterocarp forests had been removed in just 30 years[12]. Most logging practices in the Philippines are intensive and, as is common in other 'developing' countries in the Pacific Basin, there is little state enforcement of environmental laws, and corruption encourages the continuance of unsustainable practices (Bautista, 1990).

Direct effects of increasing urban population densities in East Asian cities (and those elsewhere in the Pacific Basin) are exemplified by the problems associated with water supply, which have caused rapid subsidence in places (Table 15.7).

---

[12] Cruder though no less compelling figures indicate that in AD 1521 the archipelago was perhaps 90% forested; one hundred years ago this was down to 70%; at the end of World War II, 60%; 35% by the time Ferdinand Marcos won his second presidential election in 1969; and around 20% today (Broad, 1993)

Just as many of East Asia's social problems can be attributed to high population densities, so much of the current environmental crisis can be directly ascribed to the problems associated with producing enough food while at the same time 'developing' the economy. There is perhaps no better example of the various views on this dilemma than that provided by the 175 m high Three Gorges Dam being built across the Yangtze River. When operational, this will be the world's largest hydroelectricity generating project but there will be an immense environmental cost payable in the currency of flooding, changed river discharge quality and quantity, and altered channel and floodplain morphology.

15.10 LATE HOLOCENE ENVIRONMENTS OF AUSTRALASIA

15.10.1 The Early and Middle Late Holocene in New Guinea

Although New Guinea has a long Holocene history of human impact, one instance where this is unquestionably absent involves fluctuations of ice on the highest mountains such as Mount Jaya (Plate 15.7). The Last Glacial ice sheet is now reduced to a few small ice caps here but these extended significantly during the late Holocene and present a unique record of high-altitude tropical climate change in the western Pacific (Hope and Peterson, 1976). The earliest Holocene advance occurred some 3000 BP followed by others <2400 BP, 1800–1600? BP and <1300 BP. Fluctuations occurred at similar times on Mount Wilhelm (Hope, 1976).

The middle part of the late Holocene in New Guinea may have been associated with increased lowland precipitation – a carefully constrained record from Waigani Lake shows flooding 2500–1200 BP followed by a reduction in effective precipitation (Osborne et al., 1993). As noted in Chapter 13, an absence of human influence on the greater part of the Waigani record is clear, in contrast to most other such records from New Guinea. An example of the latter comes from Lake Wanum in the Markham Valley in eastern New

Table 15.7   Subsidence caused by groundwater extraction in parts of the Pacific Rim
From sources in Goudie (1995) and Wang (1998)

Location	Amount (m)	Dates of measurement (years AD)	Rate (mm/year)
East Asia			
Tokyo, Japan	4.0	1892–1972	500
Osaka, Japan	>2.8	1935–1972	76
Shanghai, China	0.65	1921–1948	24
	0.28	1949–1956	40
	0.44	1957–1961	110
	0.21	1962–1965	69
	0.07	1966–1992	2.5
Total	1.64	1921–1992	23
Niigata, Japan	>1.5	–	–
Tanggu Harbour, China	1.1	1975–1991	68
Tianjin, China	0.83	1975–1991	52
Elsewhere in Pacific Basin			
California, Central Valley	8.53	–	–
Mexico City	7.5	–	250–300

Guinea, where precipitation increase during the late Holocene is indicated by a higher lake level (Garrett-Jones, 1979). Doubts about the correctness of this interpretation arise because of the unknown role of people and volcanic eruptions in increasing lake-floor sedimentation and thus lake-surface level.

Continuing earlier trends, humans are thought to have effected great changes to highland environments of New Guinea during the late Holocene. A prominent change from clay to soil-aggregate deposition at Kuk Swamp around 2500 BP has been interpreted as marking the start of soil tillage in the surrounding grassland soils, probably in response to the progressive depletion of their nutrients. Subsequent erosion appears to have been systematically addressed around 1200 BP by planting *Casuarina* (Gillieson et al., 1987) which had the additional benefits of providing firewood and fuelwood in areas depleted of forest, and enriching the soil through nitrogen fixation.

The cultural complexity of the Sepik–Ramu area of New Guinea has long intrigued outsiders, and may ultimately be explained by environmental change (Swadling and Hope, 1992). A combination of uplift, sea-level fall and large

volumes of fluvial sediment being deposited around the mouths of rivers led to the emergence and progradation of this area during the late Holocene. This in turn made access between coastal and highland areas much easier and may have encouraged an interchange of ideas and products which found expression in the elaborate artefacts of the Sepik–Ramu area[13].

15.10.2   The Early and Middle Late Holocene in Australia

The beginning of the late Holocene in northeast Australia marked the onset of significantly cooler and drier conditions than during the middle Holocene, which have persisted until today. This change has been interpreted as causing the rainforest canopy to open up and sclerophyllic plants to expand their range (Markgraf et al., 1992). In many parts of northern Australia, an increase in rates of coastal dune accumulation

---

[13] Decreased cultural contact with Malay peoples may also have resulted from these environmental changes and contributed to the appearance of these distinctive indigenous cultural traits

Plate 15.7   One of the less familiar ice fields in the Pacific Basin is that capping Mount Jaya (Cartensz) in New Guinea
Cartensz Glaciers Expedition 1971

2700–1800 BP has been linked to a reduction in wet-season precipitation (Lees et al., 1990). Evidence for aridity 2600–1400 BP has been reported from the Atherton Tableland (Hiscock and Kershaw, 1992).

Farther south, conditions 3500–1800 BP in the alpine zone of Mount Kosciusko in New South Wales were cooler than during the middle Holocene and marked by increased slope instability. This connection is based on the proposition that temperature fall (~3°C) decreased vegetation cover making slopes more vulnerable to subaerial erosion, a condition perhaps aggravated by increased windiness (Martin, 1986). At Barrington Tops in New South Wales, the early part of the late Holocene was drier yet a slight expansion of *Nothofagus moorei* forest over

the last two millennia may signal increased moisture (Dodson et al., 1986).

There continues to be controversy over the course of late Holocene sea level along the east Australian coast (see also Chapter 14). It has been claimed that the most influential data – which show that sea level never exceeded its present level after 6000 BP – come from 'anomalous' parts of the coast, and that a more representative scenario is one involving a higher-than-present sea-level stand, extending into the late Holocene (Bryant, 1992). Good examples of the evidence for the latter come from pollen analyses of Minnamurra estuarine sediments, showing that sea level was 0.8–2.3 m higher before 2500 BP; studies of fossil tube worms and barnacles at Nambucca Heads, which demonstrate that sea level was >1 m higher 3420–1800 BP (Flood and Frankel, 1989); and studies of emerged beach deposits which show that sea level was over 1 m higher 6000–1500 BP (Bryant et al., 1992; Baker and Haworth, 1997).

The late Holocene sea-level record around Australia bears signs of having been distorted by tectonic movements. Greater uplift in the south has elevated indicators of the Holocene sea-level maximum there while subsidence in the north has depressed their levels there (Figure 15.9A; see also Chapter 14). More recently, continued uplift in the south has reduced the observed rate of sea-level rise, whereas subsidence in the north has amplified this rate (Figure 15.9B).

15.10.3 The Early and Middle Late Holocene in New Zealand and New Caledonia

Late Holocene cooling of New Zealand began ~2500 BP. Its effects are clearest in South Island, where there was an expansion of forests of *Nothofagus*, a hardy species able to cope with

low temperatures and strong winds. These changes were associated with a northward shift of the belt of westerly wind flow which accounts for most precipitation on South Island today. Increasing wind strengths during the late Holocene, particularly along the west coasts, led to drier conditions in rain-shadow areas which experienced an increased incidence of forest fires as a consequence (Salinger, 1988). In North Island, a less conspicuous increase in *Nothofagus* was accompanied by a decline in *kauri* (*Agathis australis*) indicating increasing aridity during the late Holocene (Dodson et al., 1988). At one site in the far north, vegetation signifying warm, moist conditions dominated from ~3400 BP to at least 2000 BP (Elliott et al., 1995).

The late Holocene expansion of *Nothofagus* forests, not conducive to the activities of browsing herbivorous birds such as the *moa* (mostly of the family Dinornithidae), may have led to a decline in the numbers of such birds before humans arrived (Nunn, 1994b). Humans are nonetheless

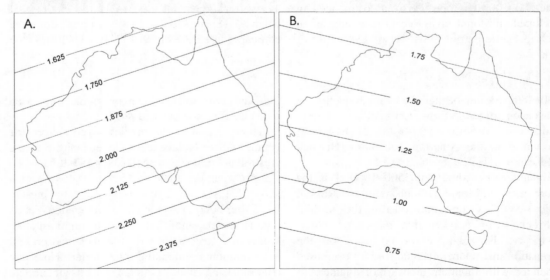

Figure 15.9   Trend-surface maps of Australian sea levels. (A) The maximum heights of sea level (relative to present mean sea level) during the Holocene suggests that subsequent uplift has been greatest in the south of the continent, including Tasmania. (B) Modern rates of sea-level rise around Australia. These are greatest in the north, suggesting that here sea-level rise is amplified by subsidence, whereas in the south its effects are reduced by uplift Reprinted from *Marine Geology*, 108, Bryant, E. 'Last interglacial and Holocene trends in sea-level maxima around Australia: implications for modern rates', 209–217. Copyright 1992, with permission from Elsevier Science

clearly implicated in the ultimate extinction of *moa* (Anderson, 1989).

As discussed in Chapter 6, the uncommonly high levels of toxins in the soils of the main island of New Caledonia gave rise to a series of unique biotic adaptations (see also Plate 11.1). These included a giant megapode (bird) which apparently became extinct shortly after humans arrived on New Caledonia some 3200 BP, and a pygmy crocodile[14] which disappeared about 1800 years ago (Flannery, 1994). Humans may be implicated in these extinctions, yet it has also been found that pre-human late Holocene sedimentation rates on the main island of New Caledonia exceeded those during the island's early post-human history (Stevenson and Dodson, 1995), once again emphasizing the need for prudence when interpreting such supposed proxies of human impact.

15.10.4   The Last Millennium in New Guinea

On Mount Wilhelm, the Little Ice Age was marked by a lowering of snowlines by ~200 m some 500–200 BP (Hope, 1976); a 26 km^2 ice cap developed on Mount Jaya. Recent recession of ice here has been interpreted as signalling the end of the Little Ice Age, which saw the ice cap shrink to about 2.5 km^2 today (see Plate 15.7).

Tephra have proved useful stratigraphic markers in sediment sequences of the New Guinea highlands, and have been used to provide comparisons of sedimentation (soil-loss proxy) rates in contrasting situations within the last millennium. The widespread Tibito tephra, for example, distinguished from others by its Rb/Sr character, was dated to around 300 BP by Russell Blong (1982). A conspicuous increase in swamp and rockshelter sedimentation rates after Tibito deposition has been interpreted as signalling increased soil erosion and forest clearance associated with a shift from swamp cultivation to the dryland cropping of the newly introduced

sweet potato (*Ipomoea batatas*; Gillieson et al., 1986). This crop was favoured as a staple by highlanders because of its resistance to frost and drought, and its short growing time compared with *taro*, the previously most common staple, particularly at high elevations. In addition, as William Clarke (1977) described, land on which *taro* could no longer be grown, because of depleted soil fertility, was brought back into production for the less demanding sweet potato. Like cassava (or manioc – *Manihot esculenta*) in many Pacific islands (see below), the spread of sweet potato through the New Guinea highlands revolutionized agricultural practices and, in many places, caused increases in the pace of environmental change; data from the Simbu area, for example, show that rates of hillslope erosion increased from less than 0.05 mm/year to as much as 1.8 mm/year as a result of the introduction of the sweet potato (Gillieson et al., 1986).

Of the agricultural innovations associated with European arrival in New Guinea, the metal axe brought 'revolutionary changes in quantitative and qualitative relations between people and land, and as well, significant shifts in the ecological balance' (Spate, 1953: 170) resulting from the comparative ease with which trees could now be felled, and thus forest cleared for planting. Changes occurred in the Pacific Islands for similar reasons (see below).

Although records of particularly damaging frosts and prolonged droughts are preserved in their oral histories, the vulnerability of highland populations to such events has risen recently as a result of increased population densities.

15.10.5   The Last Millennium in Australia

Late Holocene aridity is conspicuous in most parts of Pacific Australia yet, by the start of the last millennium, conditions became wetter than today (Harrison and Dodson, 1993). Details of climate variations during this period were gleaned from dendrochronological studies of the climatically sensitive and long-lived Huon pine (*Lagarostrobos franklinii*) in Tasmania. The Little Climatic Optimum may be reflected by the above-

---

[14] Uniquely among crocodiles, this one possessed blunt teeth to crush snails and clams to supplement its diet on resource-poor New Caledonia

average temperatures of the periods AD 940–1000 and AD 1100–1200 yet these were split by an 'anomalously cold' period in the 11th century. Evidence for the Little Ice Age is much less clear, perhaps as a result of the moderating influence of the oceans on Tasmanian climate. Of particular interest in the Tasmanian data is the observation that the most extreme temperatures of the last millennium occurred within the past century or so: the coldest conditions occurred AD 1890–1914, the warmest AD 1965–1989 (Cook et al., 1992).

Little has been discerned about the impact of Aborigines on the environments of eastern Australia prior to European settlement although their use of fire, discussed in a separate section below, may have been an important agent of change, at least locally. Environmental changes during the last millennium may have affected human lifestyles more but, again, low population densities mean that the nature of such effects remains largely conjectural. Some of the most marked such effects may have been along the coastline. For example, an important element in the subsistence economy of the inhabitants of Princess Charlotte Bay were mud-dwelling shellfish, particularly *Anadara granosa*, yet these disappeared abruptly around 600 BP, perhaps because of rapid sea-level fall (see Figure 15.10), possibly (also) overexploitation, and the people of the area began to forage on nearby reefs and to hunt dugong (Beaton, 1985).

The arrival of Europeans in Australia led to a conspicuous increase in the pace of environmental change in many areas. This impact stemmed from the naive yet still widespread assumption that the ecosystems of Australia were as European ones were, a fallacy which has led to the environmental crisis of modern Australia (Flannery, 1994). Most early European settlers assumed Australia to be an empty land – they paid scant attention to Aboriginal lifestyles – and altered its environments by replacing native with imported species of vegetation, particularly for broad-acre farming, and by introducing a range of animals for ranching (sheep, cattle), for hunting (rabbit, fox, deer), accidentally (mouse, rat), as food (pig, goat,

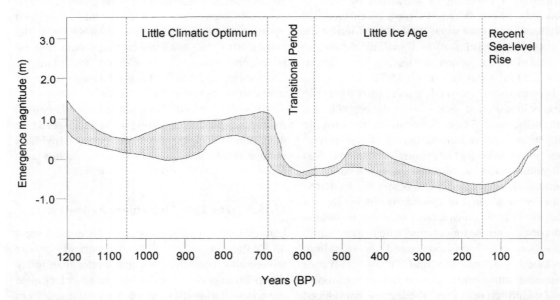

Figure 15.10   Sea-level changes around Pacific Islands over the past 1200 years cannot be identified precisely so are represented here as an envelope. Sea-level changes over this period are thought to have followed closely changes in temperature, which probably drove them. In particular, the rapid sea-level fall about 650 BP (~AD 1300) is thought to have been a response to a rapid temperature fall which had far-reaching effects in the Pacific Basin
From Nunn (1998e)

buffalo) or as draught animals (horse, donkey, camel), some of which became wild and are now having impacts on areas far from population centres (Adamson and Fox, 1982).

A good example of how European settlement transformed east Australian vegetation comes from the Monaro Plains in southern New South Wales. In early colonial times, the adjacent high country fringing the Snowy Mountains to the west was used for grazing mostly sheep, particularly during summer when droughts often affected the plains. Summer grazing in the high country coincided with the appearance of less palatable herbs so during the 1860s farmers introduced rye grasses, clovers, lucerne, prairie grass and cocksfoot which spread throughout this zone and across the Monaro Plains. This combined with regular burning of vegetation has brought about a wholesale transformation of the flora, even in an area such as this which was of only marginal interest to Europeans (Dodson et al., 1994).

The question of the use of fire by early human (generally pre-European) settlers of many parts of the Pacific is controversial. In Australia, the Aborigines have been portrayed alternatively as either cautious custodians or desperate despoilers of the environment through their use of fire. The issue is reviewed below.

15.10.6 Fire Use by Native Australians

The evidence of fires having occurred in the Australian landscape during the late Holocene (and earlier) is irrefutable. Charcoal has been found routinely in sediment cores and generally interpreted as indicating the presence of humans in the environment. Yet manifestly fires can start without humans – witness the situation in the Amazon (Saldarriaga and West, 1986) and in New Zealand, New Caledonia and Pacific Islands discussed in Chapter 14 – and it is irrational to insist that charcoal signifies a human-started fire (rather than another kind) without other compelling evidence. The debate thus hinges on this other evidence, which comes largely from studies of vegetation, climate and human lifestyles.

Many native plants which dominate(d) Australian ecosystems are those for which periodic burning signals rejuvenation rather than ruination. And there is good evidence that many such ecosystems developed long before humans could have entered the picture (Flannery, 1994). Yet there is also evidence that for at least some of the time Aborigines have occupied Australia, they used fire as part of a system of environmental management which altered pre-existing divisions between areas of fire-loving and fire-sensitive plants. The problem is how to separate changes between these divisions caused by human actions from those caused by natural climate variations.

There is abundant evidence that climate parameters, particularly precipitation, varied throughout the period of human occupation of eastern Australia. Yet, upon being confronted with a similarity in age with the first movements of people into a particular area and accelerated sedimentation in associated rockshelters and swamps, most archaeologists have opted for a simple *a priori* explanation. This may be correct yet fails to acknowledge the possibility that humans moved into the area *because* the climate had become moister and that accelerated sedimentation may have been the (principal) consequence of this, not human activity. In support of this view is recent thinking that, contrary to earlier opinions, Australia's Aboriginal society did undergo major transformations during the Holocene, probably driven by climate change. Such transformations were expressed by increasing food productivity and, less mundanely, by changed art forms and styles which refute 'the idea of Aboriginal society as conservative and basically unchanging over the last 40,000 or so years' (Bowdler, 1993: 132).

An understanding of the use of fire by native Australians, along with the earliest human colonizers of many other parts of the Pacific Basin, has commonly been premised on the belief that observations by Europeans within the last couple of hundred years or so are equally applicable to the entire period of pre-European settlement, which may be in excess of 100 000 years for Australia. There is no clear support for this extrapolation. Indeed, it has been argued that vegetation burning was considered undesirable by many Aboriginal groups because such events reduced food resources, particularly the supply of small mammals (Horton, 1982). There are opposing views and it seems likely that both may be valid for particular parts of the continent, perhaps only at particular times.

Along the Pacific Rim of Australia, owing to the more maritime climate, it seems less likely than elsewhere, particularly in the presently drier parts of the continent, that there was a major pre-European human impact on vegetation during the late Holocene (Head, 1989).

15.10.7   The Last Millennium in New Zealand and New Caledonia

An interesting study of late Holocene environmental change in New Zealand comes from Chatham Island, the largest in a small archipelago some 900 km east of South Island. Sand dunes form a belt around the north and east coasts and several distinct episodes of dune construction have been identified. Although the times of these correlate well with those on the New Zealand mainland (Table 15.8), there are important differences in cause. On the mainland, river sediment yield is important and periods of dune formation may thus signify times of increased precipitation and upland soil erosion. On Chatham Island, neither river sediment yield nor human influence have been important; dune-building episodes here were probably initiated by coastal erosion during storms (McFadgen, 1994).

From the information in Table 15.8, it is worth noting that an unstable phase predated (by several hundred years) and straddled the time of initial human arrival (~825 BP) in New Zealand implying that the environments which its first human settlers occupied were already under stress and thus not all the subsequent environmental changes should be attributed uncritically to human impact. Exactly how this stress was manifested is uncertain but it seems correct to suppose that immediately prior to human settlement, New Zealand was covered largely by rainforests dominated by podocarps and *Nothofagus*. The separation of subsequent human and non-human influences on its environment, particularly prior to the arrival of Europeans, has proved controversial and is discussed in the following section.

Evidence for the Little Ice Age in New Zealand comes from measurements of $\delta^{18}O$ from a stalagmite in a cave in northernmost South Island (Wilson et al., 1979). Yet a precise match between cooler summer temperatures, derived from tree-ring data, and times of ice expansion has proved elusive (Norton et al., 1989), perhaps because precipitation rather than temperature was occasionally the dominant control of this.

Recent warming in New Zealand has been marked by a conspicuous recession of glaciers. Since the 1850s, an overall rise of temperature comparable to that recorded for the same period in most other parts of the world has been monitored (Salinger, 1988).

The arrival of Europeans in New Caledonia had effects on its environments comparable to those

Table 15.8   Dates of dune-building episodes (unstable phases) in New Zealand and on Chatham Island
After McFadgen (1994)

New Zealand	Date (years BP)	Chatham Island	Date (years BP)
Unstable phase	150–0	Unstable phase	150–0
Stable phase	440–150	Stable phase	400–150
Unstable phase	450–440	Unstable phase	450–400
Stable phase	570–450	Stable phase	550–440
Unstable phase	1850–570	Unstable phase	1850–550

experienced in New Zealand (see below), with cattle being preferred to sheep ranching. Another important difference has been the exploitation of the nickel reserves on New Caledonia's main island. Most mining has been opencast so, in addition to the degradation of the stripped areas of mined uplands, huge volumes[15] of waste have been washed downslope filling lowland valleys and causing coastlines to prograde (Dupon, 1986).

15.10.8 Pre-European Human Impact in New Zealand

New Zealand is the place in the Pacific Basin where the debate about the principal cause of last-millennium, post-settlement environmental change has raged more fiercely than perhaps anywhere else. The reasons are many and include the islands' comparatively short history of human occupation, the conspicuous extirpation of flightless *moa* early in this period, and the prejudices of vocal groups who have been quick to 'blame' others for what they perceive as disastrous environmental changes of unprecedented proportions which could have been brought about only by humans. Following the lead of Holloway (1964), who railed in vain against the dominance of anthropogenic explanations for environmental change in the post-settlement era, there has been considerable work in the past decade or so focused on non-human causes of environmental change. Much of this work has shown that climate variations within the post-settlement era caused significant vegetation and landscape change; a good example is the Matawhero erosional period in the Ruahine Ranges which is now interpreted as resulting from an unusually stormy period in the 17th century which devastated extensive areas of forest (Pillans et al., 1992).

Recognition that natural changes during the late Holocene 'have been relatively minor' (McGlone et al., 1993: 311) compared to those of the earlier Holocene has encouraged a refocusing by New

Zealand scientists on longer-term changes rather than those of the islands' short post-settlement history which preoccupied scientists of an earlier generation. This in turn has enhanced the understanding of post-settlement changes. For example, discoveries that charcoal formed during widespread forest fires long before people arrived means that such fires may also have occurred after people arrived although, given the abrupt rise in forest-fire frequency following human arrival, it is reasonable to link the two events (Anderson and McGlone, 1992).

For whatever combination of factors, deforestation began in many parts of New Zealand soon after initial human colonization. Much of the forest was replaced by bracken which, after continued burning[16] and consequent depletion of soil nutrients, was itself replaced by grassland in places. There is also abundant evidence for increases in soil erosion following human arrival but much of this, particularly in hilly areas, cannot be demonstrated as having been the direct outcome of human activities.

The once-popular idea that the pre-European populations of New Zealand were essentially hunters and gatherers until an understanding of sweet potato (*Ipomoea batatas*) cultivation and winter storage in pits developed, has been challenged following investigations of the hitherto little-studied Maori horticulture. It is now clear that the Maori adapted the landscape by building terraces, stone walls and irrigation/drainage channels in many places to grow a variety of crops including cabbage trees (*Cordyline* spp.), *taro* and yams among others (Bulmer, 1989). What such discoveries mean for the understanding of human impact on the environments of New Zealand within the last ~800 years is unclear.

What is clear is that while Maori settlers may have adapted parts of the New Zealand landscape as their needs dictated, it has not been demonstrated that they were responsible for most

---

[15] In less than one hundred years, 110 million tonnes of nickel ore have been extracted. During this process, about five times as much waste by weight was created. The volume of this is at least 220–280 million m^3

[16] This burning may have been carried out deliberately to promote the spread and growth of the edible bracken fern *Pteridium esculentum*

of the change it experienced during the period between initial human settlement and European arrival. The effects of European settlement are better documented although not wholly free from controversy. The spread of exotic grasslands, commonly at the expense of forest in North Island and of native grasslands in South Island, was driven largely by the development of rangelands, particularly for sheep grazing. Yet it is possible that this process was facilitated in places by ecosystem stress associated with recent warming and by other processes, such as the abandonment of traditional land-management practices, the impacts of introduced (often feral) animals, and even the effects of volcanic eruptions (see Plate 15.21), which cannot be construed as direct human impact.

15.10.9   Contemporary Environmental Issues in Australasia

New Guinea is host to a range of environmental problems, ranging from those associated with mining in the wettest areas to those arising from newly introduced agricultural techniques. Gold mines are operating and planned in parts of the upper Fly River catchment where 'natural' erosion from rainfall is lowering the ground surface at rates of 3–4 mm/year – among the highest in the world (Pickup et al., 1980). Recent agricultural innovation in the New Guinea highlands, described by Eric Waddell (1972), has included the building of composted-soil mounds which allow almost continuous (rather than seasonal) cultivation. Yet this technique works best and appears sustainable only on flat lands. Where, in response to increased demand, it has been extended to upland slopes, rapid losses of soil nutrients have resulted in falling crop yields; a consequence of this is that more upland forest is being cleared to create more such mounds on these slopes which extends the environmental problems associated with this technique.

Most other parts of Australasia are experiencing problems associated with effective environmental management deriving from the imposition of alien ecological understanding

dating from the arrival of the first Europeans. Australia in particular still needs to come to terms with being a land in which cycles of change are controlled more by ENSO than by annual (seasonal) changes, a lesson its Aboriginal inhabitants learnt generations back (Flannery, 1994). Like many Pacific Islands and those in the Philippines, New Caledonia is wracked by environmental mismanagement which is increasing socio-economic problems (Dupon, 1986; David, 1994). Perhaps only New Zealand can hold its head high among its neighbours for its prescient and far-reaching legislative framework for environmental conservation, which is gradually starting to reverse many of the deleterious effects inherited from former times.

15.11   LATE HOLOCENE ENVIRONMENTS OF THE PACIFIC ISLANDS

15.11.1   The Early and Middle Late Holocene in the Pacific Islands

Many offshore islands in Papua New Guinea became colonized during the later Holocene by an Austronesian-speaking people who had different lifestyles (and consequently different environmental impacts) to the Papuan speakers who had occupied the area for several tens of thousands of years (see Table 12.2). The newcomers were the first representatives in the region of the Lapita people, discussed above (section 15.2.4). A good example of their possible environmental impact comes from the Arawe islands off southern New Britain where they constructed stilt houses out over the sea. Gosden and Webb (1994) suggested that sand accumulated around the bases of offshore house posts and that, when sea level fell during the late Holocene, the area immediately landward of these sand barriers became filled with clay washed off inland gardens. This resulted in an extension of the shoreline greater than that which might have resulted from sea-level fall alone. The authors see these 'people shaping the landscape in response to social need' (Gosden and Webb, 1994: 49) although they stop short of suggesting that the observed changes were deliberately engineered. Also, it would appear unwise to undervalue the contribution of late Holocene sea-level fall, the effects of which on other Pacific island coastlines not (significantly) affected by humans were comparable (Nunn, 1998a).

Neither remote Lord Howe Island in the southwest Pacific nor the Galápagos group in its central eastern part were settled until around 150 years ago, so the sedimentary and vegetational record of increasing precipitation found in both locations during the late Holocene can be interpreted in solely climatic terms (Colinvaux, 1972; Dodson, 1982). On most other Pacific Islands, humans were present for much of the late Holocene so contemporary climate changes are more difficult to identify, not least because it is uncertain how far one can reasonably extend conclusions from islands (like Lord Howe and the Galápagos) where humans were absent.

Humans occupied most of the main island groups in the west Pacific by the start of the late Holocene, settling most of the rest and probably reaching Central America subsequently. It has been suggested that grassland (Plate 15.8) developed on many islands only after humans arrived and allegedly burned the forests which apparently existed. This scenario has been questioned for several reasons, and it seems more likely that these grasslands originated long before humans arrived, perhaps in response to aridity around the Last Glacial Maximum (Nunn, 1997a). Yet there is no reason to suppose that Pacific Island grasslands did not spread as the result of periodic fires, including those lit deliberately by humans, particularly during the post-settlement history of Pacific islands. For Mangaia in the Cook Islands – the history of which was probably typical of many in the central Pacific – Joanna Ellison (1994) suggested that the decline in forest and spread of the fern *Dicranopteris* about 1650 BP marked the start of systematic human disturbance of the ecosystem.[17]

The early human settlers of many Pacific islands did undoubtedly cause many changes to their vegetation and landscape, but it is important not to overstate the extent of these simply because of an absence of information to the contrary. For example, people are clearly implicated in the loss of *Pritchardia* forest on Oahu island in Hawaii (Athens and Ward, 1993) and in the decimation of many Pacific bird populations (Steadman, 1989) but were not responsible for the shoreline progradation and aggradation which had transformed the coast of Aitutaki island in the Cook Islands by AD 900–1200 (Allen, 1998).

Aside from vegetation change, the dependability of other potential indicators of early human impact in the Pacific islands has been questioned from time to time. The significance of charcoal (see above) is hotly debated, some authors regarding its sustained presence in sediments as indicative of human burning, even though artefactual evidence may be absent (Kirch and Ellison, 1994), others approaching the topic more circumspectly (Spriggs and Anderson, 1993). Lowland sedimentation, particularly infilling of middle Holocene swamps, is another observation which has been interpreted widely in the Pacific Islands as an outcome of human disturbance of upland areas. Yet many such changes can be explained equally well by the effects of falling late Holocene sea level (and uplift); human impact is often unnecessary (Nunn 1994a).

Independent evidence for late Holocene sea-level fall is widespread in the Pacific (Pirazzoli and Montaggioni, 1988; Nunn, 1990a, 1994a, 1995). This evidence includes middle Holocene emerged reefs, beaches, benches and notches at levels typically 1.5–2.0 m above their modern equivalents which were abandoned as sea level fell (see Figure 14.3). There are also many signs of significant late Holocene tectonic change, discussed towards the end of this chapter, which contributed to coastal changes on Pacific Islands.

---

[17] While plausible in this example, it is important not to forget that such ideas are inferential. Pre-human vegetation disturbances have been noted on Mangaia and attributed to the effects of a severe El Niño event (Chapter 14); on New Caledonia, temporary collapses of *Nothofagus* forests in the absence of humans were ascribed to disease or the replacement of even-aged stands (Hope, 1996)

Plate 15.8 Changing land use in Pacific Island grasslands illustrated by a view of the upper Ba valley, Vitilevu island, Fiji. The grasslands which cover most slopes in leeward areas of Pacific Islands in the tradewind belts have been widely regarded as anthropogenic, but there is increasing evidence that such grasslands may have persisted in these places since the Last Glacial Maximum when the climate was generally more arid than today. Many grassland areas are now being planted with pine (as shown on the left), a process which is leading to depleted nutrient levels in nearby valley soils and reductions in stream flow which have serious consequences for agriculture in these places

15.11.2 The Last Millennium in the Pacific Islands

The temporal pattern of long-distance human colonization of Pacific islands (see Table 12.2) provides an unusual proxy of climate change. Much of the longest-distance colonization was accomplished during the Little Climatic Optimum when reduced storm/cyclonic activity, persistent tradewinds, and clear skies important for navigation may have prevailed (Bridgman, 1983). Anomalous westerly winds may have been more common, maybe propelling vessels unexpectedly eastwards (Finney, 1985). Yet during the Little Ice Age, successful long-distance voyaging ceased. Even interisland contact within archipelagoes like the Marquesas ceased abruptly, perhaps in large part because of deteriorating weather conditions associated with a cooler climate (Nunn, 1997c); a record of increased tropical-cyclone frequency

450–250 BP from Tahiti (Flenley et al., 1991) may be an indicator of this.

The comparatively low-amplitude climate changes of the last 1200 years or so in the Pacific were associated with similarly low-amplitude changes in sea level, the record of which can be subdivided into four (Figure 15.10). Sea-level rise during the Little Climatic Optimum culminated around 700 BP in a sea level that may have stood almost 1 m higher than today in parts of the Pacific. There followed a rapid fall which saw sea level dropping below its present mean level at the start of the Little Ice Age. Within this period there were minor fluctuations but lasting change did not begin until around 150 BP when sea level began to rise, a trend which has been monitored in the last century (Nunn, 1997c, 1999d).

Pre-European settlers on many Pacific Islands made substantial changes to their environments in places. Terraces were cut into hillsides otherwise too steep for planting crops without unacceptable soil loss. Artificial channels were dug to distribute water more effectively for both agricultural and domestic purposes. And huge creations, whose meaning is essentially lost to us today, were fashioned using crude implements. Perhaps the most famous of these are the statues (*moai*) of Easter Island, which were carved, transported and erected kilometres away during the period AD 1400–1680. Elsewhere, huge amounts of fill were transported by hand to places where it was required to create or extend land (Plate 15.9). The wholly artificial islands of the Langa Langa lagoon off Malaita in Solomon Islands are examples of the ingenuity and effort involved in such projects, which have also been described from Fiji (Nunn, 1999a).

On many Pacific Islands, the Little Climatic Optimum was evidently a time of plenty. Most settlements were coastal and unfortified. In response to an increasingly dry warm climate in the low-latitude Pacific Ocean, water-conservatory strategies, notably agricultural terracing, appeared on many islands. Then, around AD 1300, there was a rapid climate change which devastated the economies of many islands, resulted in intertribal conflict, and drove people from coasts and river valleys into fortified hilltop settlements. The extent of this event and its significance are discussed in a separate section near the end of this chapter. On many islands, the fallout of this event lasted throughout the Little Ice Age and beyond, many hilltop settlements not being abandoned until the arrival of Christian

Plate 15.9   Off the east coast of Kosrae island in Micronesia lies the 70 ha island Lelu. The east side of Lelu (on the right in the photo) consists of the volcanic hill Finol Poro, while the remainder (some 27 ha) is entirely artificial. Beginning around 750 years ago, lagoonal sediments were piled onto the reef flat to create this island, which became the centre of an influential regional chiefdom
J. Stephen Athens

missionaries in the 19th century. The deterioration of weather conditions during the Little Ice Age exacerbated the problems of resource depletion which humans endured as a consequence of the ~AD 1300 event.

Wittingly or not, accompanying humans on many interisland journeys were a number of plants and animals which have had distinct recognizable effects on the environments of those Pacific Islands to which they were introduced. A particularly intriguing example is the explanation given by John Flenley (1993) for the disappearance of an extinct palm (*Jubaea* sp.) from Easter Island. Carbonized roots and nuts of this palm have been discovered on Easter Island. Teeth marks on the nuts have been identified as those of the small Polynesian rat (*Rattus concolor*) whose bones have also been unearthed on the island. Flenley suggests that it was this rat, rather than the people who inadvertently introduced it to Easter Island, that was responsible for the major drop in numbers of this unique *Jubaea*, the decline of which was carried to extinction by sheep and goats introduced in the last 200 years.

As elsewhere in the Pacific Basin, plants and animals introduced to Pacific Islands by humans have had major effects on native floras and faunas. Weeds, such as crab grass (*Digitaria setigera*), were inadvertently spread throughout the Pacific Islands by their first settlers through spores in the guts of pigs or in the mud packed around corms of *taro* to keep it viable on long voyages (Kirch, 1982). The problems caused by various guavas (*Psidium* spp.), introduced by Europeans to Tubuai island in French Polynesia, are typical of such species 'which have the ability to crowd out native vegetation with their monospecific stands' (Paulay, 1994: 140). Many species of rat reached plague proportions shortly after their introduction to certain Pacific Islands, and may have been largely responsible for the extinction of many native bird species. A similar consequence resulted from the accidental introduction of the snake *Boiga irregularis* to Guam (Savidge, 1987). Owing to their comparative vulnerability to predators, the land snails of many Pacific Islands have sustained some of the most dramatic losses

since human occupation, a process which is still continuing; only 25–35% of the 1461 species of Hawaiian land snail which existed at the time of initial human colonization about 1300 BP have survived (Solem, 1990). Reviews of the topic of the impacts of exotic fauna and flora on Pacific Island ecosystems were given by Steadman (1989), Nunn (1994a) and Paulay (1994).

It was not only the introduced biota which caused an acceleration of environmental changes on many islands but also the introduction of new techniques of land management. As in New Guinea (see above), the arrival of the metal axe and, later, the cane knife (machete) led to more efficient and faster vegetation clearance; such introductions were primarily responsible for the decimation of mangrove forests along many Pacific Island coasts during the 20th century. As for Australia (see above), many new techniques failed to consider how the Pacific Island environment might differ from those for which they had been originally developed, and this led to some serious problems; soil erosion, for example, came to affect large areas of Tahiti and Chuuk (Truk) following the spread of new methods of shifting agriculture from East Asia designed to raise crop yields.

Few reliable indicators of last-millennium palaeoclimates are available for the Pacific Islands. On Nuku Hiva in the Marquesas, there is a suggestion of cooler wetter conditions some 500 BP (Sabels, 1966). Data quoted by Sanchez and Kutzbach (1974) suggest that the Galápagos Islands were also wetter 300–100 BP than either before or after. The increasing use of oxygen-isotope ratios from recent corals to determine palaeotemperatures is leading to a much clearer picture of climate changes during the past few hundred years. For example, work on Galápagos corals allowed the recognition of El Niño events that had been missing from records compiled from historical sources, and provided a temperature record from AD 1586 to AD 1953 (Dunbar et al., 1994).

Few precise palaeotemperature series are available for the Pacific Islands during the period of recent warming shown in Figure 15.10; that for sea-surface temperatures from Santo island in Vanuatu (Quinn et al., 1993) could conceivably be

Table 15.9   Rates of sea-level rise in the Pacific
After Nunn (1997c) and sources therein

Site, island group	Rate of sea-level rise (mm/year)	Comments
Honolulu, Oahu, Hawaii	1.5	Island effectively stable
Hilo, Hawai'i, Hawaii	3.8	Island sinking
Kwajalein, Marshall Islands	0.9	Island stable, short record
Nalato, Vitilevu, Fiji	2.5	Site stable
Pago Pago, Tutuila, American Samoa	1.4	Island effectively stable
Suva, Vitilevu, Fiji	No trend	Site unstable
Chuuk, Federated States of Micronesia	0.6	Island stable
Wellington, North Island, New Zealand	1.6	Tectonic component subtracted
Pacific mean	1.0	
Global mean	1.4	

interpreted as a rising trend[18] compatible with directly monitored series within the last hundred years which show a rise of around 0.6°C/100 years (Nunn, 1997c). It is plausible to suppose temperature rise to have been the principal cause of the sea-level rise and, less certainly, the rise in tropical-cyclone (hurricane) frequency, over a similar time period (Nunn, 1994a, 1997c). Owing to their disproportionately long coastlines (compared to total land areas), the effects of recent sea-level rise have registered more extensively on Pacific Islands than elsewhere in the Pacific Basin. Many island coasts have been progressively inundated and/or eroded significantly in recent decades although it is not always easy to separate natural effects from those arising from human modifications such as mangrove clearance. Representative rates of what is considered natural sea-level rise are given in Table 15.9. Variations which cannot be readily attributed to tectonic influences are likely to be the product of the varying degree of precision of the methods used in different places.

The arrival of Europeans[19] in the Pacific Islands led to many environmental changes

including the clearance of forest for plantation agriculture in places (see Plate 15.14). Yet European impact was less abrupt than elsewhere in the Pacific Basin both because their arrival was more gradual and because the archipelagic character of the region defied rapid mass colonization. Even today, it is mostly the centrally placed, often only the largest, islands which bear signs of environmental impact attributable directly to European settlement.

Europeans also introduced exotic plants and animals, the environmental impact of some of which were considerable. New diseases entered the Pacific Islands at the time of European contact and took a massive toll of indigenous peoples: the native population of Hawaii fell from about 300 000 in AD 1779 to just 55 000 in AD 1875 largely in consequence of fatalities from introduced diseases such as smallpox, measles and typhoid which the people had never experienced and to which their bodies had never developed any resistance (Schmitt, 1961). Such occurrences led to abandonment of many settlements and their associated food-production systems which, as in posthispanic Central America, then became vulnerable to invasion by exotic weeds resulting in the effective replacement of pre-European ecosystems on some islands by the end of the 19th century.

The upsurge of plantation agriculture on many islands was associated with an increase in forest

[18] Notwithstanding the notable cooling AD 1832–1866
[19] Including implicitly those people from other parts of the world whose lifestyles and demands on Pacific Islands generally differed markedly from those of indigenous Pacific Islanders

clearance in many island groups except, as is (was) the case on many islands remote from commercial centres, where coconuts are (were) the principal commercial crop.

Coming closer to the present, the Pacific War (1941–1945) took a toll of island environments which is still manifest on many. Direct changes arose from the occupation of islands by various armies, which not only were preceded by massive changes[20] but also resulted in unprecedented increases in the numbers of people which in turn produced unprecedented environmental stress (Plate 15.10); for example, on Puluwat Atoll ($\sim3\,km^2$) in the Caroline Islands, 6000 soldiers were living off the land in 1946 and had already consumed all the *taro* and all the pigs (quoted in Farrell, 1972). The detritus of battle is still visible on those islands which became sites of conflict and now, ironically, attracts overseas visitors who have become an important source of revenue to some island communities.

15.11.3   Contemporary Environmental Issues in the Pacific Ocean and Islands

Some environmental changes experienced on Pacific Islands in the last few decades are more likely to have been the result of contemporary non-human changes (particularly an increased tropical-cyclone frequency) than of direct human impact. This interpretation derives from repeated observations of landscape change occurring in the absence of significantly altered human impact (Plate 15.11). The frustration of local people at being blamed for environmental changes in which they did not consider themselves to be implicated was summed up by the Fijian man who told the writer, '*Keimami sa vakila na liga ni Kalou*[21]'.

---

[20] As much as 75% of the coconut palms – by far the dominant vegetation – were felled by the Japanese during their occupation of some islands in the Marshall Islands
[21] 'We feel the hand of God', paraphrased and incorporated into the title of the author's 1991 monograph (3rd edition, Nunn, 1997c). Similar sentiments were expressed to Simon (1997: 35) repeatedly in the degraded highlands of southern Mexico, '*porque la tierra ya no da*' ('because the earth no longer gives')

A range of non-human environmental changes is currently causing concern in the Pacific Islands. Among these, the effects of sea-level rise and temperature rise, of the increased frequency of tropical cyclones and other extreme events, including those linked to El Niño, are being aggravated by the heightened vulnerability of many island environments associated with their 'development' and with rapid population growth (Plate 15.12).

The most extreme impacts of humans on modern Pacific Island environments cannot be simply overlooked or dismissed however insignificant they may be with respect to the region's longer-term history on which this book focuses. The pace of 'development' in many small island countries has exacted a high environmental price.

Some of the most serious environmental problems in the Pacific Islands are associated with mining; one example comes from the island Bougainville in the southwest Pacific, where a massive copper mine opened at Panguna in April 1972 (Plate 15.13). In May 1989, the mine was closed because of the violent opposition of some island people to the widespread environmental pollution associated with its activities, and a catalogue of grievances concerning the distribution of benefits and compensation within Bougainville (Oliver, 1991). The mine has not reopened.

On many Pacific Islands, as in many 'developing countries' along the Pacific Rim, the gulf between theoretical and practical environmental management is wide and deep. Aside from the difficulties of effectively patrolling environmental practice in large archipelagoes where transportation and communication are often difficult at the best of times, many such problems result from a lack of national control over land, which is commonly both customarily and often legally under control of local inhabitants. This situation may be viewed as commendable when the contemporary plight of indigenous peoples elsewhere in the Pacific Basin is considered, yet it brings its own problems, not least of which is the vulnerability of many landowning groups to the persuasions of potential 'developers'. Throughout

Plate 15.10 The spread of the Second World War to the Pacific in 1941 led to massive impacts on some island environments. The abrupt arrival of huge numbers of people and armament on many atolls in the central western Pacific left scars on their environments which are still visible today. This view is of Kwajalein – the world's largest atoll – just after its occupation by United States forces in March 1944. Note the concentration of dwellings on what was previously unoccupied land, and the presence of the supply fleet offshore
National Archives (NWDNS-90-G-400949)

Plate 15.11   The upper photo shows the bridge (looking upstream) over the Wainimala river in interior Vitilevu island in Fiji. Constructed in 1979 to carry the transinsular road, this bridge was washed away in the flood which accompanied Tropical Cyclone Kina in January 1993. The lower photo shows the remains of the bridge (looking downstream) two months later. It has been claimed that the magnitude of this flood was increased by the effects of forest clearance in the upstream catchment but surveys have shown that such clearance had been negligible. It is more likely that the increased frequency of tropical cyclones, particularly in the 1980s and 1990s, was responsible (largely through landsliding) for filling river channels such as these and thereby amplifying flood levels

Plate 15.12   Few parts of the Pacific Basin have been immune from human impact over the last few centuries. The village of Lutu lies in the upper valley of the Wainimala river in the interior of Vitilevu island in Fiji. The village has grown and prospered since the transinsular road reached it in 1979. Improved access to markets has stimulated the clearance of more of the steep slopes around the village for the planting of crops like *taro* (*Colocasia esculenta*) and *yaqona* (*Piper methysticum*). A large dam upstream has regulated the discharge of the Wainimala leading to the almost permanent abandonment of channels which once carried water regularly

many Pacific Island groups, particularly in the southwest Pacific, unscrupulous developers have stripped islands of their forests and mineral resources, compensating only a fraction of their value to local people and/or governments who are left to manage and often subsist on severely degraded landscapes. It is surprising that so few parallels have been drawn between the comprehensive exploitation of Pacific 'phosphate islands' like Angaur, Banaba (Ocean Island), Makatea and Nauru during their colonial history and modern postcolonial stripping of forests on many islands, particularly in the larger archipelagoes of the southwest Pacific.

Mangrove forests continue to be cleared or polluted at alarming rates along Pacific Island coasts (David, 1994) but another casualty of 20th century development has been coral reefs. As in the Philippines, and to a lesser extent in Papua New Guinea and off northeast Australia, coral reef ecosystems throughout the Pacific Islands have become increasingly stressed in recent decades. Much of this stress is thought to have been associated with human impact; overexploitation of reef resources, and physical damage to reef structure are two common causes. But in recent years, it has become clear that stress to reef ecosystems is also coming from non-human sources. Among these are the incidents of coral death associated with prolonged above-sea exposure during low sea levels associated with El Niño, and those of comprehensive reef death through 'coral bleaching' resulting from high ocean-water temperatures (Nunn, 1997b).

Plate 15.13   The Panguna copper mine, Bougainville island, Papua New Guinea, in 1988, one year before it was
closed because of mounting protests from islanders
Douglas Oliver

Problems associated with mangroves and reefs have caught the attention of many Pacific Island leaders in the last decade who have become convinced of the heightened vulnerability of their island environments to future sea-level rise if the protective function of mangroves and reefs is reduced. Yet fine words have so far been translated into little action and even the implementation of plans designed to mitigate the effects of future sea-level rise (and climate change) in this region appear to have been left largely in the hands of regional or international organizations. Many short-term responses to problems associated with coastal land-use changes and sea-level rise have been ill-conceived and ineffective (Mimura and Nunn, 1998).

More insidious forms of environmental change are appearing on many islands, particularly as a consequence of economic, social or political pressure. One example is provided by cassava (manioc), a South American staple introduced to many island groups early in their colonial histories. Since cassava is so fast-growing and less demanding to grow than traditional staple crops, many communities have abandoned their former gardens and horticultural practices. Another example is provided by sugar cane, the main agricultural export of Fiji, now being grown unsustainably on marginal lands to take advantage of favourable though only temporary trade agreements (Plate 15.14). In one of the few quantitative assessments to have been made of this situation, William Clarke and John Morrison (1987) found that soil losses of 90 tonnes/hectare/year were occurring on 18–22° slopes under sugar cane. These rates are greatly in excess of the tropical soil-loss tolerance level of 13.5 tonnes/hectare/year and, if unchecked, will lead to widespread degradation of many upland canelands in Fiji within the next decade or so.

Although commonly perceived as a benefit, tourism is contributing to environmental change on many Pacific Islands both through the construction of the requisite infrastructure and through direct impacts by visitors on the environment, particularly coral reefs.

The question of how much (potential) environmental damage – broadly defined – is associated with nuclear weapons and their testing in the Pacific Islands is passionately debated. The French government, which conducted underground tests on atolls in French Polynesia well into the 1990s, claims the tests were perfectly safe; in which case, say the islanders, why not conduct them in France? The suspicion that recent tests may have damaged atoll structures, which are therefore at greater risk than is widely believed among scientists of having radioactive residues seep out into the ocean, derives from studies of long-term atoll evolution and reports of earlier tests and similar events, albeit mostly surficial not deep, particularly on atolls of the western Pacific which found themselves under the control of the United States at the end of the Second World War (Keating, 1999). The first test on Bikini Atoll was illustrated in Plate 2.5; many of its consequences, particularly on the human inhabitants of the area and the ecosystems on which they subsisted, were clearly unanticipated.

Much of the environmental change associated with such occurrences has been cultural rather than physical. The interfingering of the two becomes more complex as the present day is approached. Take the situation on the atoll Kwajalein (see Plate 15.10) in the Marshall Islands, for example, and that of its displaced inhabitants, recounted by Andrew Mitchell (1989: 212–213). Kwajalein

> is used by the US military as a target for intercontinental ballistic missiles fired from Vanderberg Airforce Base in California. The Marshallese who once lived in solitary splendour on the atoll now inhabit the worst slums in the Pacific, on the tiny islet of Ebeye in the giant atoll's ring. The lagoon opposite the settlement, which has a higher concentration of buildings than Hong Kong, has a bacteria count 25,000 times higher than the US Public Health Service demands. Beer cans and 'disposable' nappies float in the bays. Most of the native population now lives on US welfare; before, they needed only to harvest the land and fish, skills now largely abandoned.

The issue of how cultural environments have changed recently is beyond the scope of this book, although almost-deserted villages on outer islands

A

B

Plate 15.14   Sugar cane covers most of the drier lowland areas on the main islands of Fiji. Introduced when the islands were a British colony, sugar cane has spread into more marginal, upland areas during the last decade or so of post-colonial government as a consequence of the artificially high price paid for Fiji sugar on the world market under the Lomé Convention. Lowland sugar cane growing is currently under threat from the effects of sea-level rise, both direct inundation and increasing groundwater salinization, while upland cane growing is not sustainable in the long term and is already leading to degradation

in the tropical Pacific provide images of comparable poignancy to those of native people clustered on reserves of marginal land, of abandoned rural farmsteads, huge cheerless squatter settlements and deteriorating inner cities throughout the Pacific Basin.

The most severe oceanic changes have been concentrated along the ocean fringes of the Pacific

Rim, particularly near large urban–industrial complexes. Traditional marine ecosystems have been devastated in the Pacific off East Asia by industrial and other effluents and by the effects of oil spillage; periodic plankton blooms known as 'red tides', nine of which occurred in Tolo Harbour, Hong Kong in 1994, cause mass death of marine organisms. Other examples were given above. In many Pacific island waters, overfishing is rife, and much of the effluent reaching them is untreated.

15.12   SYNTHESIS

Given the variations in late Holocene palaeo-climate records from the Pacific Basin, even from within particular constituent regions, it is difficult to produce a meaningful synthesis, except in general terms. Indeed, various uncommonly detailed records reported above for particular times during the late Holocene hint at the kind of detailed synthesis which might be possible were many more records of comparable complexity available.

In the absence of these, the general picture of palaeoclimate change during the late Holocene in the Pacific Basin is summarized in Table 15.10. The overall cooling trend of the late Holocene is apparent in all parts of the region, this cooling

being accompanied most commonly by increasing moisture levels particularly along the eastern Pacific Rim. Elsewhere the record of changing precipitation is more variable except in Australasia and on some Pacific Islands where there are indications of increasing aridity during the late Holocene.

The cooling trend is no surprise, the result of the astronomical forcing of global climate. Moisture changes are less easy to explain although are likely to be linked along the east Pacific Rim to the strengthening of coast-parallel ocean currents (and related processes) resulting from increased equator–pole temperature gradients.

For the last millennium, the Little Climatic Optimum, the Little Ice Age, and the period of recent warming (and sea-level rise) are well marked in most parts of the Pacific Basin with the conspicuous exception of the Antarctic Peninsula–southernmost South America region.

Table 15.10   Generalized trends of late Holocene climate in the constituent regions of the Pacific Basin

Region	3 ka	2 ka	1 ka	0
Antarctica				
Pacific coast	cooling ...		warm ...	cool...warming
Antarctic Peninsula	cooling, dry ...		warm, moist...cool...warm...cooling	
South America				
South	cooling, moist ...		variable ...	
Central and north	cooling, moist		warm, dry... cool, wet...warming	
Central America	cooling, moist ...		variable.. warm, wet...cool ... warming	
North America				
United States	cooling, variable moisture ...		warm, dry... cool, wet... warming	
Canada	cooling, moist ...		(warm, dry) ...cool, wet ...warming	
Beringia	cooling, moist		(warm, dry) ...cool, wet ...warming	
East Asia				
Mainland	cooling (variable moisture) ...		warm, dry ... cool, wet... warming	
Islands	warm, moist ... (variable)		... warm, dry ... cool, wet.. warming	
Australasia				
New Guinea	cooling, dry ... cooling, wet (warm) ... cool, wet ...warm, dry			
Australia	cooling, dry ...		warm, wet... variable	
New Zealand	cooling ... drier...		warm, dry... cool	warm
Pacific Islands	cooling (variable moisture)		warm, dry ... cool, dry	warming

15.13   KEY ISSUES

Three issues of key importance to an understanding of late Holocene environmental change in the Pacific Basin are discussed in the three subsections below.

The first discusses El Niño, to the periodic influence of which anomalous, often extreme weather conditions throughout the Pacific Basin are becoming increasingly attributed. An appreciation of the role of El Niño in 20th century climate change has emerged only recently; its potential role in earlier changes is not yet well understood.

The second reviews evidence by region for a period of rapid climate change around AD 1300 which produced abrupt responses in many Pacific Basin environments. It seems clear that this was a regional phenomenon demanding a regional explanation rather than one explicable only by local-area changes.

The third discusses the question of whether latest Holocene environmental changes may have been influenced by variations in sunspot activity.

The final section of this chapter considers the influence of tectonics and volcanism on Pacific Basin environments during the entire Holocene, a topic deferred from the rest of Part E for largely organizational reasons.

15.13.1   Effects of El Niño on Pacific Basin Environments

El Niño (see also Chapter 1) is but one phase of the ENSO (El Niño–Southern Oscillation) phenomenon which presently recurs every three to seven years, centred around the southern hemisphere summer[22]. El Niño (negative-ENSO) events have had the most extreme effects on environmental change in the Pacific Basin and are the only type discussed here.

---

[22] The final draft of this book was produced in Seattle in the northwest United States during the 1997–1998 El Niño event. The winter was uncommonly mild here although the Pacific Rim farther south in California was battered by storm after storm; floods created havoc, cliffs collapsed, and a man called Al Nino was inundated with angry phone calls

An El Niño event off the Peru–Chile coast occurs when the southeast Pacific anticyclone weakens and this area is invaded by warm equatorial waters which shut down upwelling and cause a massive increase in precipitation throughout the commonly arid lands onshore. Major El Niño events have had severe impacts on coastal fisheries here which in turn often led to social disruption among the communities which depended on them; these range from reduced mollusc populations – the main source of protein for prehispanic lowland dwellers – to the collapse of the commercial anchoveta fishery in the 1970s (see above).

Teleconnections with El Niño occurrences in Pacific South America have been identified in most parts of the world but are especially well established in the low- and mid-latitude parts of the Pacific Basin. Winter precipitation increases in northern Mexico and the southwest United States, the northwest United States and western Canada are commonly drier, as is much of Australia, northern China, and most of the extra-equatorial Pacific Islands (summarized in Table 1.15). During El Niño events, equatorial Pacific Islands tend to experience anomalously high rainfall, which registers in the chemical composition of lagoonal corals (Cole et al., 1992), and which bring about uncommonly rapid environmental changes on islands in these places. On many equatorial atolls, for instance, these periods of prolonged heavy rainfall are associated with rapid terrestrial erosion, increased nearshore ocean-water turbidity and associated reef stress, and the undermining of artificial shoreline-protection structures.

During an El Niño event, a body of warm water moves across the low-latitude Pacific Ocean from west to east. One consequence of this is that abnormally low sea levels occur for prolonged periods around western Pacific islands. For example, the sea level off Pohnpei (Ponape) island in Micronesia remained 25 cm below its long-term mean level for several weeks during the 1982–1983 El Niño (Philander, 1990). Such events cause prolonged exposure of coral reefs which may combine with other sources of stress to bring about coral death and reef degradation. Reef flats around many atolls in the Marshall Islands may have become barren only in recent decades, perhaps in response to the cumulative effects of 20th century El Niño events.

The El Niño phenomenon may have begun only during the Holocene (DeVries, 1987); thus an understanding of the timing and magnitude of its occurrences has much light to shed on the controls of Holocene environmental change. Yet, despite a growing body of research, the nature of Holocene El Niño events and the relationship between their periodicity and recent climate changes such as the Little Climatic Optimum and Little Ice Age are still far from clear. One reason for this is the

difficulty of establishing a complete chronology. The longest proxy record of El Niño events comes from various indicators including, for the earliest part, Nile River level data (Figure 15.11). Changes in the frequency of occurrence of these events since AD 1700 appear to be linked to sunspot variability (Anderson, 1992a; see below).

The relationship between ENSO and the climate deterioration marking the transition ~AD 1300 (see also following section) between the Little Climatic Optimum and the Little Ice Age in the Pacific Basin is intriguing yet unclear. By far the lowest frequency of El Niño events within the past 1370 years occurs on three occasions between AD 1100 and AD 1300. It is tempting to suppose

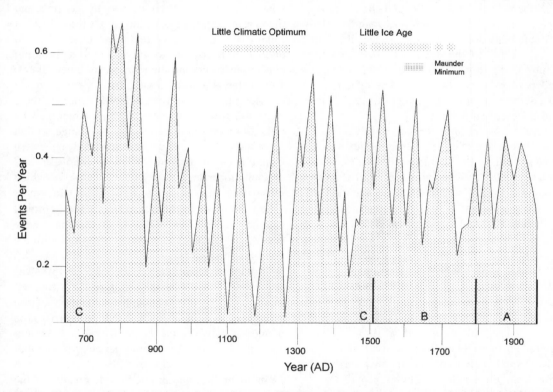

Figure 15.11   Incidence of El Niño events since AD 622. Note in particular the comparatively low incidence of El Niño events during the Little Climatic Optimum (~AD 1000–1300) which may have been associated with the dry, warm conditions during this interval in the tropical Pacific which facilitated successful long-distance voyaging. Similarly, the heightened incidence of El Niño events during the Little Ice Age (~AD 1300–1750) may express the changed weather conditions which evidently restricted interisland communication at this time and placed stress on island environments which caused conflict among their inhabitants
Time series based on data recognized in Nile streamflow data (C), strong events only (B) and all events (A). After Anderson (1992b) with permission of Cambridge University Press

that a low incidence of El Niño may have been associated with prolonged aridity in the central part of Pacific South America, for example, thus rendering the area more vulnerable than it might otherwise have been to the mighty El Niño event of ~AD 1330. In other words, the juxtaposition of periods of prolonged dryness and excessive rainfall would have been a recipe for maximizing environmental change.

15.13.2   Rapid Climate Change Around AD 1300 (~650 BP)?

Some palaeoclimate records show that after the end of the Little Climatic Optimum about AD 1300, there was an abrupt and rapid fall of temperature to a level which was maintained for the first few hundred years of the succeeding Little Ice Age (refer to the sea-level proxy for temperature in Figure 15.10). The actual transition lasted perhaps only 100 years. The evidence for associated environmental changes in the Pacific Basin is described here.

There is no clear information for this time available for Antarctica. That from South America is widespread. In the southernmost Andes, a cold interval AD 1270–1340 marks the rapid transition between the Little Climatic Optimum and the Little Ice Age (Villalba, 1990). Farther north, the transition is dated by a peat formed 650 BP (~AD 1300) which was overridden by a glacier in the Peruvian Andes (Seltzer, 1994). Glacier advance here was associated with a lowering of snowlines which has been linked to a contemporary fall in maize production. Around AD 1330, a very strong El Niño event (see Figure 15.11) left permanent marks on the coast. This produced a flood of such magnitude in coastal Peru that it caused the collapse of the Nyamlap dynasty and an invasion of the region from the south by Chimu people (Wells, 1990). It is possible that rapid cooling and increased moisture had already set the grounds for this invasion (as elsewhere in the Pacific) and that the strong El Niño simply provided the *coup de grâce*.

There is an inverse relationship between precipitation in the Quelccaya ice-core record (in which the start of the Little Ice Age is apparently

not registered until around AD 1450) and that along the northern Peruvian coast. Thus the decline in the rate of ice accumulation, controlled largely by precipitation, at Quelccaya AD 1300–1400 (not shown in Figure 15.3) was probably linked to a rise in lowland, coastal precipitation which brought about widespread and rapid environmental change of the kind comparable to that in other parts of the Pacific Basin around AD 1300. The precision of the Quelccaya record allows the conclusion that the onset of the Little Ice Age was abrupt in tropical South America (Thompson and Mosley-Thompson, 1987).

For Central America, an increase in effective precipitation (precipitation minus evaporation), which could have been caused by cooling and increased rainfall here, occurred AD 1380–1522 at Lake Pátzcuaro in central Mexico (O'Hara, 1993) and would appear to have been a regional phenomenon perhaps beginning as early as AD 1325 (Metcalfe, 1987).

Along the Pacific Rim in North America, there were some environmental changes which can be linked to a period of rapid climate change around AD 1300. This event is best marked in western Canada by a period of rapid cooling and precipitation rise, the majority of dates for which fall within the range AD 1214–1350 (Luckman, 1993). In southern Alaska, a widespread cultural change ~720–670 BP (Mills, 1994) may have been related to the ~AD 1300 (~650 BP) climate deterioration.

In parts of Japan, the time around AD 1300 was one marked by 'catastrophic forest destruction' (Yasuda 1976: 56), although recognition of the effects of rapid climate change across the boundary between the Little Climatic Optimum and the Little Ice Age are not so clear from other records (Lamb, 1977).

In Australia, the abrupt lifestyle change of the late Holocene inhabitants of Princess Charlotte Bay (see above) about 600 BP has been ascribed to 'dramatic climate events' (Hiscock and Kershaw, 1992: 66). For New Zealand, the Nelson stalagmite record shows an abrupt fall of ~1.3°C in some 75 years beginning around AD 1380 (derived graphically from Wilson et al., 1979).

Throughout northern New Zealand, by the 15th century the Maori had begun building forts (*pa*) to protect food stocks, perhaps (partly) in response to an abrupt and profound reduction in the resource base associated with a rapid climate change around AD 1300. Similar events characterize human history in many parts of the Pacific islands. In the Fiji group, for example, coastal settlements were abruptly abandoned around AD 1300–1400 in favour of fortified hilltop settlements (Plate 15.15). At the same time, weapons of war began to be manufactured and cannibalism, the practice of which may have developed primarily to intimidate rival groups, spread. All such changes point to rapidly increased intertribal conflict throughout the region which cannot be satisfactorily explained by local circumstances in every instance. Rather they point to the effects of a rapid, region-wide climate change which produced

a rapid decline in human-resource availability and, consequently, (increased) competition for a diminished pool of available resources on many islands (Nunn, 1994b).

In many parts of the tropical Pacific, the effects of a rapid climate change around AD 1300 would have been manifested in several arenas simultaneously. Coral reefs on which most coastal people depended for food would have become exposed during the associated sea-level fall (illustrated in Figure 15.10). The result would have been a massive prolonged drop in reef productivity having grave consequences for the populations which had depended on this. Many food crops would also have become stressed by the temperature fall and, particularly on low coasts and low islands, by the drop in the water-table which accompanied sea-level fall. It has also been suggested that agricultural terraces and other

Plate 15.15   View along the mountain ridge of Totoya island, southeast Fiji. Settlements were established along the coast during the Little Climatic Optimum but the transition to the Little Ice Age about AD 1300 so reduced the marine and terrestrial resource bases that coastal settlements were abandoned in favour of fortified mountain-top sites. Most of the peaks along the spine of Totoya are surrounded by deep ditches and walls; there is an oral tradition which tells of groups on adjoining hilltops lobbing firebombs at each other, a sign of the conflict which is believed to have lasted for most of the Little Ice Age

artificial structures developed to utilize available water to the full during the dry conditions of the Little Climatic Optimum would have been devastated by a short-lived though massive increase in precipitation across much of the low- and mid-latitude Pacific at this time, evidence for which is found in proxy records from California and China (Nunn, 1997c).

A classic example of the environmental outcome of the ~AD 1300 climate deterioration in the Pacific Islands is provided by Easter Island (Rapanui) where, before this time, many agricultural terrace systems were operating, and clearly doing so with such efficiency that the people found the ample time they needed to carve, excavate, transport and erect the massive statues (*moai*) for which the island is renowned. After about AD 1300, the situation changed to one of *moai*-toppling during which many statues were pulled down and smashed, and the resource base on the island shrank rapidly. Most writers envisage this change as the outcome of population having reached an unsustainable level on this remote island (Bahn and Flenley, 1992) although others suggest that climate change played a key role (McCall, 1993; Nunn, 1994a).

There is enough evidence from many parts of the Pacific Basin to suppose that the ~AD 1300 event affected a vast region, transforming environments in ways which had profound implications for their human inhabitants. Clearly more investigations are needed before the cause of this event can be reliably determined although a link with what appears to be a phase change in El Niño recurrence (shown in Figure 15.11) provides a promising start.

The identity and Pacific-wide character of the ~AD 1300 change has been obscured for decades by a widespread insistence that most environmental change within the later post-settlement history of much of the Pacific Basin was brought about almost solely by people. The dominance of this view also explains, at least in part, why a relationship between sunspots and Pacific climate change, discussed in the next section, has been undervalued by some writers.

15.13.3   Climate Control by Sunspot Minima: Evidence in the Pacific Basin

Scientists have connected variations in sunspot activity to climate changes, particularly extreme events, for a long time but this has tended to focus on Europe and North America. Of those studies which have proposed correlations between sunspot cycles and climate changes earlier than the last millennium in the Pacific Basin, that by Scuderi (1993) for the last 2000 years in the Sierra Nevada is notable.

Two sunspot minima – the Spörer and the Maunder – mark the coldest times of the Little Ice Age in Europe and North America. According to tree-ring data, the Spörer Minimum (AD 1400–1510) may have been responsible for the extreme cold AD 1445–1469 in Tasmania (Cook et al., 1992). There is some evidence for effects of the Maunder Minimum (AD 1645–1715) in Alaska (Eddy, 1977).

More recent events have been linked to 11-year and 22-year solar cycles (Currie, 1987; Anderson, 1992b; Meko, 1992). In particular, the incidence of droughts in North America and China has been linked to the 11-year solar cycle. In both places, drought incidence is out of phase with this solar cycle although the phase shifts. More interestingly, for the Pacific Basin, the drought cycles in China and western North America are also out of phase, implying that a standing wave pattern exists in the atmosphere above the North Pacific (Currie and Fairbridge, 1985).

15.14   TECTONICS AND VOLCANISM IN THE PACIFIC BASIN DURING THE HOLOCENE

In earlier parts of this book, when longer time spans were discussed, the effects of tectonics and volcanism were often those which, with hindsight, appeared most important to an understanding of environmental change. The principal reason for this was that, unlike most other causes of environmental change in the distant past, the effects of tectonics and volcanism persisted through to the present. Of course, most effects of

Holocene tectonism and volcanism have also persisted to the present but the difference is that, because of an abundance of other, clearly discernible reasons for environmental change within this epoch, we can see how relatively unimportant the effects of tectonics and volcanism are in a regional sense over such short time periods. This is true particularly of the Holocene during which humans became a progressively more important agent of environmental change.

Within the Holocene, most environmental changes associated with tectonics and volcanism have been highly localized; particular parts of a shoreline have abruptly risen or sunk, for instance; particular eruptions clothed surrounding areas with lava and ash. While not intending to underrate either the extreme effects or often serious consequences of such events, it is recognized that these are not of great regional importance to Holocene environmental change in the Pacific Basin.

Those tectonic changes to have had the greatest effects on Pacific environments during the Holocene fall into two broad categories: aseismic and coseismic. Aseismic uplift has been most marked in formerly ice-covered areas particularly around the Antarctic periphery and along the western seaboard of Canada. The effects of coseismic uplift and subsidence are generally more localized, being associated with single earthquake events, but have often had more profound environmental effects, examples of which abound above convergent plate boundaries in Pacific South and Central America, in the island arcs of the northwest Pacific including Japan, and in New Guinea and New Zealand, and most southwest Pacific Island groups; representative examples are given in Table 15.11. Tsunami affect most Pacific Basin coasts.

The effects of Holocene volcanism on Pacific Basin environments have been more diverse; no attempt is made to classify them in this book (although see Ollier, 1988). The most profound environmental changes associated with volcanism are brought about not by lava flows or even pyroclastic eruptions in many places because these affect mostly areas formed from the products of earlier eruptions. More important are the long-range effects of volcanic eruption, which include ash (tephra) clouds and lahars.

This section reviews the evidence for tectonic and volcanic environmental changes in the Pacific Basin during the Holocene. The discussion proceeds by region recounting representative case studies where appropriate.

15.14.1 Holocene Tectonics and Volcanism in Antarctica

The withdrawal of ice from its late Last Glacial limits led to the isostatic uplift of the Antarctic continental periphery causing the emergence of early Holocene shorelines. No records of coseismic uplift are known.

A complete picture of Holocene volcanism in Antarctica is prohibited by the environment: cloud formations and windblown snow can be mistaken for smoke plumes, windblown ash can emulate a true volcanic ashfall, and the effects of most eruptions may be quickly obliterated by blizzards (LeMasurier, 1990). The greatest concentration of volcanoes active during the Holocene in Antarctica is found along the Antarctic Peninsula and on nearby island groups where they are associated with remnants of a convergent plate boundary. There has also been Holocene activity at some of the volcanoes along the Pacific coast in Marie Byrd Land and along the west side of the Ross Sea. Most of these eruptions are associated with fracture zones along which earlier volcanism had occurred.

15.14.2 Holocene Tectonism and Volcanism in South America

Tectonics have continued to play a role in environmental change in South America during the Holocene especially along its Pacific coast. The rate of coupling between the South American Plate and the Nazca Plate responsible for the Cenozoic uplift of the Pacific margins of South America has varied through time. For example, a weakening of coupling has been cited as the reason for the comparative late Holocene stability

Table 15.11   Magnitudes of selected Holocene coseismic uplift and subsidence events in the Pacific Basin. For Quaternary (longer-term) aseismic rates, see Table 9.2. For rates of ground subsidence caused by groundwater extraction, see Table 15.7

Location	Uplift/subsidence magnitude (m) and year	Source of information
UPLIFT		
South America		
Guamblin Island, Chile	5.7 (AD 1960)	Plafker (1972)
Central America		
Sonora, Mexico	6.1 (AD 1887)	Goudie (1995)
North America		
California	0.91 (AD 1906)	Goudie (1995)
Beringia		
Montague Island, Alaska	11.3 (AD 1964)	Plafker (1972)
East Asia		
Honshu, Japan	3.0–6.0 (AD 1703)	Kayanne and Yoshikawa (1986)
Okinawa, Ryukyu Islands	2.5 (late Holocene)	Kawana and Pirazzoli (1985)
Australasia		
New Zealand	4.5 (AD 1929)	Soons and Selby (1982)
Pacific Islands		
Beqa, Fiji	1.32 (mean of five events)	Nunn (1990b)
Guadacanal, Solomon Islands	1.5 (AD 1961)	Grover (1965)
Malakula, Vanuatu	1.2 (AD 1965)	Taylor et al. (1980)
SUBSIDENCE		
Beringia		
Kodiak Island, Alaska	2.3 (AD 1964)	Plafker (1972)
East Asia		
Honshu, Japan	0.45 (AD 1964)	Rothé (1969)
Pacific Islands		
Santo, Vanuatu	0.15–0.2 (AD 1971)	Prevot and Chatelain (1983)

of north-central Chile, an area which was tectonically very active before the middle Pleistocene (Ota et al., 1995).

The Holocene sea-level record in parts of South America has been distorted by the effects of uplift. For example, three levels of emerged beaches occur along the sides of the Beagle Channel in southernmost South America at 8–10 m, 4–6 m and 1.5–3 m above present mean sea level. The highest levels have yielded dates of 8240 ± 60 BP and 7518 ± 58 BP which demonstrate that the channel was open (ice-free) by this time, and that there has been significant uplift since this time. The lack of any significant east–west variation in beach level led Gordillo et al. (1992) to favour the cumulative effect of abrupt coseismic uplift events (associated with plate convergence) rather than glacio-isostatic recovery as the cause of this uplift.

Coseismic movements have been important elsewhere. The great earthquakes which struck southern Chile in May 1960 were accompanied by both uplift and subsidence (Plafker and Savage, 1970) exemplifying the complex legacy of shorter-term coseismic change in this region. Parts of the south-central Chile coast were uplifted 6 m in this event, others nearby sank 2 m (compare with Figure 15.13). Yet some areas that sank in 1960 have undergone net late Holocene emergence, probably as a result of regional aseismic deformation (Atwater et al., 1992). Nelson and Manley (1992) reached a similar conclusion about Isla Mocha in the same part of Chile. They suggested the emerged shorelines had been raised largely by aseismic uplift at a rate of ~70 mm/year although particular shorelines were elevated coseismically during large earthquakes. As elsewhere in the Pacific Basin, it is likely that the cumulative effects of coseismic uplift are subdued because of the relaxation, manifested as interseismic subsidence, which takes place between coseismic-uplift events.

Active volcanism continues to affect many parts of the Andes, commonly at long-established stratovolcanoes. Notable eruptions within the last millennium include the AD 1660 eruption of Huaynaputina in Peru, the AD 1877 eruption of Cotopaxi in Ecuador, and the 1967 eruption of Quizapu in Chile. The amount of change accomplished locally by these volcanoes can be very great, particularly when eruptions occur beneath snow and/or ice and produce large volumes of meltwater, or when unconsolidated materials are later moved in heavy rains. The resulting flows (lahars) can be very destructive: more than 23 000 Colombians died for this reason when Nevado del Ruiz volcano erupted on 13 November 1985[23].

***

[23] The initial eruption occurred at 3.15 pm. Attempts to evacuate the nearby town of Armero were frustrated by heavy rain. At 9.05 pm a strong tremor on the 5400 m volcano was followed by a rain of hot pumice and ash which melted part of its ice cap, caused the Guali River to flood, and a natural dam to burst, releasing a 70 km/h torrent and mudflow which enveloped Armero (Siegel and Witham, 1991)

15.14.3 Holocene Tectonics and Volcanism in Central America

Parts of the Pacific Rim in Central America adjoin a line of plate convergence and have been affected by coseismic movements throughout the Holocene as a consequence.

Volcanic eruptions have been common throughout most of central America during the Holocene. Owing to seasonally high rainfalls, lahars are important agents of environmental change here. For example, a large area of coastal Guatemala was inundated by lahars in 1902; boulders weighing as much as 200 tonnes were moved in lahars with velocities as high as 36 km/h during the 1963–1964 eruption of Irazú Volcano, Costa Rica (Blong, 1984).

Most Holocene volcanic eruptions in the Pacific Basin occurred at sites of earlier ones so that their environmental consequences commonly duplicated those. One exception was the 1943 appearance in a Mexican corn field of the volcano that became Parícutin (Plate 15.16). An indication of the effects on the surrounding area is given in Table 15.12.

15.14.4 Holocene Tectonics and Volcanism in North America

Studies of the effects of large-magnitude earthquakes along the Pacific margin of the United States have helped elucidate its Holocene history. Uplift is a common product of these events; the record of 7 m of shoreline uplift about 1000 BP in southern Puget Sound provides a reminder of the threat posed to these environments by movements along what have been assumed to be inactive faults (Bucknam et al., 1992). Yet subsidence appears even more widespread and is the most important cause of environmental change associated with the 'earthquake deformation' cycle, discussed in the following section. Of great concern in many parts of California are the effects of large earthquakes associated with the San Andreas fault zone which threaten some of the most complex human environments in the Pacific Basin (Dolan et al., 1997; Plate 15.17).

Plate 15.16   On the afternoon of 20 February 1943, a hole emitting dust appeared in a corn field a few kilometres south of San Juan Parangaricutiro. Within 10 years, it had grown into Parícutin Volcano, pictured here in 1972, one of the most recent reminders of the eruptive potential of the trans-Mexico volcanic belt and associated fractures
Mary Lee Nolan

Table 15.12   Areas affected by the 1943–1952 eruption of Parícutin Volcano, Mexico
After Rees (1979)

Zones	Areas prior to eruption (ha)			
	Forest	Cleared[1]	Settlements	Total
Total-kill zone[2]	975	1374	131	2 480
Nearly total-kill zone[3]	3 435	940	0	4 375
Partial survival zone A[4]	3 866	1379	0	5 245
Partial survival zone B[5]	17 363	8695	52	26 110

[1] Lands under cultivation or in fallow
[2] Coincident with the cinder cone and lava flows
[3] Most vegetation eliminated, average ash depth 1.5 m
[4] Tree damage and heavy kill of shrubs and herbs, 0.5–1.5 m of ash deposition
[5] Slight tree damage and partial survival of shrubs and herbs, 0.15–0.5 m of ash deposition

Plate 15.17   The great San Francisco earthquake of 18 April 1906 was produced by slip of the northern section of the San Andreas Fault. Associated ground rupture occurred along a 330 km trace and may have extended as much as 130 km offshore (Niemi and Hall, 1992). Fires broke out across much of the city after the main tremors; soldiers helped evacuate its inhabitants and prevent looting
Library of Congress

Tectonic change probably affected all Pacific Canada during the Holocene, largely as an outcome of isostatic readjustment and/or because of plate convergence. Precise details are comparatively poorly known. The greatest amount of uplift was aseismic yet coseismic uplift played an important role, especially along the coasts. The three large-magnitude earthquakes to have affected Vancouver Island this century were the outcome of the overriding of the Pacific Plate by the North America Plate. From their observation of a buried soil overlain by tsunami-deposited sand on western Vancouver Island, Clague and Bobrowsky (1994) argued that one such earthquake occurred here about 300 years ago.

Sea-level history is a good indicator of regional uplift during the late Holocene in this region. Figure 15.12 shows the generalized pattern of sea-level change along the coast of British Columbia. The Queen Charlotte Islands, which were not covered by the continental ice sheet during the Last Glacial, exhibit the common picture of Holocene (eustatic) sea-level change in the Pacific Basin but on the mainland, from which the massive Cordilleran ice sheet disappeared during the Holocene, the land is rebounding upwards, so the trend is reversed.

Holocene volcanism occurred along many mountain ranges in the western United States, notably in the Cascade region (see Figure 9.3).

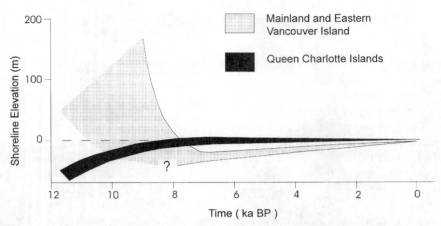

Figure 15.12   Sea-level changes over the past ~12 ka along the British Columbia coast. Note the contrast between areas which are rising isostatically and those which are not
After Clague (1989b) with permission of John J. Clague and *Episodes*

Four belts of volcanoes were active in Pacific Canada during the Holocene (see Chapter 9). Although the active nature of volcanoes in this region had been known for generations, nothing had prepared its inhabitants for the scale of the 18 May 1980 eruption of Mount St Helens. An avalanche from the volcano's flanks buried much of the surrounding landscape to depths as great as 200 m; the accompanying lateral blast flattened forests across an area of 400 km^2 (compare Plate 15.21). Ash fell over an area of 80 000 km^2 within 24 hours.

15.14.5   The Earthquake Deformation Cycle Along the Western Continental Margin of North America

During large-magnitude earthquakes associated with nearby subduction, such as occur off the northwest coast of the United States, abrupt coseismic uplift and subsidence sometimes occur. Geological evidence for subsidence can be less conspicuous than that for uplift, yet soils buried by estuarine mud provide strong evidence for coseismic subsidence in this area. Brian Atwater's (1987) explanation involved a soil-covered coastal lowland being abruptly submerged by tides after subsiding during an earthquake. The former lowland now becomes

the shallow floor of an estuary, and tidal mud then rebuilds land on which a new soil develops. Then another earthquake occurs and this dry land becomes submerged once again, resetting the cycle.

Abundant rhythmic bedding of this kind, some associated with tsunami deposits, has been found in coastal estuaries and salt marshes, and has been interpreted as the product of repeated large earthquakes causing the simultaneous subsidence of hundreds of kilometres of the northwest coast of the United States (Long and Shennan, 1994).

Data other than those from coastal sediments support the idea that abrupt subsidence events recurred here. For example, tree-ring studies have been employed in this manner, incomplete rings being explained by sudden saltwater inundation consistent with rapid subsidence (Atwater and Yamaguchi, 1991).

Farther north on west Vancouver Island, coseismic subsidence has occurred within the past 400 years in association with a large earthquake. Interestingly, this subsidence occurred in an area which had hitherto been characterized by emergence. Clague and Bobrowsky (1994) highlighted the similarity between this situation and that in south-central Chile where aseismic uplift is also occasionally interrupted by coseismic subsidence.

15.14.6 Holocene Tectonics and Volcanism in Beringia

As in Pacific Canada, the principal cause of Holocene tectonics has been difficult to identify in parts of Beringia where isostatic rebound and uplift associated with plate convergence have both been active. The issue is well illustrated by the historical uplift of Glacier Bay in southeast Alaska (Adams and Clague, 1993).

In terms of the area affected by surface deformation, the 1964 Prince William Sound (Alaska) earthquake was the greatest such event of the 20th century. It occurred at the eastern end of the Aleutian arc where the Pacific Plate is being thrust beneath the North American Plate at ~6.3 cm/year. The area affected by deformation was around $140\,000\,km^2$. Maximum uplift reached 11.3 m on Montague Island, maximum subsidence was 2.3 m on Kodiak Island (Figure 15.13). In addition to ground deformation (Plate 15.18), the entire region between Anchorage and the Gulf of Alaska shifted seaward at least 20 m (Plafker, 1972). Tsunami generated by the abrupt movement of the ocean floor affected most Pacific coasts although they were not as large as those produced by the 1946 Aleutian Islands earthquake.

Little effort has been directed to assessing the cumulative effects of such earthquakes on long-term environmental change yet it is reasonable to suppose that they are a major cause of change in such places, affecting a whole range of landforms.

Volcanism also had major effects on some Beringian environments during the Holocene. Some of the best documented followed the 1912 eruption of Katmai Volcano which produced $20\,km^3$ of ash and a *nuée ardente* which filled a valley 20 km long and 3 km wide, now called the Valley of Ten Thousand Smokes.

An intriguing connection between major volcanic eruptions during the later part of the middle Holocene and climate was proposed by Begét et al. (1992). The Alaskan volcano Aniakchak erupted around 3435 BP spreading tephra across distances of more than 1500 km. At the same time, there were major eruptions of Mount St Helens in the northwest United States,

Vesuvius in Italy and Santorini in the Aegean. The combined effect of these eruptions may have been to cause a short-lived period of abrupt cooling of possible global extent.

15.14.7 Holocene Tectonics and Volcanism in East Asia

Earthquakes threaten most of China; the most devastating in recent history was the Tangshan earthquake on 28 July 1976 which claimed more than 240 000 lives. Studies of associated ground-surface deformation, such as that illustrated in Plate 15.19, have illuminated the role of such events in long-term environmental change. Throughout the country, movement along fault zones is being monitored increasingly with the intention of predicting similar events; measurements from Guangxi illustrate the complexity of such records (Table 15.13).

Tectonic movements have been common throughout Japan during the Holocene. It has been suggested that periods of particularly intense movement in the Ryukyu Islands were associated with periods of acceleration in the rate of subduction of the Pacific Plate beneath the Philippine Plate (Pirazzoli and Kawana, 1986). Farther north in the rest of Japan, uplift has been a conspicuous product of large-magnitude earthquakes – the 1995 Hanshin (Kobe) earthquake provides a recent example. Around the Ashizuri Peninsula in southern Shikoku, there were three or four coseismic-uplift events during the late Holocene, each represented by a distinct step in the landscape (Ota and Odagiri, 1994). A comparable situation is found along many parts of the Taiwan coast (Plate 15.20) where spasmodic uplift is thought to account for rates of late Holocene uplift in excess of 8 mm/year in places (Liew et al., 1993).

Tsunami affect the Japanese islands comparatively frequently owing to both their proximity to ocean trenches and the exposure of their eastern coasts to the Pacific, across which tsunami generated as far away as South America travel. The environmental impacts of tsunami are less well known than their human toll. In AD 1771,

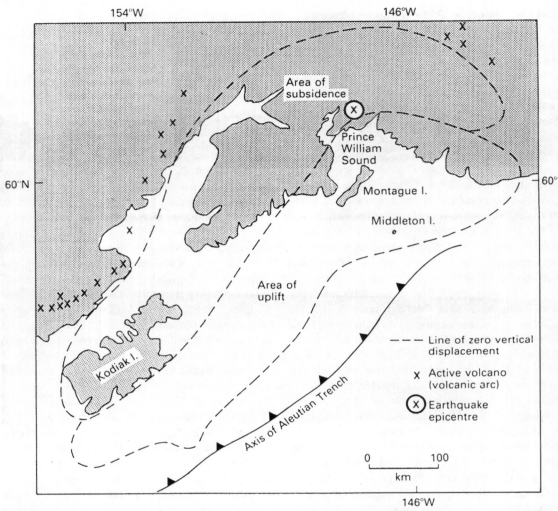

Figure 15.13   Deformation associated with the 1964 Prince William Sound (Alaska) earthquake. The zone of uplift runs parallel to the trench axis; uplift averaged 4–8 m reaching 11.3 m on Montague Island. The zone of subsidence occurred farther back from the trench close to the volcanic arc; a maximum of 2.3 m of subsidence was recorded on Kodiak Island

From Nunn (1994a)

for instance, the Meiwa Tsunami swept across the southern Ryukyus killing some 12 000 people. Such records have been used in conjunction with dates from tsunamigenic deposits to compile a chronology of tsunami (Kawana and Nakata, 1994). The Philippines are similarly tsunami-prone; that in the Moro Gulf on 17 August 1976 killed some 8000 people and brought about profound changes to the area onshore.

It is also Japan and the Philippines which have experienced the greatest effects of volcanism within this region. For example, the eruption of Mount Pinatubo on Luzon island in the Philippines on 12–15 June 1991 destroyed huge areas of surrounding forest, and deposited vast amounts of material across a wide area. Although not as well documented for recent centuries as along the eastern Pacific Rim, it is plausible to suppose that

Plate 15.18   A street in Anchorage, Alaska, following the massive Prince William Sound earthquake in March 1964. See also Figure 15.13
Special Collections Division, University of Washington Libraries. Negative no. UW 15404

lahars have been important agents of environmental change in many parts of monsoonal East Asia (Blong, 1984).

15.14.8   Holocene Tectonics and Volcanism in Australasia

Most of New Guinea has been tectonically active during the Holocene, some of the clearest manifestations of this being coseismic-uplift events along its reef-fringed coasts; along the Huon Peninsula, these events recur approximately every 1000–1300 years with an average amplitude of about 3 m (Chappell et al., 1996). As agents of environmental change, large earthquakes are particularly effective in New Guinea on account of steep slopes and heavy rainfall (Peart, 1991).

Although generally perceived as being seismically inactive on account of its intraplate location, East Australia is still occasionally subject to large earthquake events. The distortion of rates of Holocene sea-level change in Australia has also been attributed to tectonic change (see Figure 15.9).

Many parts of New Zealand are moving either up or down at rates fast enough to have influenced recent coastal landform development. For example, four marine terraces along parts of the northeast coast of North Island emerged only during the late Holocene (Ota et al., 1992).

Plate 15.19   Railway tracks on a bridge across the Jiyunhe river in eastern China were compressed when the river banks moved 2.37 m closer together during the 1976 Tangshan earthquake. The cumulative observed effects of such compression events during the Holocene have been substantial in parts of the Pacific Basin
Reprinted from Chen et al., *The Great Tangshan Earthquake of 1976*. Copyright (1988), with permission from Elsevier Science

Table 15.13   Recent movements along the Dongxing–Lingshan and Xiaodong–Nanxiao faults, Guangxi, southern China. Note how the direction as well as the magnitude of movement along particular fault zones varies through time, an illustration of the difficulties in interpreting such data meaningfully
After Huang et al. (1984)

Fault	Movement of west wall relative to east wall (mm)				Net movement (mm)
	1956–1966	1966–1972	1972–1976	1976–1977	1956–1977
Jiping (Taishu)	−7.2	+1.7	+3.1	−2.9	−5.3
Louzhuping	+5.0	−2.7	+2.7	−1.2	+3.8
Taitong (Lingshan)	−5.0	+5.3	−3.1	+0.75	−2.05
Xiaodong	−1.0	+3.3	−1.86	+1.3	+1.74

Plate 15.20   A stepped coastline in andesite breccia along the coast of Taiwan island, western Pacific, is interpreted
as evidence of repeated coseismic-uplift events during the late Holocene
Paolo Pirazzoli

Both New Guinea and New Zealand are especially prone to the effects of volcanic eruptions. The account by Russell Blong (1982) of the oral-historical and geological evidence for eruptions of New Guinea volcanoes during the late Holocene demonstrated the ways in which people were affected by the associated changes which both destroyed and enriched highland environments[24]. Most Holocene eruptions in New Zealand have been confined to North Island, their environmental effects being best marked in the Taupo Volcanic Zone. Some of the forest destruction associated with the AD 1886 eruption of Mount Tarawera is shown in Plate 15.21.

Recent eruptions of Mount Ruapehu have led to ash being deposited across much of North Island.

15.14.9   Holocene Tectonics and Volcanism in the Pacific Ocean and Islands

Tectonic changes on many Pacific Islands during the Holocene have had significant effects on their environments. Those changes having the greatest effects have been coseismic; examples abound on islands close to the convergent plate boundaries in the south and west Pacific. For example, on Tongatapu island in Tonga, a rate of net Holocene uplift of 1.6 mm/year has been reported (Nunn and Finau, 1995) which is representative of rates from similar situations elsewhere in this region.

Pacific island intraplate volcanism is generally not explosive and its products are slow-moving. Much seminal work has been done on Kilauea, the

---

[24] Blong recognizes this suggestion as esoteric but points to the effect of a tephra fall smothering the ground yet also replenishing nutrients in (overworked) agricultural soils

Plate 15.21   The area around Tikitapu Lake, near Rotorua, North Island, New Zealand, shortly after the 10 June 1886 eruption of Mount Tarawera. Most of the large trees have been toppled, and the whole area buried under ash
Photo C.10724 from Burton Brothers Collection, Museum of New Zealand, Te Papa Tangarewa

most active eruptive centre on Mauna Loa Volcano on Hawai'i – one of the world's largest. The environmental effects of such volcanism are better illustrated by the island Savaii in Samoa, much of the northern part of which has become covered by lavas (and evacuated by humans) in the last few hundred years.

More explosive eruptions in the Pacific Islands are associated with plate convergence but these generally affect only single islands and produce similar landscapes to those already in existence. One exception is Rabaul Volcano on New Britain island in Papua New Guinea where a large city has developed around the drowned caldera of what is now realized to be a moderately active volcano; precursors of eruptions such as earthquake activity and ground deformation are routinely monitored in the area (de Saint Ours et al., 1991).

PART F

Conclusions

CHAPTER 16

Afterword

... at my back I always hear
Time's wingéd chariot hurrying near,
And yonder all before us lie
Deserts of vast eternity.
Andrew Marvell, *To His Coy Mistress*

16.1 INTRODUCTION

In an age of fierce competition for a dwindling pool of research funds, every field of scientific endeavour is being forced to justify itself. In a perfect world, the study of environmental change would have to make little effort to achieve this; both the need and the wish are clear. Knowledge of the past is invaluable to understanding the current state of the Earth's environments and to planning their future. Increasing stress on Earth systems has been accompanied by increasing calls for optimum management of its environments in the future. Several themes emerge from this book, some of which are flagged below as being particularly worthy of continued attention in the future.

16.2 NEOCATASTROPHISM AND GRADUALISM

The first of these refers to the current state of the old controversy between those (gradualists) who regard the greatest amount of environmental change as resulting from the cumulative effect of processes operating under 'normal' conditions, and others (neocatastrophists) who conclude that such change is dwarfed even over long time periods by that which occurs during short-lived large-magnitude events. The controversy suffers from being polarized in this way yet the evidence from the Pacific Basin generally favours the latter view. It is

likely that an increased focus on catastrophic rather than gradual change will yield a clearer picture of how and why its environments have changed in the past and how they might do so in future.

The main reason for taking this view for the Pacific Basin is that most of it is prone to catastrophic changes associated with tectonics and volcanism. Another example is provided by the likely effects of extraterrestrial bodies colliding with the Earth. After initial opposition, the likelihood of the K–T extinctions having been caused by a huge bolide impacting Central America is now widely accepted (Chapter 5). Yet there is caution surrounding the possible effects of bolide impacts during the Cenozoic although evidence for these is widespread (Kyte, 1988). Bolide impacts in the Pacific Ocean would have generated huge waves, the effects of which would be similar to those generated by earthquakes or produced by massive landslides.

16.3 HUMAN–ENVIRONMENT RELATIONS

In an age when the effects of humans on many of the environments they inhabit is overwhelmingly manifest, it has become unfashionable to talk of the ways in which environmental change has 'determined' human actions. Yet a recurring theme through much of this book is precisely that, whether it be human communities responding to sea-level change, prolonged drought or increased precipitation levels by altering their lifestyles. Of course humans have changed the environments they occupy, but changes in these can still unquestionably cause lifestyle changes.

Having said that, there is some way to go before the reluctance of many people to accept that physiological adaptations to environment, principally mean annual temperature and diurnal temperature regime, have occurred since the first appearance of modern humans. There is some evidence to suggest that this is so in the Pacific Basin. For example, Bergmann's Rule as applied to humans anticipates that larger, more muscular people will be found in cold climates while smaller or at least more linear people should inhabit hotter climates. This appears broadly correct for the indigenous peoples of various parts of the Pacific Basin (Table 16.1) with the conspicuous exception of some Pacific Islanders whose large bodies may be adaptations to cold wet conditions experienced during lengthy ocean journeys (Houghton, 1996). Such provocative ideas may stimulate the future re-evaluation of the role of environments and environmental change in shaping human bodies as well as human lifestyles.

16.4 UNCERTAINTIES

However positive one is about the nature of modern environments and the ways in which they have changed, it would be misleading to imply that the uncertainties are not still considerable.

Many problems refer to dating. Every radiometric dating technique involves uncertainties that persist despite efforts to resolve them. Yet in an age when modellers demand ever greater precision, such uncertainties frustrate the understanding of environmental change. Witness the problem with identifying the timing and spatial extent of the Younger Dryas cooling event, reviewed in Chapter 11, because it occurred at a time of a radiocarbon plateau.

Still other difficulties refer to the materials selected for dating, and the possibilities that they

Table 16.1   Stature-mass ratios for selected groups of indigenous Pacific peoples. Bergmann's Rule suggests that people living in colder places will exhibit lower stature-mass ratios than those living in hotter places
After Houghton (1996)

Region of Pacific Basin	Group	Female stature–mass ratio	Male stature–mass ratio
North America	United States	2.85	2.47
Beringia	Aleut	2.71	2.56
	Russia	2.70	
East Asia	China (Canton)	2.74	
	Japan	3.04	2.65
	Korea		2.90
Australasia	New Zealand (Maori)		2.27
Pacific Islands	Cook Islands (Pukapuka)	2.39	2.39
	Fiji (coastal)	2.61	2.41
	Fiji (Lau group)		2.24
	Hawaii	2.34	2.22
	Papua New Guinea (Manus)	3.13	2.71
	Samoa	2.47	2.26
	Solomon Islands (Ontong Java)	2.62	2.56
	Tokelau	2.22	2.24
	Tonga (Foa)	2.28	2.28

are actually poor indicators of the event they were selected to represent because of problems which have been recognized only subsequently. Examples include the possibility that 'old' radiocarbon at ocean-upwelling sites or in long river estuaries could have rendered dates from marine shells far older than they actually are.

Many proxies of environmental change have also been used in ways which, with the benefit of their improved understanding, some researchers today might shy away from. A major discovery in pollen analysis, for example, is that species respond individually to climate change; there is no community response, implying that some such changes inferred in this way may be erroneous because 'wrong' indicator species were unwit-

Plate 16.1   The Three Congruent Glaciers in Yukon Territory, Canada. The glacier on the left is advancing, that in the centre is receding, and that on the right has advanced recently but may now be stable. This situation captures well the difficulties which dog the palaeoclimatic interpretation of such features. If two of these glaciers were unknown, unreachable, or exhibited an uncertain record, a scientist could end up with any one of three contrasting palaeoclimate interpretations for this site. With no other records from the surrounding region, the chances of an erroneous conclusion being extended across a wide area are high. There are undoubtedly fallacies of similar pedigree buried in the data reviewed and accepted for this book and many others
Austin Post, University of Washington

tingly used. A related problem derives from the likelihood that changing tree-line positions in tropical montane regions, many instances of which have been interpreted as indications of changing temperature, may rather have been an outcome of vegetation response to changed carbon-dioxide levels in the atmosphere (Street-Perrott et al., 1997). An excellent illustration of the problems with using glacier changes as an indicator of climate change is shown in Plate 16.1.

16.5.   CONCLUSION

This book was written with the view of informing its readers about the long history of environmental change which has led to the creation of the modern Pacific Basin. One of its major themes has been to emphasize that human impact has not brought about the greatest environmental changes in the history of the region. Yet it is evident that in most parts of the Pacific Basin, humans have either generated or exacerbated environmental problems, the potential severity of which threaten human survival in some areas. This book seeks to raise awareness of these not simply by recounting some of them but also by putting them into an appropriate long-term context. This is premised on the idea that optimally effective management practices derive not only from observations of what is currently happening but also from understanding as much as possible about what happened before.

It is clear that the solution to many of the Pacific Basin's environmental ills lies in education, and this is a major goal of this book. Education about the environment is needed today more urgently than ever before. We may forgive T.H. Huxley for believing in 1883 that the ocean's resources were inexhaustible; we may even dismiss as unfortunately timed C.W. Thornthwaite's assertion in 1956 that humans were 'incapable of making any significant change in the climatic pattern on the earth' (both quoted in Cushing, 1975). But what should we make of gung-ho statements like that of Henry Pratt, executive assistant to a former Governor of Alaska in 1969: 'Hell, this country's so goddamn big that even if industry ran wild we could never wreck it. We can have our cake and eat it, too' (quoted by Pollak, 1971)?

There is a growing and commendable tendency to look to the past for analogies of the current environmental crises facing certain parts of the Pacific Basin (e.g. Bahn and Flenley, 1992; Dunham, 1994). Understanding how human–environment relationships changed at such times has the potential to aid future environmental management. Yet many such investigations have focused on either the human or the non-human elements of past environmental crises whereas a proper understanding depends on a realistic balance between the two. Much of this book has strived for such a balance but this is still a long way from being achieved in many instances.

References

Abbott, M.B., Seltzer, G.O., Kelts, K.R. and Southon, J. 1997. Holocene paleohydrology of the tropical Andes from lake records. *Quaternary Research*, 47, 70–80.

Adam, D.P. and West, G.J. 1983. Temperature and precipitation estimates through the last glacial cycle from Clear Lake, California, pollen data. *Science*, 219, 168–170.

Adams, C.G., Lee, D.E. and Rosen, B.R. 1990. Conflicting isotopic and biotic evidence for tropical sea-surface temperatures during the Tertiary. *Palaeogeography, Palaeoclimatology, Palaeoecology*, 77, 289–313.

Adams, J. and Clague, J.J. 1993. Neotectonics and large-scale geomorphology of Canada. *Progress in Physical Geography*, 17, 248–264.

Adamson, D.M. and Fox, M.D. 1982. Change in Australasian vegetation since European settlement. *In:* Smith, J.M.B. (editor). *A History of Australasian Vegetation*. Sydney: McGraw Hill, 109–149.

Adovasio, J.M., Boldurian, A.T. and Carlisle, R.C. 1988. Who are those guys? Some biased thoughts on the initial peopling of the New World. *In:* Carlisle, R.C. (editor). *Americans before Columbus: Ice-Age Origins*. Pittsburgh: Department of Anthropology, University of Pittsburgh, 45–61.

Agenbroad, L.D. 1988. Clovis people: the human factor in the Pleistocene megafaunal extinction equation. *In:* Carlisle, R.C. (editor). *Americans before Columbus: Ice-Age Origins*. Pittsburgh: Department of Anthropology, University of Pittsburgh, 63–74.

Ager, T.A. 1982. Vegetational history of western Alaska during the Wisconsin glacial interval and the Holocene. *In:* Hopkins, D.M., Matthews, J.V., Schweger, C.E. and Young, S.B. (editors). *Paleoecology of Beringia*. New York: Academic Press, 75–94.

Ahagon, N., Tanaka, Y. and Ujiie, H. 1993. *Florisphaera profunda*, a possible nannoplankton indicator of late Quaternary changes in seawater turbidity at the northwestern margin of the Pacific. *Marine Micropaleontology*, 22, 255–273.

Aharon, P., Chappell, J. and Compston, W. 1980. Stable isotope and sea level data from New Guinea support Antarctic ice-surge theory of ice ages. *Nature*, 283, 649–651.

Aharon, P., Goldstein, S.L., Wheeler, C.W. and Jacobson, G. 1993. Sea-level events in the South Pacific linked with the Messinian salinity crisis. *Geology*, 21, 771–775.

Aitchison, J.C., Clarke, G.L., Meffre, S. and Cluzel, D. 1995. Eocene arc-continent collision in New Caledonia and implications for regional southwest Pacific tectonic evolution. *Geology*, 23, 161–164.

Allègre, C. 1988. *The Behavior of the Earth*. Cambridge, Mass.: Harvard University Press, 272 p.

Allen, M.S. 1998. Holocene sea-level change on Aitutaki, Cook Islands: landscape change and human response. *Journal of Coastal Research*, 14, 10–22.

Alpers, C.N. and Brimhall, G.H. 1988. Middle Miocene climatic change in the Atacama Desert, northern Chile: evidence from supergene mineralization at La Escondida. *Geological Society of America, Bulletin*, 100, 1640–1656.

Alvarez, L.W., Alvarez, W., Asaro, F. and Michel, H.V. 1980. Extraterrestrial cause for the Cretaceous–Tertiary extinction. *Science*, 208, 1095–1108.

An, Z., Wu, X., Wang, P., Wang, S., Dong, G., Sun, X., Zhang, D., Lu, Y., Zheng, S. and Zhao, S. 1991. Changes in the monsoon and associated environmental changes in China since the Last Interglacial. *In:* Tungsheng, L., Zhongli, D. and Zhengtang, G. (editors). *Loess, Environment and Global Change*. Beijing: Science Press, 1–29.

An, Z., Porter, S.C., Zhou, W., Lu, Y., Donahue, D.J., Head, M.J., Wu, X., Ren, J. and Zheng, H. 1993. Episode of strengthened summer monsoon climate of Younger Dryas age on the Loess Plateau of central China. *Quaternary Research*, 39, 45–54.

Anderson, A. 1989. *Prodigious Birds*. Cambridge University Press, 238 p.

Anderson, A. 1994. Comment on J. Peter White's paper 'Site 820 and the evidence for early occupation in Australia'. *Quaternary Australasia*, 12, 30–33.

Anderson, A. and McGlone, M. 1992. Living on the edge – prehistoric land and people in New Zealand.

In: Dodson, J. (editor). *The Naive Lands: prehistory and environmental change in Australia and the south-west Pacific.* Melbourne: Longman Cheshire, 199–241.

Anderson, J.B. and Bartek, L.R. 1992. Cenozoic glacial history of the Ross Sea revealed in intermediate resolution seismic reflection data combined with drill site information. *In:* Kennett, J.P. and Warnke, D.A. (editors). *The Antarctic Paleoenvironment: A Perspective on Global Change, Part One.* Washington: American Geophysical Union, Antarctic Research Series 56, 231–263.

Anderson, P.M. and Brubaker, L.B. 1993. Holocene vegetation and climate histories of Alaska. *In:* Wright, H.E., Kutzbach, J.E., Webb, T., Ruddiman, W.F., Street-Perrott, F.A. and Bartleim, P.J. (editors). *Global Climates since the Last Glacial Maximum.* Minneapolis: University of Minnesota, 386–400.

Anderson, P.M. and Brubaker, L.B. 1994. Vegetation history of northcentral Alaska: a mapped summary of late-Quaternary pollen data. *Quaternary Science Reviews*, 13, 71–92.

Anderson, P.M. and Lozhkin, A.V. 1999. The stage 3 interstadial complex (Karginskii/middle Wisconsinian interval) of Beringia: variations in paleoenvironments and implications for paleoclimatic interpretations. *Quaternary Science Reviews*, in press.

Anderson, P.M., Bartlein, P.J. and Brubaker, L.B. 1994. Late Quaternary history of tundra vegetation in northwestern Alaska. *Quaternary Research*, 41, 306–315.

Anderson, R.Y. 1992a. Possible connection between surface winds, solar activity and the earth's magnetic field. *Nature*, 358, 51–53.

Anderson, R.Y. 1992b. Long-term changes in the frequency of occurrence of El Niño events. *In:* Diaz, H.F. and Markgraf, V. (editors). *El Niño: historical and paleoclimate aspects of the Southern Oscillation.* New York: Cambridge University Press, 193–200.

Andres, M.S., McKenzie, J.A., Vasconcelos, C. and Members of International GBR Drilling Consortium. 1997. Influence of Late-Quaternary paleoclimatology on the evolution of the Great Barrier Reef, north east Australia. *Supplement to EOS, Transactions American Geophysical Union*, 78, F52.

Arakawa, H. 1957. Climatic change as revealed by the data from the Far East. *Weather*, 12, 46–51.

Aristarain, A.J., Jouzel, J. and Lorius, C. 1990. A 400 years isotope record of the Antarctic Peninsula climate. *Geophysical Research Letters*, 17, 2369–2372.

Arkhipov, S.A., Isayeva, L.L., Bespaly, V.G. and

Glushkova, O. 1986. Glaciation of Siberia and north-east USSR. *Quaternary Science Reviews*, 5, 463–474.

Arldt, T. 1918. Die frage der permanenz der ozeane und kontinente. *Geographischer Anzeiger*, 19, 2–12.

Arnold, J.E. and Tissot, B. 1993. Measurement of significant marine paleotemperature variation using black abalone shells from prehistoric middens. *Quaternary Research*, 39, 390–394.

Askin, R.A. 1983. Tithonian (uppermost Jurassic)–Barremian (Lower Cretaceous) spores, pollen, and microplankton from the South Shetland Islands, Antarctica. *In:* Oliver, R.L., James, P.R. and Jago, J.B. (editors). *Antarctic Earth Science*. Canberra: Australian Academy of Science, 295–297.

Athens, J.S. and Ward, J.V. 1993. Environmental change and prehistoric Polynesian settlement in Hawai'i. *Asian Perspectives*, 32, 205–223.

Athens, J.S. and Ward, J.V. 1998. *Paleoenvironment and Prehistoric Landscape Change: a sediment record from Lake Hagoi, Tinian, CNMI.* Honolulu: International Archaeological Research Institute, 128 p.

Atwater, B.F. 1987. Evidence for great Holocene earthquakes along the outer coast of Washington State. *Science*, 236, 942–944.

Atwater, B.F. and Yamaguchi, D.K. 1991. Sudden, probably coseismic submergence of Holocene trees and grass in coastal Washington State. *Geology*, 19, 706–709.

Atwater, B.F., Núñez, H.J. and Vita-Finzi, C. 1992. Net late Holocene emergence despite earthquake-induced submergence, south-central Chile. *Quaternary International*, 15/16, 77–85.

Aubry, M.P. 1992. Late Paleogene calcareous nannoplankton evolution: a tale of climatic deterioration. *In:* Prothero, D.R. and Berggren, W.A. (editors). *Eocene–Oligocene Climatic and Biotic Evolution.* Princeton University Press, 272–309.

Bachmann, K. 1987. Speciation in *Microseris*. *In:* Hovenkamp, P. (editor). *Systematics and Evolution: a matter of diversity.* Utrecht: Utrecht University Press, 67–80.

Bahlburg, H. 1993. Hypothetical southeast Pacific continent revisited: new evidence from the middle Palaeozoic basins of northern Chile. *Geology*, 21, 909–912.

Bahn, P.G. 1993. 50,000-year-old Americans of Pedra Furada. *Nature*, 362, 114–115.

Bahn, P.G. and Flenley, J. 1992. *Easter Island, Earth Island.* London: Thames and Hudson, 240 p.

Baker, R.G.V. and Haworth, R.J. 1997. Further evidence from relic shellcrust sequences for a late Holocene higher sea level for eastern Australia. *Marine Geology*, 141, 1–9.

Barbetti, M. and Allen, H. 1972. Prehistoric man at Lake Mungo, Australia, by 32,000 years BP. *Nature*, 240, 46–48.

Barnosky, A.D. 1989. The late Pleistocene event as a paradigm for widespread mammal extinction. *In:* Donovan, S.K. (editor). *Mass Extinctions: processes and evidence.* London: Belhaven, 235–254.

Baroni, C. and Orombelli, G. 1991. Holocene raised beaches at Terra Nova Bay, Victoria Land, Antarctica. *Quaternary Research*, 36, 157–177.

Baroni, C. and Orombelli, G. 1994. Abandoned penguin rookeries as Holocene paleoclimatic indicators in Antarctica. *Geology*, 22, 23–26.

Barrash, W. and Venkatakrishnan, R. 1982. Timing of late Cenozoic volcanic and tectonic events along the western margin of the North American plate. *Geological Society of America, Bulletin*, 93, 977–989.

Barrera, E. 1994. Global environmental changes preceding the Cretaceous–Tertiary boundary: early–late Maastrichian transition. *Geology*, 22, 877–880.

Barrett, P.J., Adams, C.J., McIntosh, W.C., Swisher, C.C. and Wilson, G.S. 1992. Geochronological evidence supporting Antarctic deglaciation three million years ago. *Nature*, 359, 816–818.

Barron, E.J. and Peterson, W.H. 1989. Model simulation of the Cretaceous Ocean circulation. *Science*, 244, 684–686.

Barron, J.A. and Baldauf, J.G. 1990. Development of biosiliceous sedimentation in the North Pacific during the Miocene and early Pliocene. *In:* Tsuchi, R. (editor). *Pacific Neogene Events: their timing, nature and interrelationships.* Tokyo: University of Tokyo Press, 43–63.

Barry, R.G. and Chorley, R.J. 1982. *Atmosphere, Weather and Climate.* London: Methuen, 4th edition, 407 p.

Bartlett, A.S. and Barghoorn, E.S. 1973. Phytogeographic history of the Isthmus of Panama during the last 12,000 years. *In:* Graham, A. (editor). *Vegetation and Vegetational History of Northern Latin America.* Amsterdam: Elsevier, 203–299.

Bautista, G.M. 1990. The forestry crisis in the Philippines: nature, causes, and issues. *The Developing Economies*, 28, 67–94.

Beaton, J.M. 1985. Evidence for a coastal occupation time-lag at Princess Charlotte Bay, north Queensland, and the implications for coastal colonisation and population growth theories for Aboriginal Australia. *Archaeology in Oceania*, 20, 1–20.

Beck, J.W., Edwards, R.L., Ito, E., Taylor, F.W., Recy, J., Rougerie, F., Joannot, P. and Henin, C. 1992. Sea-surface temperature from coral skeletal strontium/calcium ratios. *Science*, 257, 644–647.

Begét, J., Mason, O. and Anderson, P. 1992. Age, extent and climatic significance of the c. 3400 BP Aniakchak tephra, western Alaska. *The Holocene*, 2, 51–56.

Belasky, P. and Runnegar, B. 1994. Permian longitudes of Wrangellia, Stikinia, and eastern Klamath terranes based on coral biogeography. *Geology*, 22, 1095–1098.

Benson, L.V., Lund, S.P., Burdett, J.W., Kashgarian, M., Rose, T.P., Smoot, J.P. and Schwartz, M. 1998. Correlation of late-Pleistocene lake-level oscillations in Mono Lake, California, with North Atlantic climatic events. *Quaternary Research*, 49, 1–10.

Berger, W.H. and Labeyrie, L.D. (editors). 1987. *Abrupt Climatic Change.* Dordrecht: Reidel, 425 p.

Berggren, W.A. and Prothero, D.R. 1992. Eocene–Oligocene climatic and biotic evolution. *In:* Prothero, D.R. and Berggren, W.A. (editors). *Eocene–Oligocene Climatic and Biotic Evolution.* Princeton University Press, 29–45.

Berggren, W.A., Hilgen, F.J., Langereis, C.G., Kent, D.V., Obradovich, J.D., Raffi, I., Raymo, M.E. and Shackleton, N.J. 1995. Late Neogene chronology: new perspectives in high-resolution stratigraphy. *Geological Society of America, Bulletin*, 107, 1272–1287.

Berkman, P.A. 1992. Circumpolar distribution of Holocene marine fossils in Antarctic beaches. *Quaternary Research*, 37, 256–260.

Bespaly, V.G., Glushkova, O.Y., Ivanov, V.F., Kolpakov, V.V. and Prokhorova, T.P. 1993. North-east of Asia section: stratigraphy and paleogeography of the late Pleistocene. *In:* Velichko, A.A. (editor). *Evolution of Landscapes and Climate of Northern Eurasia.* Moscow: Nauka, 59–62.

Bestland, E.A., Retallack, G.J. and Swisher, C.C. 1997. Stepwise climate change recorded in Eocene–Oligocene paleosol sequences from central Oregon. *Journal of Geology*, 105, 153–172.

Binford, M.W., Kolata, A.L., Brenner, M., Janusek, J., Seddon, M., Abbott, M. and Curtis, J. 1997. Climate variation and the rise and fall of an Andean civilization. *Quaternary Research*, 47, 235–248.

Birkenmajer, K. and Zastawniak, E. 1989. Late Cretaceous–Tertiary floras of King George Island, West Antarctica: their stratigraphic distribution and paleoclimatic significance. *In:* Crame, J. (editor). *Origin and Evolution of the Antarctic Biota.* Geological Society of London, Special Publication 47, 227–240.

Bishop, P. 1994. Late Quaternary environments in Australasia. *In: Quaternary Stratigraphy of Asia and the Pacific, IGCP 296.* ESCAP, Mineral Resources Development Series 63, 9–27.

Björck, S., Olsson, S., Ellis-Evans, C., Hökansson, H., Humlum, O. and de Lirio, J.M. 1996. Late Holocene palaeoclimatic records from lake sediments on James

Ross Island, Antarctica. *Palaeogeography, Palaeoclimatology, Palaeoecology*, 121, 195–220.

Black, R.F. 1980. Late Quaternary climatic changes in the Aleutian islands. *In:* Mahaney, W.C. (editor). *Quaternary Paleoclimate*. Norwich: GeoAbstracts, 47–62.

Blong, R.J. 1982. *The Time of Darkness: local legends and volcanic reality in Papua New Guinea*. Canberra: Australian National University, 257 p.

Blong, R.J. 1984. *Volcanic Hazards*. Sydney: Academic Press, 424 p.

Bloom, A.L. 1970. Holocene submergence in Micronesia as the standard for eustatic sea-level changes. *Quaternaria*, 145–154.

Bloom, A.L., Broecker, W.S., Chappell, J.M.A., Matthews, R.K. and Mesolella, K.J. 1974. Quaternary sea level fluctuations on a tectonic coast: new ^{230}Th/^{234}U dates from the Huon Peninsula, New Guinea. *Quaternary Research*, 4, 185–205.

Blunier, T., Chappellaz, J., Schwander, J., Stauffer, B. and Raynaud, D. 1995. Variations in atmospheric methane concentration during the Holocene epoch. *Nature*, 374, 46–49.

Bockheim, J.G., Wilson, S.C., Denton, G.H., Andersen, B.G. and Stuiver, M. 1989. Late Quaternary ice-surface fluctuations of Hatherton Glacier, Transantarctic Mountains. *Quaternary Research*, 31, 229–254.

Bonavia, D. and Grobman, A. 1989. Andean maize: its origins and domestication. *In:* Harris, D.R. and Hillman, G.C. (editors). *Foraging and Farming: the evolution of plant exploitation*. London: Unwin Hyman, 456–470.

Boninsegna, J.A. 1992. South American dendrochronological records. *In:* Bradley, R.S. and Jones, P.D. (editors). *Climate since AD 1500*. London: Routledge, 446–462.

Bonnichsen, R. and Turnmire, K.L. (editors). 1991. *Clovis: origins and adaptations*. Corvallis: Center for the Study of the First Americans, 344 p.

Bourgois, J., Lagabrielle, Y., De Wever, P., Suess, E. and NAUTIPERC Team. 1993. Tectonic history of the northern Peru convergent margin during the past 400 ka. *Geology*, 21, 531–534.

Bourrouilh-Le Jan, F.G. 1989. The oceanic karsts: modern bauxite and phosphate ore deposits on the high carbonate islands (so-called 'uplifted atolls') of the Pacific Ocean. *In:* Bosák, P., Ford, D.C., Glazek,. J. and Horácek, I. (editors). *Paleokarst: a systematic and regional review*. Amsterdam: Elsevier, 443–471.

Bowdler, S. 1993. Views of the past in Australian prehistory. *In:* Spriggs, M., Yen, D.E., Ambrose, W., Jones, R., Thorne, A. and Andrews, A. (editors). *A Community of Culture: the people and prehistory of the Pacific*. Canberra: Australian National University, 123–138.

Bowen, D.Q., Richmond, G.M., Fullerton, D.S., Sibrava, V., Fulton, R.J. and Velichko, A.A. 1986. Correlation of Quaternary glaciations in the northern hemisphere. *Quaternary Science Reviews*, 5, 509–510.

Bowler, J. 1971. Pleistocene salinities and climatic change: evidence from lakes and lunettes in southeastern Australia. *In:* Mulvaney, D.J. and Golson, J. (editors). *Aboriginal Man and Environment in Australia*. Canberra: The Australian National University, 47–65.

Bowler, J.M., Hope, G.S., Jennings, J.N., Singh, G. and Walker, D. 1976. Late Quaternary climates of Australia and New Guinea. *Quaternary Research*, 6, 359–394.

Bradbury, J.P. 1982. Holocene chronostratigraphy of Mexico and Central America. *Striae*, 16, 46–48.

Bradbury, J.P. 1989. Late Quaternary lacustrine paleoenvironments in the Cuenca de México. *Quaternary Science Reviews*, 8, 75–100.

Bradbury, J.P. 1992. Late Cenozoic lacustrine and climatic environments at Tule Lake, northern Great Basin, USA. *Climate Dynamics*, 6, 275–285.

Bradbury, J.P. 1997. A diatom record of climate and hydrology for the past 200 ka from Owens Lake, California with comparison to other Great Basin records. *Quaternary Science Reviews*, 16, 203–219.

Bradley, R.S. and Jones, P.D. 1993. 'Little Ice Age' summer temperature variations: their nature and relevance to recent global warming trends. *The Holocene*, 3/4, 367–376.

Bridgman, H.A. 1983. Could climatic change have had an influence on the Polynesian migrations? *Palaeogeography, Palaeoclimatology, Palaeo-ecology*, 41, 193–206.

Briggs, J.C. 1987. *Biogeography and Plate Tectonics*. Amsterdam: Elsevier, 204 p.

Brigham-Grette, J. and Carter, L.D. 1992. Pliocene marine transgressions of northern Alaska: circum-arctic correlations and paleoclimatic interpretations. *Arctic*, 45, 74–89.

Broad, R. 1993. *Plundering Paradise: the struggle for the environment in the Philippines*. Berkeley: University of California Press, 197 p.

Broecker, W.S. 1997. Thermohaline circulation, the Achilles heel of our climate system: will man-made CO_2 upset the current balance? *Science*, 278, 1582–1588.

Broecker, W.S. and Denton, G.G.H. 1990. The role of ocean–atmosphere reorganisations in glacial cycles. *Quaternary Science Reviews*, 9, 305–341.

Broecker, W.S., Andree, M., Klas, M., Bonani, G., Wolfli, W. and Oeschger, H. 1988. New evidence from the South China Sea for an abrupt termination of the last glacial period. *Nature*, 333, 156–158.

Broecker, W.S., Kennett, J.P., Flower, B.P., Teller, J.T.,

Trumbore, S., Bonani, G. and Wolfli, W. 1989. Routing of meltwater from the Laurentide ice sheet during the Younger Dryas cold episode. *Nature*, 341, 318–321.

Brook, E.J., Kurz, M.D., Ackert, R.P., Denton, G.H., Brown, E.T., Raisbeck, G.M. and Yiou, F. 1993. Chronology of Taylor Glacier advances in Arena Valley, Antarctica, using *in situ* cosmogenic [3]He and [10]Be. *Quaternary Research*, 39, 11–23.

Brothers, R.N. and Lillie, A.R. 1985. Regional geology of New Caledonia. *In:* Nairn, A.E.M., Stehli, F.G. and Uyeda, S. (editors). *The Ocean Basins and Margins, Volume 7A. The Pacific Ocean.* New York: Plenum Press, 325–374.

Browman, D.L. 1983. Tectonic movement and agrarian collapse in prehispanic Peru. *Nature*, 302, 568–569.

Bruno, L.A., Baur, H., Graf, T., Schlüchter, C., Signer, P. and Wieler, R. 1997. Dating of Sirius Group tillites in the Antarctic Dry Valleys with cosmogenic [3]He and [21]Ne. *Earth and Planetary Science Letters*, 147, 37–54.

Bryant, E. 1992. Last interglacial and Holocene trends in sea-level maxima around Australia: implications for modern rates. *Marine Geology*, 108, 209–217.

Bryant, E.A., Young, R.W., Price, D.M. and Short, S.A. 1992. Evidence for Pleistocene and Holocene raised marine deposits, Sandon Point, New South Wales. *Australian Journal of Earth Sciences*, 39, 481–493.

Bucknam, R.C., Hemphill-Haley, E. and Leopold, E.B. 1992. Abrupt uplift within the past 1700 years at southern Puget Sound, Washington. *Science*, 258, 1612–1617.

Bull, W.B. and Cooper, A.F. 1986. Uplifted marine terraces along the Alpine Fault, New Zealand. *Science*, 234, 1225–1228.

Bulmer, S. 1989. Gardens in the south: diversity and change in prehistoric Maaori agriculture. *In:* Harris, D.R. and Hillman, G.C. (editors). *Foraging and Farming: the evolution of plant exploitation.* London: Unwin Hyman, 688–705.

Bunting, R., 1995. The environment and settler society in western Oregon. *Pacific Historical Review*, 64, 413–432.

Burbank, D.W. 1981. A chronology of late Holocene glacier fluctuations on Mount Rainier, Washington. *Arctic and Alpine Research*, 13, 369–386.

Burckhardt, C. 1902. Traces géologiques d'un ancien continente Pacifique. *Revista del Museo La Plata*, 10, 179–192.

Burke, R.M. and Birkeland, P.W. 1984. Holocene glaciation in the mountain ranges of the western United States. *In:* Wright, H.E. (editor). *Late-Quaternary Environments of the United States, Volume 2. The Holocene.* London: Longman, 3–11.

Burke, W.H., Denison, R.E., Hetherington, E.A., Koepnick, R.B., Nelson, H.F. and Otto, J.B. 1982. Variation of [87]Sr/[86]Sr throughout Phanerozoic time. *Geology*, 10, 516–519.

Burn, C.R. 1997. Cryostratigraphy, paleogeography, and climate change during the early Holocene warm interval, western Arctic coast, Canada. *Canadian Journal of Earth Sciences*, 34, 912–925.

Burrett, C., Duhig, N., Berry, R. and Varne, R. 1991. Asian and south-western Pacific continental terranes derived from Gondwana, and their biogeographic significance. *Australian Systematic Botany*, 4, 13–24.

Burrows, C.J. and Russell, J.B. 1990. Aranuian vegetation history of the Arrowsmith Range, Canterbury I. Pollen diagrams, plant macrofossils, and buried soils from Prospect Hill. *New Zealand Journal of Botany*, 28, 323–345.

Burt, R.L. and Williams, W.T. 1988. Plant introduction in Australia. *In:* Home, R.W. (editor). *Australian Science in the Making.* Cambridge: Cambridge University Press, 252–276.

Bush, A.B.G. and Philander, S.G.H. 1998. The role of ocean–atmosphere interactions in tropical cooling during the Last Glacial Maximum. *Science*, 279, 1341–1344.

Cabioch, G., Thomassin, B.A. and Lecolle, J.F. 1989. Age d'émersion des récifs frangeants holocènes autour de la 'Grande Terre' de Nouvelle-Calédonie (SO Pacifique); nouvelle interprétation de la courbe des niveaux marins depuis 8 000 ans B.P. *Comptes Rendus de l'Academie des Sciences, Paris*, 308, 419–425.

Calder, N. 1974. *The Weather Machine.* London: BBC, 143 p.

Calvert, S.E. 1964. Factors affecting distribution of laminated diatomaceous sediments in Gulf of California. *American Association of Petroleum Geologists, Memoir*, 3, 311–330.

Cambray, H. and Cadet, J.P. 1994. Testing global synchronism in peri-Pacific arc volcanism. *Journal of Volcanology and Geothermal Research*, 63, 145–164.

Campa U., M.F. 1985. The Mexican thrust belt. *In:* Howell, D.G. (editor). *Tectonostratigraphic Terranes of the Circum-Pacific Region.* Houston: Circum-Pacific Council for Energy and Mineral Resources, Earth Science Series, No. 1, 299–313.

Cande, S.C. and Kent, D.V. 1992. A new geomagnetic polarity time scale for the late Cretaceous and Cenozoic. *Journal of Geophysical Research*, 97, 13917–13951.

Cann, R.L. 1996. Hybrids, mothers, and clades: who is right? *In:* Akasawa, T. and Szathmáry, E.J.E. (editors). *Prehistoric Mongoloid Dispersals.* Oxford: Oxford University Press, 41–51.

Cantrell, C. and Rodgers, K.A. 1989. Australia and the Funafuti connection, 2. An all Australian assault. *Search*, 20, 27–30.

Caputo, M.V. and Crowell, J.C. 1985. Migration of glacial centers across Gondwana during Palaeozoic era. *Geological Society of America, Bulletin*, 96, 1020–1036.

Carey, S.W. 1976. *The Expanding Earth*. Amsterdam: Elsevier, 488 p.

Carter, L.D., Brigham-Grette, J., Marincovich, L., Pease, V.L. and Hillhouse, J.W. 1986. Late Cenozoic Arctic Ocean sea ice and terrestrial paleoclimate. *Geology*, 14, 675–678.

Casamiquela, R.M. 1980. Considérations écologiques et zoogéographiques sur les vertébrés de la zone littorale de la mer du Maestrichien dans le Nord de la Patagonie. *Mémoires de la Société Géologique de France*, 139, 53–55.

Cawood, P.A. and Leitch, E.C. 1985. Accretion and dispersal tectonics of the southern New England Fold Belt, eastern Australia. *In:* Howell, D.G. (editor). *Tectonostratigraphic Terranes of the Circum-Pacific Region*. Houston: Circum-Pacific Council for Energy and Mineral Resources, Earth Science Series, No. 1, 481–492.

Chang, T.T. 1976. The origin, evolution, cultivation, dissemination, and diversification of Asian and African rices. *Euphytica*, 25, 425–441.

Chang, T.T. 1989. Domestication and spread of the cultivated rices. *In:* Harris, D.R. and Hillman, G.C. (editors). *Foraging and Farming: the evolution of plant exploitation*. London: Unwin Hyman, 408–417.

Chappell, J. 1974. Geology of coral terraces, Huon Peninsula, New Guinea: a study of Quaternary tectonic movements and sea level changes. *Geological Society of America, Bulletin*, 85, 553–570.

Chappell, J. 1983. A revised sea-level record for the last 300,000 years from Papua New Guinea. *Search*, 14, 99–101.

Chappell, J. and Polach, H. 1991. Post-glacial sea-level rise from a coral record at Huon Peninsula, Papua New Guinea. *Nature*, 349, 147–149.

Chappell, J. and Shackleton, N.J. 1986. Oxygen isotopes and sea level. *Nature*, 324, 137–140.

Chappell, J. and Veeh, H.H. 1978. ^{238}Th/^{234}U age support of an interstadial sea level of −40 m at 30,000 yr BP. *Nature*, 276, 602–603.

Chappell, J., Rhodes, E.G., Thom, B.G. and Wallensky, E. 1982. Hydro-isostasy and the sea-level isobase of 5500 BP in north Queensland, Australia. *Marine Geology*, 49, 81–90.

Chappell, J., Ota, Y. and Berryman, K. 1996. Late Quaternary coseismic uplift history of Huon Peninsula, Papua New Guinea. *Quaternary Science Reviews*, 15, 7–22.

Chatwin, B. 1987. *The Songlines*. New York: Viking Penguin, 293 p.

Chen, C. and Olsen, J.W. 1990. China at the Last Glacial Maximum. *In:* Soffer, O. and Gamble, C. (editors). *The World at 18 000 BP, Volume 1*. London: Unwin Hyman, 276–295.

Chen, F. 1979. The Holocene strata of Beijing area and the change of its natural environment. *Scientia Sinica*, 9, 900–907.

Chen, S. 1994. Early urbanization in the eastern Zhou in China (770–221 BC): an archaeological view. *Antiquity*, 68, 724–744.

Chen, Y., Kam-ling, T., Chen, F., Gao, Z., Zou, Q. and Chen, Z. 1988. *The Great Tangshan Earthquake of 1976*. Oxford: Pergamon, 153 p.

Chi, C-C. 1994. Growth with pollution: unsustainable development in Taiwan and its consequences. *Studies in Comparative International Development*, 29, 23–47.

Chinzei, K., Fujioka, K., Kitazato, H., Koizumi, I., Oba, T., Oda, M., Okada, H., Sakai, T. and Tanimura, Y. 1987. Postglacial environmental change of the Pacific Ocean off the coast of central Japan. *Marine Micropaleontology*, 11, 273–291.

Chorley, R.J. 1963. Diastrophic background to twentieth-century geomorphological thought. *Geological Society of America, Bulletin*, 74, 953–970.

Christophel, D.C. 1990. The impact of mid-Tertiary climatic changes on the development of the modern Australian flora. *Geological Society of America, Abstracts with Programs* 22, A77.

Chuey, J.M., Rea, D.K. and Pisias, N.G. 1987. Late Pleistocene paleoclimatology of the central equatorial Pacific: a quantitative record of eolian and carbonate deposition. *Quaternary Research*, 28, 323–339.

Churchward, J. 1959. *The Lost Continent of Mu*. London: Spearman, 335 p.

Churkin, M., Whitney, J.W. and Rogers, J.F. 1985. The North American-Siberian connection, a mosaic of craton fragments in a matrix of oceanic terranes. *In:* Howell, D.G. (editor). *Tectonostratigraphic Terranes of the Circum-Pacific Region*. Houston: Circum-Pacific Council for Energy and Mineral Resources, Earth Science Series, No. 1, 79–84.

Ciesielski, P.F. and Weaver, F.M. 1974. Early Pliocene temperature changes in the Antarctic seas. *Geology*, 2, 511–515.

Cinq-Mars, J. 1990. La place des Grottes du Poisson-Bleu dans la prehistoire Beringienne. *Revista de Arquelogia Americana*, 1, 9–32.

Clague, J.J. 1989a. Quaternary geology of the Queen Charlotte Islands. *In:* Scudder, G.G.E. and Gessler, N. (editors). *The Outer Shores*. University of British Columbia (Proceedings of the Queen Charlotte Islands First International Symposium, August 1984), 65–74.

Clague, J.J. 1989b. Sea levels on Canada's Pacific coast: past and future trends. *Episodes*, 12, 29–33.

Clague, J.J. 1991. Quaternary glaciation and sedimentation. *In:* Gabrielse, H. and Yorath, C.J. (editors). *Geology of the Cordilleran Orogen in Canada*. Geological Survey of Canada, Geology of Canada, No. 4, 419–434.

Clague, J.J. and Bobrowsky, P.T. 1994. Evidence for a large earthquake and tsunami 100–400 years ago on western Vancouver Island, British Columbia. *Quaternary Research*, 41, 176–184.

Clague, J.J. and Evans, S.G. 1993. Historic retreat of Grand Pacific and Melbern Glaciers, Saint Elias Mountains, Canada: an analogue for decay of the Cordilleran ice sheet at the end of the Pleistocene? *Journal of Glaciology*, 39, 619–624.

Clapperton, C.M. 1983. The glaciation of the Andes. *Quaternary Science Reviews*, 2, 83–155.

Clapperton, C.M. 1990a. Quaternary glaciations in the southern hemisphere: an overview. *Quaternary Science Reviews*, 9, 299–304.

Clapperton, C.M. 1990b. Quaternary glaciations in the southern ocean and Antarctic Peninsula area. *Quaternary Science Reviews*, 9, 229–252.

Clapperton, C.M. 1993a. *Quaternary Geology and Geomorphology of South America*. Amsterdam: Elsevier, 779 p.

Clapperton, C.M. 1993b. Nature of environmental changes in South America at the last glacial maximum. *Palaeogeography, Palaeoclimatology, Palaeoecology*, 101, 189–208.

Clapperton, C.M. and Sugden, D.E. 1982. Late Quaternary glacial history of George VI Sound area, West Antarctica. *Quaternary Research*, 18, 243–267.

Clapperton, C.M., Hall, M., Mothes, P., Hole, M.J., Still, J.W., Helmens, K.F., Kuhry, P. and Gemmell, A.M.D. 1997. A Younger Dryas icecap in the equatorial Andes. *Quaternary Research*, 47, 13–28.

Clarke, W.C. 1977. A change of subsistence staple in prehistoric New Guinea. *In:* Leakey, C.L.A. (editor). *Proceedings of the Third Symposium of the International Society for Root Crops*. Ibadan, Nigeria: International Society for Tropical Root Crops, 159–163.

Clarke, W.C. 1995. Lands rich in thought. *South Pacific Journal of Natural Science*, 14, 55–67.

Clarke, W.C. and Morrison, J. 1987. Land mismanagement and the development imperative in Fiji. *In:* Blaikie, P. and Brookfield, H. (editors). *Land Degradation and Society*. New York: Methuen, 76–85.

CLIMAP Project Members. 1976. The surface of the ice-age earth. *Science*, 191, 1131–1137.

CLIMAP Project Members. 1984. The last interglacial ocean. *Quaternary Research*, 21, 123–224.

Coccioni, R. and Galeotti, S. 1994. K-T boundary extinction: geological instantaneous or gradual event? Evidence from deep-sea benthic foraminifera. *Geology*, 22, 779–782.

COHMAP Members, 1988. Climatic changes of the last 18,000 years: observations and model simulations. *Science*, 241, 1043–1052.

Colbert, E.H. 1982. Mesozoic vertebrates of Antarctica. *In:* Craddock, C. (editor). *Antarctic Geoscience*. Madison: University of Wisconsin, 619–627.

Cole, J.E., Shen, G.T., Fairbanks, R.G. and Moore, M. 1992. Coral monitors of El Niño/Southern Oscillation dynamics across the equatorial Pacific. *In:* Diaz, H. and Markgraf, V. (editors). *El Niño: historical and paleoclimatic aspects of the Southern Oscillation*. Cambridge: Cambridge University Press, 349–375.

Cole, J.W. 1985. Distribution and tectonic setting of late Cenozoic volcanism in New Zealand. *Royal Society of New Zealand, Bulletin* 23, 7–20.

Cole, K.L. and Liu, G-W. 1994. Holocene paleoecology of an estuary on Santa Rosa island, California. *Quaternary Research*, 41, 326–335.

Colhoun, E.A. 1991. *Climate during the Last Glacial maximum in Australia and New Guinea*. Australia and New Zealand Geomorphology Group Special Publication 2, 71 p.

Colinvaux, P.A. 1972. Climate and the Galapagos islands. *Nature*, 240, 17–20.

Colinvaux, P.A. 1984. The Galápagos climate: present and past. *In:* Perry, R. (editor). *Galapagos*. Oxford: Pergamon, 55–69.

Colinvaux, P.A. and Liu, K. 1987. The late-Quaternary climate of the western Amazon Basin. *In:* Berger, W.H. and Labeyrie, L.D. (editors). *Abrupt Climatic Change*. Dordrecht: Reidel, 113–122.

Collins, L.S., Coates, A.G., Berggren, W.A., Aubry, M-P. and Zhang, J. 1996. The late Miocene Panama isthmian strait. *Geology*, 24, 687–690.

Colten, R.H. 1994. Prehistoric animal exploitation, environmental change, and emergent complexity on Santa Cruz island, California. Unpublished paper, *Fourth California Island Symposium*, Santa Barbara.

Coltrinari, L. 1993. Global Quaternary changes in South America. *Global and Planetary Change*, 7, 11–23.

Coney, P.J. 1989. The North American cordillera. *In:* Ben-Avraham, Z. (editor). *The Evolution of the Pacific Ocean Margins*. New York: Oxford University Press, 43–52.

Cook, E., Bird, T., Peterson, M., Barbetti, M., Buckley, B., D'Arrigo, R. and Francey, R. 1992. Climatic change over the last millennium in Tasmania reconstructed from tree-rings. *The Holocene*, 2,3, 205–217.

Cooke, R.G., Norr, L. and Piperno, D.R. 1996. Native Americans and the Panamanian landscape. *In:* Reitz,

E.J., Newsom, L.A. and Scudder, S.J. (editors). *Case Studies in Environmental Archaeology*. New York: Plenum, 103–126.

Cosgrove, R. 1989. Thirty thousand years of human colonization in Tasmania: new Pleistocene dates. *Science*, 243, 1706–1708.

Costin, A.B. 1972. Carbon 14 dates from the Snowy Mountains area, southeastern Australia and their interpretation. *Quaternary Research*, 2, 579–590.

Coulson, F.I. 1985. Solomon Islands. *In:* Nairn, A.E.M., Stehli, F.G. and Uyeda, S. (editors). *The Ocean Basins and Margins, Volume 7A. The Pacific Ocean*. New York: Plenum Press, 607–682.

Cox, A., Debiche, M.G. and Engebretson, D.C. 1989. Terrane trajectories and plate interactions along continental margins in the North Pacific Basin. *In:* Ben-Avraham, Z. (editor). *The Evolution of the Pacific Ocean Margins*. New York: Oxford University Press, 20–35.

Cox, C.B. 1990. New geological theories and old biogeographical problems. *Journal of Biogeography*, 17, 117–130.

Cox, P.A. and Elmqvist, T. 1997. Ecocolonialism and indigenous-controlled rainforest preserves in Samoa. *Ambio*, 26, 84–89.

Creber, G.T. and Chaloner, W.G. 1985. Tree growth in the Mesozoic and early Tertiary and the reconstruction of palaeoclimates. *Palaeogeography, Palaeoclimatology, Palaeoecology*, 52, 35–60.

Cronin, T.M. 1992. Quantitative reconstruction of Pliocene and Quaternary oceanography, Sea of Japan, based on marine ostracoda. *In: ICP IV, Kiel, Abstracts*, 90–91.

Csejtey, B., Cox, D.P., Evarts, R.C., Stricker, G.D. and Foster, H.L. 1982. The Cenozoic Denali fault system and the Cretaceous accretionary development of southern Alaska. *Journal of Geophysical Research*, 87, 3741–3754.

Curray, R.P. 1966. Glaciation about 3,000,000 years ago in the Sierra Nevada. *Science*, 154, 770–771.

Currie, R.G. 1987. Examples and implications of 18.6- and 11-yr terms in world weather records. *In:* Rampino, M.R., Sanders, J.E., Newman, W.S. and Königsson, L.K. (editors). *Climate History, Periodicity, and Predictability*. New York: Van Nostrand Reinhold, 378–403.

Currie, R.G. and Fairbridge, R.W. 1985. Periodic 18.6-year and cyclic 11-year induced drought and flood in northeastern China. *Quaternary Science Reviews*, 4, 109–134.

Cushing, D. 1975. *Fisheries Resources of the Sea and their Management*. Oxford: Oxford University Press, 87 p.

Dalrymple, G.B. and Clague, D.A. 1976. Age of the Hawaiian-Emperor bend. *Earth and Planetary Science Letters*, 31, 313–329.

Dalziel, I.W.D. 1991. Pacific margins of Laurentia and East-Antarctica-Australia as a conjugate rift pair: evidence and implications for an Eocambrian supercontinent. *Geology*, 19, 598–601.

Dalziel, I.W.D. and Forsythe. R.D. 1985. Andean evolution and the terrane concept. *In:* Howell, D.G. (editor). *Tectonostratigraphic Terranes of the Circum-Pacific Region*. Houston: Circum-Pacific Council for Energy and Mineral Resources, Earth Science Series, No. 1, 565–581.

Dalziel, I.W.D., Dalla Salda, L.H. and Gahagan, L.M. 1994. Palaeozoic Laurentia–Gondwana interaction and the origin of the Appalachian–Andean mountain system. *Geological Society of America, Bulletin*, 106, 243–252.

Dansgaard, W. 1987. Ice core evidence of abrupt climatic changes. *In:* Berger, W.H. and Labeyrie, L.D. (editors). *Abrupt Climatic Change*. Dordrecht: Reidel, 223–233.

Dansgaard, W., Johnsen, S.J., Clausen, H.B. and Langway, C.C. 1971. Climatic record revealed by the Camp Century ice core. *In:* Turekian, K.K. (editor). *The Late Cenozoic Glacial Ages*. New Haven: Yale University Press, 37–56.

Darwin, C.R. 1842. *Structure and Distribution of Coral Reefs*. London: Smith, Elder, 214 p.

Darwin, G. 1879. The precession of a viscous spheroid and the remote history of the Earth. *Philosophical Transactions of the Royal Society of London*, 170, 447–538.

David, G. 1994. Dynamics of the coastal zone in the high islands of Oceania: management implications and options. *In:* Waddell, E. and Nunn, P.D. (editors). *The Margin Fades: geographical itineraries in a world of islands*. Suva: Institute of Pacific Studies, The University of the South Pacific, 189–213.

Davis, O.K. 1992. Rapid climatic change in coastal southern California inferred from pollen analysis of San Joaquin marsh. *Quaternary Research*, 37, 89–100.

Davis, O.K. and Sellers, W.D. 1987. Contrasting climatic histories for western north America during the early Holocene. *Current Research in the Pleistocene*, 4, 87–89.

Debiche, M.G., Cox, A. and Engebretson, D.C. 1987. The motion of an allochthonous terranes across the North Pacific Basin. *Geological Society of America, Special Paper* 207, 1–49.

DeDeckker, P., Correge, T. and Head, J. 1991. Late Pleistocene record of cyclic eolian activity from tropical Australia suggesting the Younger Dryas is not an unusual climatic event. *Geology*, 19, 602–605.

Dengo, G. 1985. Mid America: tectonic setting for the Pacific margin from southern Mexico to northwestern Colombia. *In:* Nairn, A.E.M., Stehli,

F.G. and Uyeda, S. (editors). *The Ocean Basins and Margins, Volume 7A. The Pacific Ocean.* New York: Plenum Press, 123–180.

Denton, G.H. and Hendy, C.H. 1994. Younger Dryas age advance of Franz Josef Glacier in the Southern Alps of New Zealand. *Science*, 264, 1434–1437.

Denton, G.H., Bockheim, J.G., Wilson, S.C. and Stuiver, M. 1989a. Late Wisconsin and early Holocene glacial history, inner Ross Embayment, Antarctica. *Quaternary Research*, 31, 151–182.

Denton, G.H., Bockheim, J.G., Wilson, S.C., Leide, J.E. and Andersen, B.G. 1989b. Late Quaternary ice-surface fluctuations of Beardmore Glacier, Transantarctic Mountains. *Quaternary Research*, 31, 183–209.

Denton, G.H., Prentice, M.L. and Burckle, L.H. 1991. Cainozoic history of the Antarctic ice sheet. *In:* Tingey, R.J. (editor). *The Geology of Antarctica.* Oxford: Clarendon, 365–433.

Dersch, M. and Stein, R. 1994. Late Cenozoic records of eolian quartz flux in the Sea of Japan (ODP Leg 128, Sites 798 and 799) and paleoclimate in Asia. *Palaeogeography, Palaeoclimatology, Palaeoecology*, 108, 523–535.

de Saint Ours, P., Talai, B., Mori, J., McKee, C. and Itikarai, I. 1991. Coastal and seafloor changes at an active volcano: example of Rabaul caldera, Papua New Guinea. *SOPAC Technical Bulletin*, 7, 1–13.

DeVries, T.J. 1987. A review of geological evidence for ancient El Niño activity in Peru. *Journal of Geophysical Research*, 92, 14471–14479.

Dewar, G.J. 1967a. Some aspects of the topography and glacierization of Adelaide Island. *British Antarctic Survey, Bulletin*, 11, 37–47.

Dewar, G.J. 1967b. Block terraces in the Adelaide Island area. *British Antarctic Survey, Bulletin*, 11, 97–100.

Diamond, J. 1988. Express train to Polynesia. *Nature*, 336, 307–308.

Diamond, J. 1992. *The Third Chimpanzee.* New York: Harper, 407 p.

Dickinson, W.R. 1995. The times are always changing: the Holocene saga. *Geological Society of America, Bulletin*, 107, 1–7.

Dickinson, W.R. 1999. Holocene sea-level record on Funafuti and potential impact of global warming on central Pacific atolls. *Quaternary Research*, in press.

Dickinson, W.R. and Snyder, W.S. 1978. Plate tectonics of the Laramide orogeny. *Geological Society of America, Memoir* 151, 355–366.

Dietz, R.S. 1961. Continent and ocean basin evolution by spreading of the sea floor. *Nature*, 190, 854–857.

Dillehay, T. (editor). 1989. *Monte Verde: A Late Pleistocene Settlement in Chile.* Washington: Smithsonian Institution, 306 p.

Dixon, E.J. 1993. *Quest for the Origins of the First Americans.* Albuquerque: University of New Mexico Press, 156 p.

Dobson, J.E. 1992. Spatial logic in paleogeography and the explanation of continental drift. *Annals of the Association of American Geographers*, 82, 187–206.

Dodson, J.R. 1982. Modern pollen rain and recent vegetation history on Lord Howe Island: evidence of human impact. *Review of Palaeobotany and Palynology*, 38, 1–21.

Dodson, J.R., Greenwood, P.W. and Jones, R.L. 1986. Holocene forest and wetland dynamics at Barrington Tops, New South Wales. *Journal of Biogeography*, 13, 561–585.

Dodson, J.R., Enright, N.J. and McLean, R.F. 1988. A late Quaternary vegetation history for far northern New Zealand. *Journal of Biogeography*, 15, 647–656.

Dodson, J.R., de Salis, T., Myers, C.A. and Sharp, A.J. 1994. A thousand years of environmental change and human impact in the alpine zone at Mt Kosciusko, New South Wales. *Australian Geographer*, 25, 77–87.

Dolan, J.F., Sieh, K., Rockwell, T.K., Guptill, T.K. and Miller, G. 1997. Active tectonics, paleoseismology, and seismic hazards of the Hollywood Fault, northern Los Angeles Basin, California. *Geological Society of America, Bulletin*, 109, 1595–1616.

Domning, D.P. 1987. Sea cow family reunion. *Natural History*, 4, 64–71.

Douglas, J.G. 1994. Cretaceous vegetation: the macrofossil record. *In:* Hill, R.S. (editor). *History of the Australian Vegetation: Cretaceous to Recent.* Cambridge: Cambridge University Press, 171–188.

Dowdeswell, J.A. and White, J.W.C. 1995. Greenland ice core records and rapid climate change. *Philosophical Transactions of the Royal Society of London*, A352, 359–371.

Dowsett, H., Thompson, R., Barron, J., Cronin, T., Fleming, F., Ishman, S., Poore, R., Willard, D. and Holtz, T. 1994. Joint investigations of the Middle Pliocene climate I: PRISM paleoenvironmental reconstructions. *Global and Planetary Change*, 9, 169–195.

Dubois, J., Launay, J. and Recy, J. 1974. Uplift movements in New Caledonia–Loyalty Islands area and their plate tectonics interpretation. *Tectonophysics*, 24, 133–150.

Dunbar, R.B., Wellington, G.M., Colgan, M.W. and Glynn, P.W. 1994. Eastern Pacific sea surface temperatures since 1600 A.D.: the $\delta^{18}O$ record of climatic variability in Galapagos corals. *Paleoceanography*, 9, 291–315.

Dunham, P.S. 1994. Into a mirror darkly: the ancient Maya collapse and modern world environmental policy. *In:* Hickey, J.E. and Longmire, L.A. (editors). *The Environment: global problems, local solutions.* Westport: Greenwood Press, 441–468.

Duplessy, J-C., Arnold, M., Bard, E., Juillet-Leclerc, A., Kallel, N. and Labeyrie, L. 1989. AMS ^{14}C study of transient events and of the ventilation rate of the Pacific intermediate water during the last deglaciation. *Radiocarbon*, 31, 493–502.

Dupon, J.F. 1986. The effects of mining on the environments of high islands: a case study of nickel mining in New Caledonia. *South Pacific Commission, SPREP Environmental Case Studies* 1.

Dupré, W.R., Morrison, R.B., Clifton, H.E., Lajoie, K.R., Ponti, D.J., Powell, C.L., Mathieson, S.A., Sarna-Wojcicki, A.M., Leithold, E.L., Lettis, W.R., McDowell, P.F., Rockwell, T.K., Unruh, J.R. and Yeats, R.S. 1991. Quaternary geology of the Pacific margin. *In:* Morrison, R.B. (editor). *Quaternary Nonglacial Geology; Conterminous U.S.* Boulder: Geological Society of America, 141–214.

Eddy, J.A. 1977. Climate and the changing sun. *Climatic Change*, 1, 173–179.

Edelman, M. 1995. Rethinking the hamburger thesis: deforestation and the crisis of Central America's beef exports. *In:* Painter, M. and Durham, W.H. (editors). *The Social Causes of Environmental Destruction in Latin America.* Ann Arbor: The University of Michigan Press, 25–62.

Edwards, M.E. and Barker, E.D. 1994. Climate and vegetation in northeastern Alaska 18,000 yr B.P.-Present. *Palaeogeography, Palaeoclimatology, Palaeoecology*, 109, 127–135.

Edwards, R.L., Beck, J.W., Burr, G.S., Donahue, D.J., Chappell, J.M.A., Bloom, A.L., Druffel, E.R.M. and Taylor, F.W. 1993. A large drop in atmospheric ^{14}C/^{12}C and reduced melting in the Younger Dryas, documented with ^{230}Th ages of corals. *Science*, 260, 962–968.

Elias, S.A., Short, S.K. and Birks, H.H. 1997. Late Wisconsin environments of the Bering land bridge. *Palaeogeography, Palaeoclimatology, Palaeoecology*, 136, 293–308.

Elliot, D.H. and Larsen, D. 1993. Mesozoic volcanism in the central Transantarctic mountains, Antarctica: depositional environment and tectonic setting. *In:* Findlay, R.H., Unrug, R., Banks, M.R. and Veevers, J.J. (editors). *Gondwana Eight: assembly, evolution and dispersal.* Rotterdam: Balkema, 397–410.

Elliott, M.B., Striewski, B., Flenley, J.R. and Sutton, D.G. 1995. Palynological and sedimentological evidence for a radiocarbon chronology of environmental change and Polynesian deforestation from Lake Taumatawhana, Northland, New Zealand. *Radiocarbon*, 37, 899–916.

Ellison, J.C. 1994. Palaeo-lake and swamp stratigraphic records of Holocene vegetation and sea-level changes, Mangaia, Cook Islands. *Pacific Science*, 48, 1–15.

Engstrom, D.R., Hansen, B.S.C. and Wright, H.E. 1990. A possible younger Dryas record in southeastern Alaska. *Science*, 250, 1383–1385.

Ernst, W.G., Ho, C.S. and Liou, J.G. 1985. Rifting, drifting, and crustal accretion in the Taiwan sector of the Asiatic continental margin. *In:* Howell, D.G. (editor). *Tectonostratigraphic Terranes of the Circum-Pacific Region.* Houston: Circum-Pacific Council for Energy and Mineral Resources, Earth Science Series, No. 1, 375–389.

Evans, J.C. 1992. *Tea in China: the history of China's national drink.* New York: Greenwood, 169 p.

Evans, J.W. 1925. Anniversary address of the President. *Quarterly Journal of the Geological Society of London*, 31, 72.

Fairbridge, R.W. 1961. Eustatic changes in sea level. *Physics and Chemistry of the Earth*, 4, 99–185.

Fairbridge, R.W. 1992. Holocene marine coastal evolution of the United States. *In:* Fletcher, C.H. and Wehmiller, J.F. (editors). *Quaternary Coasts of the United States: marine and lacustrine systems.* Tulsa: Society for Sedimentary Geology, 9–20.

Fang, J-Q. 1993. Lake evolution during the last 3000 years in China and its implications for environmental change. *Quaternary Research*, 39, 175–185.

Farrell, B.H. 1972. The alien and the land of Oceania. *In:* Ward, R.G. (editor). *Man in the Pacific Islands.* Oxford: Clarendon, 34–73.

Farrell, J.W. and Prell, W.L. 1991. Pacific CaCO$_3$ preservation and δ^{18}O since 4 Ma: paleoceanic and paleoclimatic implications. *Paleoceanography*, 6, 485–498.

Feary, D.A., Davies, P.J., Pigram, C.J. and Symonds, P.A. 1991. Climatic evolution and control on carbonate deposition in northeast Australia. *Palaeogeography, Palaeoclimatology, Palaeoecology*, 89, 341–361.

Feng, X. and Epstein, S. 1994. Climatic implications of an 8000-year hydrogen isotope time series from bristlecone pine trees. *Science*, 265, 1079–1081.

Feng, Z., Thompson, L.G., Mosley-Thompson, E. and Yao, T. 1993. Temporal and spatial variations of climate in China during the last 10 000 years. *The Holocene*, 3, 174–180.

Filippelli, G.M. 1997. Intensification of the Asian monsoon and a chemical weathering event in the late Miocene–early Pliocene: implications for late Neogene climate change. *Geology*, 25, 27–30.

Finney, B. 1985. Anomalous westerlies, El Niño, and the colonization of Polynesia. *American Anthropologist*, 87, 9–26.

Fitzgerald, P.G., Sandiford, M., Barrett, P.J. and Gleadow, A.J.W. 1986. Asymmetric extension associated with uplift and subsidence in the Transantarctic Mountains and the Ross Embayment. *Earth and Planetary Science Letters*, 81, 67–78.

Flannery, T. 1994. *The Future Eaters*. Port Melbourne: Reed Books, 423 p.

Fleming, C.A. 1979. *The Geological History of New Zealand and its Life*. Auckland: Auckland University Press, 141 p.

Fleming, D. 1964. Science in Australia, Canada and the United States: some comparative remarks. *Proceedings of the 10th International Congress of the History of Science*, Ithaca, New York, 179–196.

Fleming, R.F. and Barron, J.A. 1996. Evidence of Pliocene *Nothofagus* in Antarctica from Pliocene marine sedimentary deposits (DSDP Site 274). *Marine Micropaleontology*, 27, 227–236.

Flenley, J.R. 1979. The Late Quaternary vegetational history of the equatorial mountains. *Progress in Physical Geography*, 3, 488–509.

Flenley, J.R. 1993. The palaeoecology of Easter Island, and its ecological disaster. *In:* Fischer, S.R. (editor). *Easter Island Studies: contributions to the history of Rapanui in memory of William T. Mulloy*. Oxford: Oxbow, 27–45.

Flenley, J.R. and King, S.M. 1984. Late Quaternary pollen records from Easter Island. *Nature*, 307, 47–50.

Flenley, J., Parkes, A. and Teller, J.T. 1991. A 500-year climate change record from Tahiti. *In:* Hay, J.E. (editor). *South Pacific Environments: interactions with weather and climate*. Auckland University, Environmental Science Occasional Publication 6, 31.

Fletcher, C.H. and Sherman, C.E. 1995. Submerged shorelines on O'ahu, Hawai'i: archive of episodic transgression during the deglaciation? *Journal of Coastal Research, Special Issue*, 17, 141–152.

Flood, P.G. and Frankel, E. 1989. Late Holocene higher sea level indicators from eastern Australia. *Marine Geology*, 90, 193–195.

Flöttmann, T., Kleinschmidt, G. and Funk, T. 1993. Thrust patterns of the Ross/Delamerian orogens in northern Victoria Land (Antarctica) and southeastern Australia and their implications for Gondwana reconstructions. *In:* Findlay, R.H., Unrug, R., Banks, M.R. and Veevers, J.J. (editors). *Gondwana Eight: assembly, evolution and dispersal*. Rotterdam: Balkema, 131–139.

Foley, R. 1987. *Another Unique Species: patterns in human evolutionary ecology*. Harlow: Longman. 313 p.

Fordyce, R.E. 1992. Cetacean evolution and Eocene–Oligocene environments. *In:* Prothero, D.R. and Berggren, W.A. (editors). *Eocene–Oligocene Climatic and Biotic Evolution*. Princeton: Princeton University Press, 368–381.

Frakes, L.A., Francis, J.E. and Syktus, J.I. 1992. *Climate Modes of the Phanerozoic*. Cambridge: Cambridge University Press, 274 p.

Francis, J.E. 1986. Growth rings in Cretaceous and Tertiary wood from Antarctica and its palaeoclimatic implications. *Palaeontology*, 48, 285–307.

Friele, P.A. and Hutchinson, J. 1993. Holocene sea-level change on the central west coast of Vancouver Island, British Columbia. *Canadian Journal of Earth Sciences*, 30, 832–840.

Fritts, H.C. and Lough, J.M. 1985. An estimate of average annual temperature variations for North America, 1602 to 1961. *Climatic Change*, 7, 203–224.

Frost, A. 1988. Science for political purposes: European explorations of the Pacific Ocean, 1764–1806. *In:* MacLeod, R. and Rehbock, P.F. (editors). *Nature in its Greatest Extent: Western Science in the Pacific*. Honolulu: University of Hawaii Press, 27–44.

Fuji, N. 1988. Palaeovegetation and palaeoclimatic changes around Lake Biwa, Japan during the last ca. 3 million years. *Quaternary Science Reviews*, 7, 21–28.

Fujita, K. and Newberry, J.T. 1982. Tectonic evolution of northeastern Siberia and adjacent regions. *Tectonophysics*, 89, 337–357.

Fukui, E. 1977. *The Climate of Japan*. Amsterdam: Elsevier, 317 p.

Fung, C. 1994. The beginnings of settled life. *In:* Murowchick, R.E. (editor). *China*. Norman: University of Oklahoma Press, 51–59.

Gardner, G.J., Mortlock, A.J., Price, D.M., Readhead, M.L. and Wasson, R.J. 1987. Thermoluminescence and radiocarbon dating of Australian desert dunes. *Australian Journal of Earth Science*, 34, 343–357.

Garrett-Jones, S. 1979. Evidence for changes in Holocene vegetation and lake sedimentation in the Markham Valley, Papua New Guinea. *Unpublished PhD thesis, Australian National University*, Canberra.

Gasperi, J.T. and Kennett, J.P. 1993. Vertical thermal structure evolution of Miocene surface waters: western equatorial Pacific DSDP site 289. *Marine Micropaleontology*, 22, 235–254.

Gaven, C.M. and Bourrouilh-Le Jan, F.G. 1981. Géochronologie (^{230}Th–^{234}U–^{238}U), sédimentologie et néotectonique des faciès récifaux pléistocènes a Maré, archipel des Loyauté, SW Pacifique. *Oceanis*, 7, 347–365.

Gavenda, R.T. 1992. Hawaiian Quaternary paleo-environments: a review of geological, pedological, and botanical evidence. *Pacific Science*, 46, 295–307.

Geist, E.L., Vallier, T.L. and Scholl, D.W. 1994. Origin, transport, and emplacement of an exotic island-arc terrane exposed in eastern Kamchatka, Russia. *Geological Society of America, Bulletin*, 106, 1182–1194.

Gernet, J. 1996. *A History of Chinese Civilization*.

Cambridge: Cambridge University Press, 2nd edition, 801 p.

Gibb, J. 1986. A New Zealand regional holocene eustatic sea-level curve and its application to determination of vertical tectonic movements. *Royal Society of New Zealand, Bulletin*, 24, 377–395.

Gibbons, J.R.H. and Clunie, F.G.A.U. 1986. Sea level changes and Pacific prehistory. *Journal of Pacific History*, 21, 58–82.

Gillieson, D., Oldfield, F. and Krawiecki, A. 1986. Records of prehistoric soil erosion from rock-shelter sites in Papua New Guinea. *Mountain Research and Development*, 6, 315–324.

Gillieson, D., Gorecki, P., Head, J. and Hope, G. 1987. Soil erosion and agricultural history in the central highlands of Papua New Guinea. *In:* Gardiner, V. (editor). *International Geomorphology 1986, Part II*. Chichester: Wiley, 507–522.

Gingerich, P.D. 1977. Patterns of evolution in the mammalian fossil record. *In:* Hallam, A. (editor). *Patterns of Evolution as Illustrated by the Fossil Record*. Amsterdam: Elsevier, 469–500.

Gingerich, P.D. 1985. South American mammals in the Paleocene of North America. *In:* Stehli, F.G. and Webb, S.D. (editors). *The Great American Biotic Interchange*. New York: Plenum, 123–137.

Goebel, T. and Aksenov, M. 1995. Accelerator radiocarbon dating of the initial Upper Palaeolithic in southeast Siberia. *Antiquity*, 69, 349–357.

Gollan, C. 1984. The Australian dingo: in the shadow of man. *In:* Archer, M. and Clayton, G. (editors). *Vertebrate Zoogeography and Evolution in Australasia*. Perth: Hesperian Press, 921–927.

Golovneva, L.B. 1994. The flora of the Maastrichtian–Danian deposits of the Koryak upland, northeast Russia. *Cretaceous Research*, 15, 89–100.

Golson, J. 1982. Kuk and the history of agriculture in the New Guinea highlands. *In:* May, R.J. and Nelson, H. (editors). *Melanesia: beyond diversity*. Canberra: Australian National University Press, 297–307.

Golson, J. and Hughes, P.J. 1980. The appearance of plant and animal domestication in New Guinea. *Journal de la Société des Océanistes*, 36, 294–303.

Gorbarenko, S.A. 1996. Stable isotope and lithologic evidence of Late-Glacial and Holocene oceanography of the northwestern Pacific and its marginal seas. *Quaternary Research*, 46, 230–250.

Gordillo, S., Bujalesky, G.G., Pirazzoli, P.A., Rabassa, J.O. and Saliège, J-F. 1992. Holocene raised beaches along the northern coast of the Beagle Channel, Tierra del Fuego, Argentina. *Palaeogeography, Palaeoclimatology, Palaeoecology*, 99, 41–54.

Gorecki, P.P., Horton, D.R., Stern, N. and Wright, R.V.S. 1984. Co-existence of humans and megafauna in Australia: improved stratigraphic evidence. *Archaeology in Oceania*, 19, 117–119.

Gorman, C.F. 1971. The Hoabinhian and after: subsistence patterns in southeast Asia during the late Pleistocene and early Recent periods. *World Archaeology*, 2, 300–320.

Gosden, C. and Webb, J. 1994. The creation of a Papua New Guinean landscape: archaeological and geomorphological evidence. *Journal of Field Archaeology*, 21, 29–51.

Goudie, A.S. 1995. *The Changing Earth; rates of geomorphological processes*. Oxford: Blackwell, 302 p.

Gould, S.J. 1989. *Wonderful Life: The Burgess Shale and the nature of history*. London: Hutchinson Radius, 347 p.

Goy, J.L., Macharé, J., Ortlieb, L. and Zazo, C. 1992. Quaternary shorelines in southern Peru: a record of global sea-level fluctuations and tectonic uplift in Chala Bay. *Quaternary International*, 15/16, 99–112.

Graf, K. 1981. Palynological investigation of two postglacial peat bogs near the boundary of Bolivia and Peru. *Journal of Biogeography*, 8, 353–368.

Graham, A. 1989. Late Tertiary paleoaltitudes and vegetational zonation in Mexico and Central America. *Acta Botanica Neerlandica*, 38, 417–424.

Gray, S.C., Hein, J.R., Hausmann, R. and Radtke, U. 1992. Geochronology and subsurface stratigraphy of Pukapuka and Rakahanga atolls, Cook Islands: late Quaternary reef growth and sea level history. *Palaeogeography, Palaeoclimatology, Palaeoecology*, 91, 377–394.

Grayson, D.K. 1988. Perspectives on the archaeology of the first Americans. *In:* Carlisle, R.C. (editor). *Americans before Columbus: Ice-Age Origins*. Pittsburgh: Department of Anthropology, University of Pittsburgh, 107–123.

Greenwood, D.R. 1994. Palaeobotanical evidence for Tertiary climates. *In:* Hill, R.S. (editor). *History of the Australian Vegetation, Cretaceous to Recent*. Cambridge: Cambridge University Press, 44–59.

Greenwood, D.R. and Wing, S.L. 1995. Eocene continental climates and latitudinal temperature gradients. *Geology*, 23, 1044–1048.

Grigg, R.W. 1988. Paleoceanography of coral reefs in the Hawaiian-Emperor chain. *Science*, 240, 1737–1743.

Grigg, R.W. and Hey, R. 1992. Paleoceanography of the tropical eastern Pacific Ocean. *Science*, 255, 172–178.

Grigg, R.W. and Jones, A.T. 1997. Uplift caused by lithospheric flexure in the Hawaiian archipelago as revealed by elevated coral deposits. *Marine Geology*, 141, 11–25.

GRIP (Greenland Ice-core Project) Members. 1993. Climate instability during the last interglacial period recorded in the GRIP ice core. *Nature*, 364, 203–207.

Grist, D.H. 1986. *Rice*. New York: Longman, 6th edition, 333 p.

Grosjean, M., Núñez, L., Cartajena, I. and Messerli, B. 1997. Mid-Holocene climate and culture change in the Atacama Desert, northern Chile. *Quaternary Research*, 48, 239–246.

Gross, M.G. 1972. *Oceanography: a view of the Earth*. Englewood Cliffs: Prentice-Hall, 581 p.

Groube, L., Chappell, J., Muke, J. and Price, D. 1986. A 40,000 year-old human occupation site at Huon Peninsula, Papua New Guinea. *Nature*, 324, 453–455.

Grove, J.M. 1988. *The Little Ice Age*. London: Methuen, 498 p.

Grover, J.C. 1965. Seismological and volcanological studies in the British Solomon Islands to 1961. *British Solomon Islands Geological Record*, 2, 183–188.

Gruszczynski, G., Halas, S., Hoffman, A. and Malkowski, K. 1989. A brachiopod calcite record of oceanic carbon and oxygen isotope shifts at the Permian–Triassic transition. *Nature*, 337, 64–68.

Haberle, S.G. 1998. Late Quaternary vegetation change in the Tari Basin, Papua New Guinea. *Palaeogeography, Palaeoclimatology, Palaeoecology*, 137, 1–24.

Hall, B.L., Denton, G.H., Lux, D.R. and Schlüchter, C. 1997. Pliocene paleoenvironment and Antarctic ice-sheet behavior: evidence from Wright Valley. *Journal of Geology*, 105, 285–294.

Hall, D.A. 1992. Siberian site defies theories on peopling: pebble tools are dated to 3 million years. *Mammoth Trumpet*, 8, 1, 4–5.

Hallam, A. 1981. *Facies Interpretation and the Stratigraphic Record*. San Francisco: Freeman, 291 p.

Hallam, A. 1984. Pre-Quaternary sea-level changes. *Annual Review of Earth and Planetary Science*, 12, 205–243.

Hallam, A. 1986. Evidence of displaced terranes from Permian to Jurassic faunas around the Pacific margins. *Journal of the Geological Society of London*, 143, 209–216.

Hamilton, T.D. and Brigham-Grette, J. 1991. The Last Interglaciation in Alaska: stratigraphy and paleoecology of potential sites. *Quaternary International*, 10–12, 49–71.

Hansen, B.C.S., Seltzer, G.O. and Wright, H.E. 1994. Late Quaternary vegetational change in the central Peruvian Andes. *Palaeogeography, Palaeoclimatology, Palaeoecology*, 109, 263–285.

Hanson, K.L., Lettis, W.R., Wesling, J.R., Kelson, K.I. and Mezger, L. 1992. Quaternary marine terraces, south-central coastal California: implications for crustal deformation and coastal evolution. *In:* Fletcher, C.H. and Wehmiller, J.F. (editors). *Quaternary Coasts of the United States: marine and lacustrine systems*. Tulsa: Society for Sedimentary Geology, 323–341.

Haq, B.U., Hardenbol, J. and Vail, P.R. 1987. Chronology of fluctuating sea levels since the Triassic. *Science*, 235, 1156–1167.

Harbert, W. and Cox, A. 1989. Late Neogene motion of the Pacific Plate. *Journal of Geophysical Research*, 94, 3052–3064.

Harden, C. 1988. Mesoscale estimation of soil erosion in the Rio Ambato drainage, Ecuadorian Sierra. *Mountain Research and Development*, 8, 331–341.

Harmon, R.S., Ford, D.C. and Schwarcz, H.P. 1977. Interglacial chronology of the the Rocky and Mackenzie Mountains based upon ^{230}Th–^{234}U dating of calcite speleothems. *Canadian Journal of Earth Sciences*, 14, 2543–2552.

Harrisson, S.P. 1993. Late Quaternary lake-level changes and climates of Australia. *Quaternary Science Reviews*, 12, 211–231.

Harrison, S. and Dodson, J. 1993. Climates of Australia and New Guinea since 18,000 yr B.P. *In:* Wright, H.E., Kutzbach, J.E., Webb, T., Ruddiman, W.F., Street-Perrott, F.A. and Bartleim, P.J. (editors). *Global Climates since the Last Glacial Maximum*. Minneapolis: University of Minnesota, 265–293.

Harrisson, T. 1967. Niah Caves, Sarawak. *Asian and Pacific Archaeology Series*, 1, 77–78.

Hart, M.B. 1980. A water depth model for the evolution of the planktonic Foraminiferida. *Nature*, 286, 252–254.

Hashiguchi, N. 1994. The Izu islands: their role in the historical development of ancient Japan. *Asian Perspectives*, 33, 121–149.

Hashimoto, K., Jami, C. and Skar, L. (editors). 1995. *East Asian Science: tradition and beyond*. Osaka: Kansai University Press, 568 p.

Hastenrath, S. 1981. *The Glaciation of the Ecuadorian Andes*. Rotterdam: Balkema, 159 p.

Hayami, I. and Kase, T. 1977. *A Systematic Survey of the Palaeozoic and Mesozoic Gastropods and Paleocene Bivalvia from Japan*. Tokyo: University of Tokyo Press, 154 p.

Haynes, C.V. 1991. Geoarchaeological and palaeohydrological evidence for a Clovis-age drought in North America and its bearing on extinction. *Quaternary Research*, 35, 438–450.

Hays, J.D. and Pitman, W.C. 1973. Lithospheric plate motion, sea level changes and climatic and ecological consequences. *Nature*, 246, 16–22.

Hays, J.D., Imbrie, J. and Shackleton, N.J. 1976. Variations in the earth's orbit: pacemaker of the ice ages. *Science*, 194, 1121–1132.

Head, L. 1989. Prehistoric aboriginal impacts on

Australian vegetation: an assessment of the evidence. *Australian Geographer*, 20, 37–46.

Heaton, T.H., Talbot, S.L. and Shields, G.F. 1996. An ice age refugium for large mammals in the Alexander Archipelago, southeast Alaska. *Quaternary Research*, 46, 186–192.

Heine, J.T. 1993. A reevaluation of the evidence for a Younger Dryas climatic reversal in the tropical Andes. *Quaternary Science Reviews*, 12, 769–779.

Heine, K. 1978. Neue beobachtungen zur chronostraigraphie der mittelwisconsinzeitlichen vergletscherungen und böden mexikanischer vulkane. *Eiszeitalter und Gegenwart*, 28, 139–147.

Hendy, C.H., Hendy, T.R., Rayner, E.M., Shaw, J. and Wilson, A.T. 1979. Late Pleistocene glacial chronology of the Taylor Valley, Antarctica, and the global climate. *Quaternary Research*, 11, 172–184.

Herguera, J.C. and Berger, W.H. 1994. Glacial to postglacial drop 'in productivity in the western equatorial Pacific: mixing rate vs. nutrient concentrations. *Geology*, 22, 629–632.

Hesse, P.P. 1993. A Quaternary record of the Australian environment from aeolian dust in Tasman Sea sediments. *Unpublished PhD thesis*, Australian National University, Canberra, 396 p.

Heusser, C.J. 1960. *Late Pleistocene Environments of North Pacific North America*. American Geographical Society, Special Publication 35, 305 p.

Heusser, C.J. 1974. Vegetation and climate of the southern Chilean lake district during and since the Last Interglaciation. *Quaternary Research*, 4, 290–315.

Heusser, C.J. 1977. Quaternary paleoecology of the Pacific slope of Washington. *Quaternary Research*, 8, 282–306.

Heusser, C.J. 1978. Postglacial vegetation of Adak island, Aleutian Islands, Alaska. *Bulletin of the Torrey Botanical Club*, 15, 18–23.

Heusser, C.J. 1984a. Late Glacial-Holocene climate of the Lake District of Chile. *Quaternary Research*, 22, 77–90.

Heusser, C.J. 1984b. Late Quaternary climates of Chile. *In:* Vogel, J.C. (editor). *Late Cainozoic Palaeoclimates of the Southern Hemisphere*. Rotterdam: Balkema, 59–83.

Heusser, C.J. 1989. Late Quaternary vegetation and climate of southern Tierra del Fuego. *Quaternary Research*, 31, 396–406.

Heusser, C.J. 1990. Ice age vegetation and climate of subtropical Chile. *Palaeogeography, Palaeoclimatology, Palaeoecology*, 80, 107–127.

Heusser, C.J. and Rabassa, J. 1987. Cold climatic episode of Younger Dryas age in Tierra del Fuego. *Nature*, 328, 609–611.

Heusser, C.J., Heusser, L.E. and Peteet, D.M. 1985.

Late-Quaternary climatic change on the American North Pacific coast. *Nature*, 315, 485–487.

Heusser, L.E. and Morley, J.J. 1990. Climatic change at the end of the last glaciation in Japan inferred from pollen in three cores from the northwest Pacific Ocean. *Quaternary Research*, 34, 101–110.

Heusser, L.E. and Shackleton, N.J. 1994. Tropical climatic variation on the Pacific slopes of the Ecuadorian Andes based on a 25,000-year pollen record from deep-sea sediment core Tri 163-31B. *Quaternary Research*, 42, 222–225.

Heusser, L.E. and van de Geer, G. 1994. Direct correlation of terrestrial and marine paleoclimatic records from four glacial–interglacial cycles – DSDP site 594, southwest Pacific. *Quaternary Science Reviews*, 13, 273–282.

Hickey, L.J. and Doyle, J.A. 1977. Early Cretaceous fossil evidence for angiosperm evolution. *Botanical Review*, 43, 3–104.

Hilborn, R. 1990. Marine biota. *In:* Turner, B.L., Clark, W.C., Kates, R.W., Richards, J.F., Mathews, J.T. and Meyer, W.B. (editors). *The Earth as Transformed by Human Action*. Cambridge University Press, 371–385.

Hilde, W.W., Uyeda, S. and Kroenke, L. 1977. Evolution of the west Pacific and its margin. *Tectonophysics*, 38, 145–165.

Hildebrand, A.R., Penfield, G.T., Kring, D.A., Pilkington, M., Carmargo Z.A., Jacobsen, S.B. and Boynton, W.V. 1991. Chicxulub crater: a possible Cretaceous/Tertiary boundary impact crater on the Yucatán Peninsula, Mexico. *Geology*, 19, 867–871.

Hill, R.S. and MacPhail, M.K. 1985. A fossil flora from rafted Plio-Pleistocene mudstones at Regatta Point, Tasmania. *Australian Journal of Botany*, 33, 497–517.

Hirooka, K. 1992. Paleomagnetic evidence of the deformation of Japan and its paleogeography during the Neogene. *In:* Tsuchi, R. and Ingle, J. (editors). *Pacific Neogene Environments, Evolution and Events*. Tokyo: University of Tokyo Press, 151–156.

Hiscock, P. and Kershaw, A.P. 1992. Palaeo-environments and prehistory of Australia's tropical Top End. *In:* Dodson, J. (editor). *The Naive Lands: prehistory and environmental change in Australia and the south-west Pacific*. Melbourne: Longman Cheshire, 43–75.

Ho, C-K. and Li, Z-W. 1987. Paleolithic subsistence strategies in North China. *Current Research in the Pleistocene*, 4, 7–9.

Hodell, D.A., Elmstrom, M. and Kennett, J.P. 1986. Latest Miocene benthic $\delta^{18}O$ changes, global ice volume, sea level and the 'Messinian salinity crisis'. *Nature*, 320, 411–414.

Hoffmeister, J.E. 1932. Geology of Eua, Tonga. *B.P. Bishop Museum (Honolulu), Bulletin*, 96.

Hollin, J.T. 1972. Interglacial climates and Antarctic ice surges. *Quaternary Research*, 2, 401–408.

Hollin, J.T. 1980. Climate and sea level in isotope Stage 5: an east Antarctic ice surge at ~95,000 BP? *Nature*, 283, 629–633.

Holloway, J.T. 1964. The forests of South Island: the status of the climatic change hypothesis. *New Zealand Geographer*, 20, 1–9.

Hooghiemstra, H. 1989. Quaternary and upper Pliocene glaciations and forest development in the tropical Andes: evidence from a long high-resolution pollen record from the sedimentary basin of Bogotá, Colombia. *Palaeogeography, Palaeoclimatology, Palaeoecology*, 72, 11–26.

Hooijer, D.A. 1963. Further 'Hell' mammals from Niah. *Sarawak Museum Journal*, 11, 196–200.

Hope, G.S. 1976. The vegetational history of Mt. Wilhelm, Papua New Guinea. *Journal of Ecology*, 64, 627–664.

Hope, G.S. 1983. The vegetational changes of the last 20,000 years at Telefomin, Papua New Guinea. *Singapore Journal of Tropical Geography*, 4, 25–33.

Hope, G.S. 1994. Quaternary vegetation. *In:* Hill, R.S. (editor). *History of the Australian Vegetation: Cretaceous to Recent*. Cambridge: Cambridge University Press, 368–389.

Hope, G.S. 1996. History of *Nothofagus* in New Guinea and New Caledonia. *In:* Veblen, T.T., Hill, R.S. and Read, J. (editors). *The Ecology and Biogeography of Nothofagus Forest*. New Haven: Yale University Press, 257–270.

Hope, G.S. and Golson, J. 1995. Late Quaternary change in the mountains of New Guinea. *Antiquity*, 69, 818–830.

Hope, G.S. and Peterson, J.A. 1976. Paleoenvironments. *In:* Hope, G.S., Peterson, J.A., Radok, U. and Allison, I. (editors). *The Equatorial Glaciers of New Guinea*. Rotterdam: Balkema, 173–205.

Hopkins, D.M. 1979. The Flaxman Formation of northern Alaska: record of early Wisconsinan shelf glaciation in the high Arctic? *14th Pacific Science Congress, Khabarovsk, Abstracts*, 15–16.

Hopkins, D.M. 1982. Aspects of the paleogeography of Beringia during the late Pleistocene. *In:* Hopkins, D.M., Matthews, J.V., Schweger, C.E. and Young, S.B. (editors). *Paleoecology of Beringia*. New York: Academic Press, 3–28.

Hopley, D. 1984. The Holocene 'high energy window' on the central Great Barrier Reef. *In:* Thom, B.G. (editor). *Coastal Geomorphology in Australia*. London: Academic Press, 135–150.

Horn, S.P. 1993. Postglacial vegetation and fire history in the Chirripó Páramo of Costa Rica. *Quaternary Research*, 40, 107–116.

Horn, S.P. and Sanford, R.L. 1992. Holocene fires in Costa Rica. *Biotropica*, 24, 354–361.

Hornibrook, N.D.B. 1992. New Zealand Cenozoic marine paleoclimates: a review based on the distribution of some shallow water and terrestrial biota. *In:* Tsuchi, R. and Ingle, J. (editors). *Pacific Neogene Environments, Evolution and Events*. Tokyo: University of Tokyo Press, 83–106.

Horton, D.R. 1982. The burning question: aborigines, fire and Australian ecosystems. *Mankind*, 13, 237–251.

Horton, D.R. 1984. Red kangaroos: last of the megafauna. *In:* Martin, P.S. and Klein, R.G. (editors). *Quaternary Extinctions*. Tucson: University of Arizona Press, 639–380.

Hostetler, S. and Bartlein, P.J. 1990. Simulation of lake evaporation with application to modeling lake level variations of Harney-Malheur Lake, Oregon. *Water Resources Research*, 26, 2603–2612.

Houghton, P. 1996. *People of the Great Ocean*. Cambridge: Cambridge University Press, 292 p.

Hovan, S.A., Rea, D.K., Pisias, N.G. and Shackleton, N.J. 1989. A direct link between the China loess and marine $\delta^{18}O$ records: aeolian flux to the north Pacific. *Nature*, 340, 296–298.

Howard, M.C. 1993. Small-scale mining and the environment in southeast Asia. *In:* Howard, M.C. (editor). *Asia's Environmental Crisis*. Boulder: Westview, 83–110.

Howell, B.F. 1959. *Introduction to Geophysics*. New York: McGraw Hill, 399 p.

Howell, D.G. (editor). 1985. *Tectonostratigraphic Terranes of the Circum-Pacific Region*. Houston: Circum-Pacific Council for Energy and Mineral Resources, Earth Science Series, No. 1.

Howell, D.G. and Jones, D.L. 1989. Terrane analysis: a circum-Pacific overview. *In:* Ben-Avraham, Z. (editor). *The Evolution of the Pacific Ocean Margins*. New York: Oxford University Press, 36–40.

Howell, D.G., Jones, D.L. and Schermer, E.R. 1985. Tectonostratigraphic terranes of the circum-Pacific region. *In:* Howell, D.G. (editor). *Tectonostratigraphic Terranes of the Circum-Pacific Region*. Houston: Circum-Pacific Council for Energy and Mineral Resources, Earth Science Series, No. 1, 3–30.

Hu, F.S., Brubaker, L.B. and Anderson, P.M. 1993. A 12 000 year record of vegetation change and soil development from Wien Lake, central Alaska. *Canadian Journal of Botany*, 71, 1133–1142.

Huang, Y., Xia, F., Huang, D. and Lin, H. 1984. Holocene sea level changes and recent crustal movements along the northern coast of the South China Sea. *In:* Whyte, R.O. (editor). *The Evolution of the East Asian Environment, Volume 1: Geology and Palaeoclimate*. Hong Kong: Centre of Asian Studies, University of Hong Kong, 271–287.

Hughes, B.A. and Hughes, T.J. 1994. Transgressions: rethinking Beringian glaciation. *Palaeogeography, Palaeoclimatology, Palaeoecology*, 110, 275–294.

Hughes, M.K. and Diaz, H.F. 1994. Was there a 'Medieval Warm Period', and if so, where and when? *Climatic Change*, 26, 109–142.

Hulton, N., Sugden, D., Payne, A. and Clapperton, C. 1994. Glacier modeling and the climate of Patagonia during the Last Glacial Maximum. *Quaternary Research*, 42, 1–19.

Hunter-Anderson, R.L., Thompson, G.B. and Moore, D.R. 1995. Rice as a prehistoric valuable in the Mariana Islands, Micronesia. *Asian Perspectives*, 34, 69–89.

Huntley, B., Cramer, W., Morgan, A.V., Prentice, H.C. and Allen, J.R.M. (editors). 1997. *Past and Future Rapid Environmental Changes*. Berlin: Springer-Verlag, 523 p.

Igarashi, Y. 1994. Quaternary forest and climate history of Hokkaido, Japan, from marine sediments. *Quaternary Science Reviews*, 13, 335–344.

Imamura, K. 1996. *Prehistoric Japan: new perspectives on insular East Asia*. London: UCL Press, 246 p.

Imbrie, J., Hays, J.D., Martinson, D.G., McIntyre, A., Mix, A.C., Morley, J.J., Pisias, G., Prell, W.L. and Shackleton, N.J. 1984. The orbital theory of Pleistocene climate: support from a revised chronology of the marine $\delta^{18}O$ record. *In:* Berger, A. (editor). *Milankovitch and Climate, Part 1*. Dordrecht: Reidel, 269–305.

Inoue, K. and Naruse, T. 1991. Accumulation of Asian long-range eolian dust in Japan and Korea from the late Pleistocene to the Holocene. *Catena, Supplement 20*, 25–42.

Iriondo, M. 1994. The Quaternary of Ecuador. *Quaternary International*, 21, 101–112.

Irwin, G. 1992. *The Prehistoric Exploration and Colonisation of the Pacific*. Cambridge: Cambridge University Press, 240 p.

Islebe, G.A. and Hooghiemstra, H. 1997. Vegetation and climate history of montane Costa Rica since the Last Glacial. *Quaternary Science Reviews*, 16, 589–604.

Isozaki, Y., Maruyama, S. and Furuoka, F. 1990. Accreted oceanic materials in Japan. *Tectonophysics*, 181, 179–205.

Iwauchi, A. 1994. Late Cenozoic vegetational and climatic changes in Kyushu, Japan. *Palaeogeography, Palaeoclimatology, Palaeoecology*, 108, 229–280.

Izett, G.A. and Naeser, C.W. 1976. Age of the Bishop Tuff of eastern California as determined by the fission-track method. *Geology*, 4, 587–590.

Jacob, J.S. and Hallmark, C.T. 1996. Holocene stratigraphy of Cobweb Swamp, a Maya wetland in northern Belize. *Geological Society of America, Bulletin*, 108, 883–891.

Jacoby, G.C., Cook, E.R. and Ulan, L.D. 1985. Reconstructed summer degree days in central Alaska and northwestern Canada since 1524. *Quaternary Research*, 23, 18–26.

Jennings, S.A. and Elliott-Fisk, D.L. 1993. Packrat midden evidence of late Quaternary vegetation change in the White Mountains, California-Nevada. *Quaternary Research*, 39, 214–221.

Jones, A.T. 1995. Geochronology of drowned Hawaiian coral reefs. *Sedimentary Geology*, 99, 233–242.

Jones, A.T. 1997. Late Holocene shoreline development in the Hawaiian Islands. *Journal of Coastal Research*, 14, 3–9.

Jones, B.G., Fergusson, C.L. and Zambelli, P.F. 1993. Ordovician contourites in the Lachlan Fold Belt, eastern Australia. *Sedimentary Geology*, 82, 257–270.

Jones, R. 1985. Ordering the landscape. *In:* Donaldson, I. and Donaldson, T. (editors). *Seeing the First Australians*. Sydney: George, Allen and Unwin, 181–209.

Jouzel, J., Lorius, C., Petit, J.R., Genthon, C., Barkov, N.I., Kotlyakov, V.M. and Petrov, V.M. 1987. Vostok ice core: a continuous isotope temperature record over the last climatic cycle (160,000 years). *Nature*, 329, 403–408.

Jouzel, J., Barkov, N.I., Barnola, J.M., Bender, M., Chappellaz, J., Genthon, C., Kotlyakov, V.M., Lipenkov, V., Lorius, C., Petit, J.R., Raynaud, D., Raisebeck, G., Ritz, C., Sowers, T., Stievenard, M., Yiou, F. and Yiou, P. 1993. Extending the Vostok ice-core record of palaeoclimate to the penultimate glacial period. *Nature*, 364, 407–412.

Kaizuka, S. 1980. Late Cenozoic palaeogeography of Japan. *GeoJournal*, 4, 101–109.

Kakuwa, Y. 1996. Permian–Triassic mass extinction event recorded in bedded chert sequence in southwest Japan. *Palaeogeography, Palaeoclimatology, Palaeoecology*, 121, 35–51.

Kamata, T., Fujimura, S., Kajiwara, H. and Yamada, A. 1993. The Takamori site: a possible >300,000-year-old site in Japan. *Current Research in the Pleistocene*, 10, 28–30.

Kamei, T. 1984. Fossil mammals: Lake Biwa and fossil mammals: faunal changes since the Pliocene time. *In:* Horie, S (editor). *Lake Biwa*. Dordrecht: Junk, 475–495.

Kamei, T. 1989. The Japan Sea and elephants. *The Quaternary Research (Japan)*, 29, 163–172.

Karlstrom, T.N.V. and Ball, G.E. 1969. *The Kodiak Island Refugium: its geology, flora, fauna and history*. Toronto: Ryerson, 262 p.

Kawana, T. and Nakata, T. 1994. Timing of late Holocene tsunamis originated around the southern Ryukyu islands, Japan, deduced from coralline tsunami deposits. *Journal of Geography*, 103, 352–376.

Kawana, T. and Pirazzoli, P.A. 1985. Holocene coastline changes and seismic uplift in Okinawa island, the Ryukyus, Japan. *Zeitschrift für Geomorphologie*, Supplementband 57, 11–31.

Kawana, T., Miyagi, T., Fujimoto, K. and Kikuchi, T. 1995. Late Holocene sea-level changes and mangrove development in Kosrae Island, the Carolines, Micronesia. *In:* Kikuchi, T. (editor). *Rapid Sea Level Rise and Mangrove Habitat.* Gifu University, Japan: Institute for Basin Ecosystem Studies, 1–7.

Kayanne, H. and Yoshikawa, T. 1986. Comparative study between present and emerged erosional landforms on the southeast coast of Boso Peninsula, central Japan. *Geographical Review of Japan*, 59, 18–36.

Keally, C.T. 1991. Environment and the distribution of sites in the Japanese Palaeolithic: environmental zones and cultural areas. *Bulletin of the Indo-Pacific Prehistory Association*, 10, 23–39.

Keany, J., Ledbetter, M., Watkins, N. and Ter Chien, H. 1976. Diachronous deposition of ice-rafted debris in sub-Antarctic deep-sea sediments. *Geological Society of America, Bulletin*, 87, 873–882.

Keating, B.H. 1999. Nuclear testing in the Pacific from a geologic perspective. *In:* Terry, J (editor). Suva, Fiji: The University of the South Pacific, School of Social and Economic Development, forthcoming.

Keigwin, L.D. 1980. Palaeoceanographic change in the Pacific at the Eocene–Oligocene boundary. *Nature*, 287, 722–725.

Keigwin, L.D., Jones, G.A. and Froelich, P.N. 1992. A 15,000 year paleoenvironmental record from Meiji Seamount, far northwestern Pacific. *Earth and Planetary Science Letters*, 111, 425–440.

Keller, G., D'Hondt, S., Orth, C.J., Gilmore, J.S., Oliver, P.Q., Shoemaker, E.M. and Molina, E. 1987. Late Eocene impact microspherules: stratigraphy, age and geochemistry. *Meteoritics*, 22, 25–60.

Kennedy, G.L., Wehmiller, J.F. and Rockwell, T.K. 1992. Paleoecology and paleozoogeography of late Pleistocene marine-terrace faunas of southwestern Santa Barbara County, California. *In:* Fletcher, C.H. and Wehmiller, J.F. (editors). *Quaternary coasts of the United States: marine and lacustrine systems.* Tulsa: Society for Sedimentary Geology, 343–361.

Kennedy, N. 1994. New Zealand tephro-chronology as a tool in geomorphic history of the c.140 ka Mamaku ignimbrite plateau and in relating oxygen isotope stages. *Geomorphology*, 9, 97–115.

Kennedy, W.J. and Cobban, W.A. 1976. Aspects of ammonite biology, biostratigraphy and biogeography. *Special Papers in Palaeontology*, 17, 1–94.

Kennett, J.P. 1977. Cenozoic evolution of Antarctic glaciation, the circum-Antarctic current and their impact on global paleoceanography. *Journal of Geophysical Research*, 82, 3843–3860.

Kennett, J.P. 1982. *Marine Geology.* New Jersey: Prentice-Hall, 813 p.

Kennett, J.P. and Barker, P.F. 1990. Latest Cretaceous to Cenozoic climate and oceanographic developments in the Weddell Sea, Antarctica: an ocean drilling perspective. *Proceedings of the Ocean Drilling Program*, 113, 937–960.

Kennett, J.P. and Hodell, D.A. 1993. Evidence for relative climatic stability of Antarctica during the early Pliocene: a marine perspective. *Geografiska Annaler*, 75A, 205–220.

Kennett, J.P. and Stott, L.D. 1991. Abrupt deep-sea warming, palaeoceanographic changes and benthic extinctions at the end of the Palaeocene. *Nature,* 353, 225–229.

Kennett, J.P. and von der Borch, C.C. 1985. Southwest Pacific Cenozoic paleoceanography. *In:* Kennett, J.P. and von der Borch, C.C. (editors). *Initial Reports of the Deep Sea Drilling Project, Volume XC.* Washington: US Government Printing Office, 1493–1517.

Kent, D.P., Opdyke, N.D. and Ewing, M. 1971. Climate change in the North Pacific using ice-rafted debris as a climatic indicator. *Geological Society of America, Bulletin*, 82, 2741–2754.

Kern, J.P. and Rockwell, T.K. 1992. Chronology and deformation of Quaternary marine shorelines, San Diego County, California. *In:* Fletcher, C.H. and Wehmiller, J.F. (editors). *Quaternary Coasts of the United States: marine and lacustrine systems.* Tulsa: Society for Sedimentary Geology, 377–382.

Kershaw, A.P. 1976. A late Pleistocene and Holocene pollen diagram from Lynch's Crater, north-eastern Queensland, Australia. *New Phytologist*, 77, 469–498.

Kershaw, A.P. 1986. The last two glacial–interglacial cycles in northeastern Australia: implications for climatic change and Aboriginal burning. *Nature*, 322, 47–49.

Kershaw, A.P. and Nanson, G.C. 1993. The last full glacial cycle in the Australian region. *Global and Planetary Change*, 7, 1–9.

Kershaw, A.P., McKenzie, G.M. and McMinn, A. 1993. A Quaternary vegetation history of northeastern Queensland from pollen analysis of ODP site 820. *In:* McKenzie, J.A., Davies, P.J., Palmer-Julson, A. et al. (editors). *Proceedings of the Ocean Drilling Program, Scientific Results*, 133, 107–114.

Kershaw, A.P., Stokes, T. and Bishop, P. 1994. Late Quaternary environments of tropical Australasia. *In: Quaternary Stratigraphy of Asia and the Pacific, IGCP 296.* ESCAP, Mineral Resources Development Series 63, 1–8.

King, L. 1990. *Evolution of Chumash Society*. New York: Garland, 296 p.

King, L.C. 1983. *Wandering Continents and Spreading Sea Floors on an Expanding Earth*. New York: Wiley, 232 p.

Kirch, P.V. 1982. The impact of the prehistoric Polynesians on the Hawaiian ecosystem. *Pacific Science*, 36, 1–14.

Kirch, P.V. 1997. *The Lapita Peoples: Ancestors of the Oceanic World*. Oxford: Blackwell, 353 p.

Kirch, P.V. and Ellison, J. 1994. Palaeoenvironmental evidence for human colonization of remote Oceanic islands. *Antiquity*, 68, 310–321.

Kirch, P.V., Flenley, J.R., Steadman, D.W., Lamont, F. and Dawson, S. 1992. Ancient environmental degradation: prehistoric human impacts to an island ecosystem: Mangaia, Central Polynesia. *National Geographic Research and Exploration*, 8, 166–179.

Kitagawa, H. and van der Plicht, J. 1998. Atmospheric radiocarbon calibration to 45,000 yr B.P.: Late Glacial fluctuations and cosmogenic isotope production. *Science*, 1187–1190.

Klein, R.T., Lohmann, K.C. and Kennedy, G.L. 1997. Elemental and isotopic proxies of paleotemperature and paleosalinity: climate reconstruction of the marginal northeast Pacific ca. 80 ka. *Geology*, 25, 363–366.

Knoll, A. 1991. End of the Proterozoic era. *Scientific American*, 265, 64–73.

Knoll, A. and Walter, M.R. 1992. Latest Proterozoic stratigraphy and Earth history. *Nature*, 356, 673–678.

Koba, M. 1992. Influx of the Kuroshio current into the Okinawa Trough and inauguration of Quaternary coral-reef building in the Ryukyu island arc, Japan. *The Quaternary Research (Japan)*, 31, 359–373.

Kobayashi, K. 1985. Sea of Japan and Okinawa Trough. *In:* Nairn, A.E.M., Stehli, F.G. and Uyeda, S. (editors). *The Ocean Basins and Margins, Volume 7A. The Pacific Ocean*. New York: Plenum Press, 419–458.

Konishi, K., Omura, A. and Nakamichi, O. 1974. Radiometric coral ages and sea level records from the late Quaternary reef complexes of the Ryukyu Islands. *Proceedings of the 2nd International Coral Reef Symposium*, 2, 595–613.

Korotkii, A.M. 1985. Quaternary sea-level fluctuations on the northwestern shelf of the Japan Sea. *Journal of Coastal Research*, 1, 293–298.

Korotky, A.M., Grebennikova, T.A., Razjigaeva, N.G., Volkov, V.G., Mokhova, L.M. and Bazarova, V.B. 1994. Influence of tectonic and eustatic movements to formation of Sakhalin island marine terraces. *In:* Viggósson, G. (editor). *Proceedings of the Hornafjördur International Coastal Symposium.*

Reykjavik: The Icelandic Harbour Authority, 533–552.

Kött, A., Gaupp, R. and Wörner, G. 1995. Miocene to recent history of the western Altiplano in northern Chile as revealed by lacustrine sediments of the Lauca Basin (18°5′–18°40′S/69°30′–69°05′W). *Geologische Rundschau*, 84, 770–780.

Krassilov, V.A. 1978. Araucariaceae as indicators of climate and palaeolatitudes. *Review of Palaeobotany and Palynology*, 26, 113–124.

Krymholts, G.Y., Mesezhnikov, M.S. and Westerman, G.E.G. 1988. The Jurassic ammonite zones of the Soviet Union. *Geological Society of America, Special Paper 223*, 116 p.

Ku, T.-L., Kimmel, M.A., Easton, W.H. and O'Neill, T.J. 1974. Eustatic sea level 120,000 years ago on Oahu, Hawaii. *Science*, 183, 959–962.

Kukal, Z. 1990. The rates of geological processes. *Earth-Science Reviews*, 28, 1–284.

Kukla, G. and An, Z. 1989. Loess stratigraphy in central China. *Palaeogeography, Palaeoclimatology, Palaeoecology*, 72, 203–225.

Kuzmin, Y.V. 1995. People and environment in the Russian far east from Paleolithic to Middle Ages: chronology, palaeogeography, interaction. *GeoJournal*, 35, 79–83.

Kyte, F.T. 1988. The extraterrestrial component in marine sediments: description and interpretation. *Paleoceanography*, 3, 235–247.

Labeyrie, J., Lalou, C. and Delibrias, G. 1969. Etudes des transgressions marines sur l'atoll de Mururoa par la datation des differents niveaux de corail. *Cahiers du Pacifique*, 3, 59–68.

Labeyrie, L.D., Pichon, J.J., Labracherie, M., Ippolito, P., Duprat, J. and Duplessy, J.C. 1986. Melting history of Antarctica during the past 60,000 years. *Nature*, 322, 701–706.

Labeyrie, L.D., Duplessy, J.C. and Blanc, P.L. 1987. Variations in mode of formation and temperature of oceanic deep waters over the past 125,000 years. *Nature*, 327, 477–482.

Lagoe, M.B. and Zellers, S.D. 1996. Depositional and microfaunal response to Pliocene climate change and tectonics in the eastern Gulf of Alaska. *Marine Micropaleontology*, 27, 121–140.

LaMarche, V.C. 1973. Holocene climatic variations inferred from treeline fluctuations in the White Mountains, California. *Quaternary Research*, 3, 632–660.

Lamb, H.H. 1977. *Climate, Past, Present and Future. Volume 2. Climate History and the Future*. London: Methuen.

Lambeck, K. and Nakada, M. 1992. Constraints on the age and duration of the Last Interglacial period and on sea-level variations. *Nature*, 357, 125–128.

Lambert, D. (and the Diagram Group). 1987. *The*

Cambridge Guide to Prehistoric Man. Cambridge: Cambridge University Press.

Langway, C.C., Osada, K., Clausen, H.B., Hammer, C.U., Shoji, H. and Mitani, A. 1994. New chemical stratigraphy over the last millennium for Byrd Station, Antarctica. *Tellus*, 46B, 40–51.

Lasaqa, I. 1973. Geography and geographers in the changing Pacific: an islander's view. *In:* Brookfield, H. (editor). *The Pacific in Transition: geographical perspectives on adaptation and change*. New York: St Martin's Press, 299–311.

Lauer, W. 1993. Human development and environment in the Andes: a geoecological overview. *Mountain Research and Development*, 13, 157–166.

Lecolle, J. and Bernat, M. 1985. Late Quaternary surrection history of Efate island, New Hebrides island arc (Vanuatu): Th/U dates from uplifted terraces. *Proceedings of the 5th International Coral Reef Congress*, 3, 179–84.

Lecolle, J., Bokilo, J.E. and Bernat, M. 1990. Soulèvement et tectonique de l'ile d'Efaté (Vanuatu) arc insulaire des Nouvelles-Hébrides, au cours du Quaternaire récent. Datations de terrasses soulevées par la methode U/Th. *Marine Geology*, 94, 251–270.

Lees, B.G., Yanchou, L. and Head, J. 1990. Reconnaissance thermoluminescence dating of northern Australian coastal dune systems. *Quaternary Research*, 34, 169–185.

Lees, B.G., Hayne, M. and Price, D. 1993. Marine transgression and dune initiation on western Cape York, northern Australia. *Marine Geology*, 114, 81–89.

Leg 113 Shipboard Scientific Party. 1987. Glacial history of Antarctica. *Nature*, 328, 115–116.

LeMasurier, W.E. 1990. Late Cenozoic volcanism on the Antarctic Plate. *In:* LeMasurier, W.E. and Thomson, J.W. (editors). *Volcanoes of the Antarctic Plate and Southern Oceans*. Washington DC: American Geophysical Union, Antarctic Research Series 48, 1–17.

LeMasurier, W.E. and Rex, D.C. 1983. Rates of uplift and the scale of ice level instabilities recorded by volcanic rocks in Marie Byrd Land, West Antarctica. *In:* Oliver, R.L., James, P.R. and Jago, J.B. (editors). *Antarctic Earth Science*. Canberra: Australian Academy of Science, 663–670.

LeMasurier, W.E. and Rex, D.C. 1991. Tectonic significance of linear volcanic ranges in Marie Byrd Land in late Cenozoic times (extended abstract). *In:* Thomson, M.R.A., Crame, J.A. and Thomson, J.W. (editors). *Geological Evolution of Antarctica*. Cambridge: Cambridge University Press, 531–532.

LeMasurier, W.E., Harwood, D.M. and Rex, D.C. 1994. Geology of Mount Murphy Volcano: an 8-m.y. history of interaction between a rift volcano and the West Antarctic ice sheet. *Geological Society of America, Bulletin*, 106, 265–280.

Leopold, E.B., Liu, G. and Clay-Poole, S. 1992. Low-biomass vegetation in the Oligocene? *In:* Prothero, D.R. and Berggren, W.A. (editors). *Eocene–Oligocene Climatic and Biotic Evolution*. Princeton University Press, 399–420.

Levison, M., Ward, R.G. and Webb, J.W. 1973. *The Settlement of Polynesia: a computer simulation*. Minneapolis: University of Minnesota Press, 137 p.

Leyden, B.W. 1995. Evidence of the Younger Dryas in Central America. *Quaternary Science Reviews*, 14, 833–839.

Li, J., Lowenstein, T.K., Brown, C.B., Ku, T-L and Luo, S. 1996. A 100 ka record of water tables and paleoclimates from salt cores, Death Valley, California. *Palaeogeography, Palaeoclimatology, Palaeoecology*, 123, 179–203.

Li, Y. 1988. Climatic change and the distribution of agriculture in Jiangxi Province. *In:* Zhang, J. (editor). *The Reconstruction of Climate in China for Historical Times*. Beijing: Science Press, 145–159.

Liew, P-M. 1991. Pleistocene cool stages and geological changes of western Taiwan based on palynological study. *Acta Geologica Taiwanica*, 29, 21–32.

Liew, P-M., Pirazzoli, P.A., Hsieh, M.L., Arnold, M., Barusseau, J.P., Fontugne, M. and Giresse, P. 1993. Holocene tectonic uplift deduced from elevated shorelines, eastern Coastal Range of Taiwan. *Tectonophysics*, 222, 55–68.

Lincoln, J.M. and Schlanger, S.O. 1987. Miocene sea-level falls related to the geologic history of Midway atoll. *Geology*, 15, 454–457.

Linder, S.B. 1986. *The Pacific Century: economic and political consequences of Asian-Pacific dynamism*. Stanford: Stanford University Press, 154 p.

Liu, K. 1988. Quaternary history of the temperate forests of China. *Quaternary Science Reviews*, 7, 1–20.

Löffler, E., Mackenzie, D.E. and Webb, A. 1980. Potassium argon dates from New Guinea highland volcanoes and their relevance to the geomorphic history. *Journal of the Geological Society of Australia*, 26, 387–397.

Long, A.J. and Shennan, I. 1994. Sea-level changes in Washington and Oregon and the 'earthquake deformation cycle'. *Journal of Coastal Research*, 10, 825–838.

Lorius, C. and Oeschger, H. 1994. Palaeo-perspectives: reducing uncertainty in global change? *Ambio*, 23, 30–36.

Lorius, C., Jouzel, J., Ritz, C., Melivat, L., Barkov, N.I., Korotkevich, Y.S. and Kotlyakov, V.M. 1985. A 150,000-year climatic record from Antarctic ice. *Nature*, 316, 591–596.

Lourandos, H. 1997. *Continent of Hunter-Gatherers: new perspectives in Australian prehistory*. Cambridge University Press, 390 p.

Lowenthal, D. 1958. *George Perkins Marsh: versatile Vermonter*. New York: Columbia University Press, 442 p.

Lozano-Garciá, M.S., Ortega-Guerrero, B., Caballero-Miranda, M. and Urrutia-Fucugauchi, J. 1993. Late Pleistocene and Holocene paleoclimates of Chalco Lake, central Mexico. *Quaternary Research*, 40, 332–342.

Lozhkin, A.V. 1993. Geochronology of late Quaternary events in northeastern Russia. *Radiocarbon*, 35, 429–433.

Lozhkin, A.V. and Anderson, P.M. 1995. The Last Interglaciation in northeast Siberia. *Quaternary Research*, 43, 147–158.

Lozhkin, A.V., Anderson, P.M., Eisner, W.R., Ravako, L.G., Hopkins, D.M., Brubaker, L.B., Colinvaux, P.A. and Miller, M.C. 1993. Late Quaternary lacustrine pollen records from southwestern Beringia. *Quaternary Research*, 39, 314–324.

Luckman, B.H. 1993. Glacier fluctuation and tree-ring records for the last millennium in the Canadian Rockies. *Quaternary Science Reviews*, 12, 441–450.

Luckman, B.H. 1994. Evidence for climatic conditions between ca. 900–1300 A.D. in the southern Canadian Rockies. *Climatic Change*, 26, 171–182.

Luckman, B.H., Holdsworth, G. and Osborn, G.D. 1993. Neoglacial glacial fluctuations in the Canadian Rockies. *Quaternary Research*, 39, 144–153.

Ludwig, K.R., Muhs, D.R., Simmons, K.R. and Moore, J.G. 1992. Sr-isotope record of Quaternary marine terraces on the California coast and off Hawaii. *Quaternary Research*, 37, 267–280.

Luedke, R.G. and Smith, R.L. 1991. Quaternary volcanism in the western conterminous United States. *In:* Morrison, R.B. (editor). *Quaternary Nonglacial Geology; Conterminous U.S.* Boulder: Geological Society of America, The Geology of North America, Volume K-2, 75–92.

Lumbreras, L.G. 1988. Childe y la tesis de la revolución urbana: la experiencia central andina. *In:* Manzanilla, L. (editor). *Coloquio V. Gordon Childe: estudios sobre la revolución neolítica y la revolución urbana*. México: UNAM, Instituto de Investigaciones Antropológicas, 349–366.

Lundelius, E.L. 1983. Climatic implications of late Pleistocene and Holocene associations in Australia. *Alcheringa*, 7, 125–149.

Macharé, J. and Ortlieb, L. 1992. Plio-Quaternary vertical motions and the subduction of the Nazca Ridge, central coast of Peru. *Tectonophysics*, 205, 97–108.

Magnitsky, V.A. 1961. Geophysical data and the problem of the origin of the Pacific Ocean. *10th Pacific Science Congress, Abstracts of Papers*, 361–362.

Mann, D.H. and Hamilton, T.D. 1995. Late Pleistocene and Holocene paleoenvironments of the North Pacific coast. *Quaternary Science Reviews*, 14, 449–471.

Mann, D.H. and Peteet, D.M. 1994. Extent and timing of the Last Glacial Maximum in southwestern Alaska. *Quaternary Research*, 42, 136–148.

Mannion, A.M. 1991. *Global Environmental Change*. Harlow: Longman, 404 p.

Manzanilla, L. 1997. The impact of climatic change on past civilizations: a revisionist agenda for further investigation. *Quaternary International*, 43/44, 153–159.

Marchant, D.R., Denton, G.H., Swisher, C.C. and Potter, S. 1996. Late Cenozoic Antarctic paleoclimate constructed from volcanic ashes in the Dry Valleys region of southern Victoria Land. *Geological Society of America, Bulletin*, 108, 181–194.

Markgraf, V. 1983. Late and Postglacial vegetational and palaeoclimatic changes in subantarctic, temperate, and arid environments in Argentina. *Palynology*, 7, 43–70.

Markgraf, V. 1989. Palaeoclimates in central and south America since 18,000 BP based on pollen and lake-level records. *Quaternary Science Reviews*, 8, 1–24.

Markgraf, V. 1993. Climatic history of central and south America since 18,000 yr B.P.: comparison of pollen records and model simulations. *In:* Wright, H.E., Kutzbach, J.E., Webb, T., Ruddiman, W.F., Street-Perrott, F.A. and Bartleim, P.J. (editors). *Global Climates since the Last Glacial Maximum*. Minneapolis: University of Minnesota, 357–385.

Markgraf, V., Bradbury J.P. and Busby, J.R. 1986. Paleoclimates in southwestern Tasmania during the last 13,000 years. *Palaios*, 1, 368–380.

Markgraf, V., Dodson, J.R., Kershaw, A.P., McGlone, M. and Nicholls, N. 1992. Evolution of late Pleistocene and Holocene climates in the circum-South Pacific land areas. *Climate Dynamics*, 6, 193–211.

Marshall, J.F. and Launay, J. 1978. Uplift rates of the Loyalty Islands as determined by ^{230}Th/^{234}U dating of raised coral terraces. *Quaternary Research*, 9, 186–192.

Martin, A.R.H. 1986. Late glacial and Holocene alpine pollen diagrams from the Kosciusko National Park, New South Wales, Australia. *Review of Palaeobotany and Palynology*, 47, 367–409.

Martin, H.A. 1989. Evolution of Mallee and its environment. *In:* Noble, J.C. and Bradstock, R.A. (editors). *Mediterranean Landscapes in Australia*. Melbourne: CSIRO, 83–92.

Martínez, J.I. 1994. Late Pleistocene palaeoceanography of the Tasman Sea: implications for the dynamics of the warm pool in the western Pacific. *Palaeogeography, Palaeoclimatology, Palaeoecology*, 112, 19–62.

Martinson, D.G., Pisias, N.G., Hays, J.D., Imbrie, J., Moore, T.C. and Shackleton, N.J. 1987. Age dating and the orbital theory of the Ice Ages: development of a high-resolution 0 to 300,000 yr chronostratigraphy. *Quaternary Research*, 27, 1–29.

Maruyama, S., Liou, J.G. and Seno, T. 1989. Mesozoic and Cenozoic evolution of Asia. *In:* Ben-Avraham, Z. (editor). *The Evolution of the Pacific Ocean Margins*. New York: Oxford University Press, 75–99.

Mason, O.K. and Jordan, J.W. 1993. Heightened North Pacific storminess during synchronous late Holocene erosion of northwest Alaska beach ridges. *Quaternary Research*, 40, 55–69.

Mathewes, R.W., Heusser, L.E. and Patterson, R.T. 1993. Evidence for a Younger Dryas-like cooling event on the British Columbian coast. *Geology*, 21, 101–104.

Mathews, W.H. and Rouse, G.E. 1986. An early Pleistocene proglacial succession in south-central British Columbia. *Canadian Journal of Earth Sciences*, 23, 1796–1803.

Matsuda, T. 1976. Empirical rules on sense and rate of recent crustal movements. *Journal of the Geodetic Society of Japan*, 22, 252–263.

Matsumoto, T. 1978. Japan and adjoining areas. *In:* Moullade, M. and Nairn, A.E.M. (editors). *The Phanerozoic Geology of the World II, The Mesozoic A.* Amsterdam: Elsevier, 79–144.

Matthews, J.L., Heezen, B.C., Catalano, R., Coogan, A., Tharp, M., Natland, J. and Rawson, M. 1974. Cretaceous drowning of reefs on mid-Pacific mountains and Japanese guyots. *Science*, 184, 462–464.

Matthews, J.V. and Ovenden, L.E. 1990. Late Tertiary plant macrofossils from localities in northern North America (Alaska, Yukon and Northwest Territories). *Arctic*, 43, 364–392.

Maxwell, W.D. 1989. The end Permian mass Extinction. *In:* Donovan, S.K. (editor). *Mass extinctions: processes and evidence*. New York: Columbia University Press, 152–173.

Mayewski, P.A., Lyons, W.B., Zielinski, G., Twickler, M., Whitlow, S., Dibb, J., Grootes, P., Taylor, K., Whung, P.-Y., Fosberry, L., Wake, C. and Welch, K. 1995. An ice-core-based, late Holocene history for the Transantarctic Mountains, Antarctica. *Contributions to Antarctic Research IV, Antarctic Research Series*, 67, 33–45.

Mazaud, A., Laj, C. and Bender, M. 1994. A geomagnetic chronology for antarctic ice accumulation. *Geophysical Research Letters*, 21, 337–340.

McCall, G. 1993. Little Ice Age: some speculations for Rapanui. *Rapa Nui Journal*, 7, 65–70.

McCulloch, M., Mortimer, G., Esat, T., Li, X., Pillans,

B. and Chappell, J. 1996. High resolution windows into early Holocene climate: Sr/Ca coral records from the Huon Peninsula. *Earth and Planetary Science Letters*, 138, 169–178.

McEwen Mason, J.R.C. 1991. The late Cainozoic magnetostratigraphy and preliminary palynology of Lake George, New South Wales. *In:* Williams, M.A.J., De Deckker, P. and Kershaw, A.P. (editors). *The Cainozoic in Australia: a reappraisal of the evidence*. Melbourne: Geological Society of Australia, 195–209.

McFadgen, B.G. 1994. Archaeology and holocene sand dune stratigraphy on Chatham Island. *Journal of the Royal Society of New Zealand*, 24, 17–44.

McGhee, G.R. 1996. *The Late Devonian Mass Extinction*. New York: Columbia University Press, 303 p.

McGlone, M. 1988. New Zealand. *In:* Huntley, B. and Webb, T. (editors). *Vegetation History*. Dordecht: Kluwer, 557–599.

McGlone, M. 1995. Lateglacial landscape and vegetation change and the Younger Dryas climatic oscillation in New Zealand. *Quaternary Science Reviews*, 14, 867–881.

McGlone, M.S. and Topping, W.W. 1983. Late Quaternary vegetation, Tongariro region, central North Island, New Zealand. *New Zealand Journal of Botany*, 21, 53–76.

McGlone, M., Salinger, M.J. and Moar, N.T. 1993. Paleovegetation studies of New Zealand's climate since the Last Glacial Maximum. *In:* Wright, H.E., Kutzbach, J.E., Webb, T., Ruddiman, W.F., Street-Perrott, F.A. and Bartleim, P.J. (editors). *Global Climates since the Last Glacial Maximum*. Minneapolis: University of Minnesota, 294–317.

McGowran, B. 1989. Silica burp in the Eocene ocean. *Geology*, 17, 857–860.

McMenamin, M.S. and McMenamin, D.L.S. 1990. *The Emergence of Animals: the Cambrian breakthrough*. New York: Columbia University Press, 217 p.

McWilliams, M.O. and Howell, D.G. 1982. Exotic terranes of western California. *Nature*, 297, 215–217.

Mégard, F. 1989. The evolution of the Pacific Ocean margin in South America north of the Arica elbow. *In:* Ben-Avraham, Z. (editor). *The Evolution of the Pacific Ocean Margins*. Oxford Monographs in Geology and Geophysics 8, 208–230.

Mégard, F. Noble, D.C., McKee, E.H. and Bellon, H. 1984. Multiple pulses of Neogene compressive deformation in the Ayacucho intermontane basin, Andes of central Peru. *Geological Society of America, Bulletin,* 95, 1108–1117.

Meko, D.M. 1992. Spectral properties of tree-ring data in the United States Southwest as related to El Niño/Southern Oscillation. *In:* Diaz, H.F. and Markgraf,

V. (editors). *El Niño: historical and paleoclimate aspects of the Southern Oscillation.* New York: Cambridge University Press, 227–242.

Melville, E.G.K. 1990. Environmental change and social change in the Valle del Mezquital, Mexico, 1521–1600. *Comparative Studies in Society and History*, 32, 24–53.

Menard, H.W. 1964. *Marine Geology of the Pacific.* New York: McGraw Hill, 271 p.

Menzies, N.K. 1994. *Forest and Land Management in Imperial China.* New York: St Martin's Press, 175 p.

Méon, H. and Pannetier, W. 1994. Palynological study of the late Quaternary of Loyalty Basin (SW Pacific). *Palaeogeography, Palaeoclimatology, Palaeoecology*, 111, 135–147.

Mercer, J.H. 1968. *Antarctic Ice and Sangamon Sea Level.* International Association of Scientific Hydrology, Publication 79, 217–225.

Mercer, J.H. 1976. Glacial history of southernmost South America. *Quaternary Research*, 6, 125–166.

Mercer, J.H. 1982. Holocene glacier variations in southern Patagonia. *Striae*, 18, 35–40.

Mercer, J.H. and Palacios, M.O. 1977. Radiocarbon dating of the last glaciation in Peru. *Geology*, 5, 600–604.

Mercer, J.H. and Sutter, J.F. 1982. Late Miocene – earliest Pliocene glaciation in southern Argentina: implications for global ice-sheet history. *Palaeogeography, Palaeoclimatology, Palaeoecology*, 38, 185–206.

Metcalfe, I. 1993. Southeast Asian terranes: Gondwanaland origins and evolution. *In:* Findlay, R.H., Unrug, R., Banks, M.R. and Veevers, J.J. (editors). *Gondwana Eight: assembly, evolution and dispersal.* Rotterdam: Balkema, 181–200.

Metcalfe, S.E. 1987. Historical data and climatic change in México: a review. *The Geographical Journal*, 153, 211–222.

Metcalfe, S.E. 1995. Holocene environmental change in the Zacapu Basin, Mexico: a diatom-based record. *The Holocene*, 5, 196–206.

Meyerhoff, A.A., Taner, I., Morris, A.E.L., Martin, B.D., Agocs, W.B. and Meyerhoff, H.A. 1992. Surge tectonics: a new hypothesis of Earth dynamics. *In:* Chatterjee, S. and Hotton, N. (editors). *New Concepts in Global Tectonics.* Lubbock: Texas Tech University Press, 309–409.

Milankovitch, M. 1930. Mathematische Klimalehre und astronomishe Theory der Klimaschwankungen. *In:* Köppen, W. and Geiger, R. (editors). *Handbuch der Klimatologie, band 1, Teil A.* Berlin: Borntraeger, 1–176.

Mildenhall, D.C. and Brown, L.J. 1987. An early Holocene occurrence of the mangrove Avicennia marina in Poverty Bay, North Island, New Zealand:
its climatic and geological implications. *New Zealand Journal of Botany*, 25, 281–294.

Miller, A.K. and Youngquist, W. 1949. American Permian nautiloids. *Geological Society of America, Memoir*, 41, 218 p.

Miller, C.D. 1969. Chronology of Neoglacial moraines in the Dome Peak area, north Cascade range, Washington. *Arctic and Alpine Research*, 1, 49–66.

Miller, D.C., Birkeland, P.W. and Rodbell, D.T. 1993. Evidence for Holocene stability of steep slopes, northern Peruvian Andes, based on soils and radiocarbon dates. *Catena*, 20, 1–12.

Miller K.G. 1992. Middle Eocene to Oligocene stable isotopes, climate, and deep-water history: the terminal Eocene event? *In:* Prothero, D.R. and Berggren, W.A. (editors). *Eocene–Oligocene Climatic and Biotic Evolution.* Princeton University Press, 159–177.

Miller, K.G., Fairbanks, R.G. and Mountain, G.S. 1987. Tertiary oxygen isotope synthesis, sea level history, and continental margin erosion. *Paleoceanography*, 2, 1–19.

Miller, K.G., Wright, J.D. and Fairbanks, R.G. 1991. Unlocking the ice house: Oligocene–Miocene oxygen isotopes, eustasy, and margin erosion. *Journal of Geophysical Research*, 96, 6829–6848.

Mills, R.O. 1994. Radiocarbon calibration of archaeological dates from the central Gulf of Alaska. *Arctic Anthropology*, 31, 126–149.

Mimura, N. and Nunn, P.D. 1998. Trends of beach erosion and shoreline protection in rural Fiji. *Journal of Coastal Research*, 14, 37–46.

Mitchell, A. 1989. *A Fragile Paradise.* London: Collins, 256 p.

Mochanov, I.A. 1977. *Drevnétshie Etapy Zaseeniia Chelovekom Severo Vostochnoi Azii (Ancient Stages of Human Settlement in Northeastern Asia).* Novosibirsk: Izdatrel'stvo Nauka.

Mohr, B.A.R. 1990. Eocene and Oligocene sporomorphs and dinoflagellate cysts from Leg 113 drill sites, Weddell Sea, Antarctica. *Proceedings of the Ocean Drilling Program*, 113, 595–606.

Molina-Cruz, A. 1977. The relation of the southern tradewinds to upwelling processes during the last 75,000 years. *Quaternary Research*, 8, 324–338.

Momohara, A. 1994. Floral and paleoenvironmental history from the late Pliocene to middle Pleistocene in and around central Japan. *Palaeogeography, Palaeoclimatology, Palaeoecology*, 108, 281–293.

Montaggioni, L.F., Richard, G., Bourrouilh-Le Jan, F., Gabrié, C., Humbert, L., Monteforte, M., Naim, O., Payri, C. and Salvat, B. 1985. Geology and marine biology of Makatea, an uplifted atoll, Tuamotu archipelago, central Pacific Ocean. *Journal of Coastal Research*, 1, 165–171.

Moores, E.M. 1991. Southwest U.S.–East Antarctica

(SWEAT) connection: a hypothesis. *Geology*, 19, 425–428.

Moores, E.M. 1993. Neoproterozoic oceanic crustal thinning, emergence of continents, and origin of the Phanerozoic ecosystem: a model. *Geology*, 21, 5–8.

Morgan, V.I. 1985. An oxygen isotope – climate record from the Law Dome, Antarctica. *Climatic Change*, 7, 415–426.

Mörner, N.-A. 1981. Revolution in Cretaceous sea-level analysis. *Geology*, 9, 344–346.

Morris, C. and von Hagen, A. 1993. *The Inka Empire and its Andean origins*. New York: Abbeville Press, 251 p.

Morrison, J., Geraghty, P. and Crowl, L. (editors). 1994. *Science of Pacific Island Peoples*. Suva: Institute of Pacific Studies, The University of the South Pacific, 4 volumes.

Moseley, M.E., Wagner, D. and Richardson, J.B. 1992. Space shuttle imagery of recent catastrophic change along the arid Andean coast. *In:* Johnson, L.L. (editor). *Paleoshorelines and Prehistory: an investigation of method*. Boca Raton: CRC Press, 215–235.

Mosley-Thompson, E. 1992. Paleoenvironmental conditions in Antarctica since A.D. 1500: ice core evidence. *In:* Bradley, R.S. and Jones, P.D. (editors). *Climate since A.D. 1500*. London: Routledge, 572–591.

Muhs, D.R. and Szabo, B.J. 1994. New uranium-series ages of the Waimanalo limestone, Oahu, Hawaii: implications for sea level during the last interglacial period. *Marine Geology*, 118, 315–326.

Muhs, D.R., Kelsey, H.M., Miller, G.H., Kennedy, G.L., Whelan, J.F. and McInelly, G.W. 1990. Age estimates and uplift rates for late Pleistocene marine terraces: southern Oregon portion of the Cascadia forearc. *Journal of Geophysical Research*, 95, 6685–6698.

Muhs, D.R., Rockwell, T.K. and Kennedy, G.L. 1992a. Late Quaternary uplift rates on the Pacific coast of North America, southern Oregon to Baja California Sur. *Quaternary International*, 15/16, 121–133.

Muhs, D.R., Miller, G.H., Whelan, J.F. and Kennedy, G.L. 1992b. Aminostratigraphy and oxygen isotope stratigraphy of marine-terrace deposits, Palos Verdes Hills and San Pedro areas, Los Angeles County, California. *In:* Fletcher, C.H. and Wehmiller, J.F. (editors). *Quaternary Coasts of the United States: marine and lacustrine systems*. Tulsa: Society for Sedimentary Geology, 363–382.

Muhs, D.R., Kennedy, G.L. and Rockwell, T.K. 1994. Uranium-series ages of marine terrace corals from the Pacific coast of North America and implications for Last-Interglacial sea level history. *Quaternary Research*, 42, 72–87.

Murray-Wallace, C.V. and Belperio, A.P. 1991. The last interglacial shoreline in Australia – a review. *Quaternary Science Reviews*, 10, 441–461.

Nanson, G.C., Young, R.W. and Stockton, E.D. 1987. Chronology and palaeoenvironment of the Cranebrook Terrace (near Sydney) containing artefacts more than 40,000 years old. *Archaeology in Oceania*, 22, 72–78.

Nanson, G.C., Price, D.M. and Short, S.A. 1992. Wetting and drying of Australia over the past 300 ka. *Geology*, 20, 791–794.

Neall, V.E. 1975. Climate-controlled tephra deposition on Pouakai Ring plain, Taranaki, New Zealand. *New Zealand Journal of Geology and Geophysics*, 18, 317–326.

Nelson, A.R. and Manley, W.F. 1992. Holocene coseismic and aseismic uplift of Isla Mocha, south-central Chile. *Quaternary International*, 15/16, 61–76.

Nelson, C.S., Hendy, C.H., Jarrett, G.R. and Cuthbertson, A.M. 1985. Near-synchroneity of New Zealand alpine glaciations and northern hemisphere continental glaciations during the past 750 ka. *Nature*, 318, 361–363.

Nelson, C.S., Mildenhall, D.C., Todd, A.J. and Pochnall, D.T. 1988. Subsurface stratigraphy, paleo-environments, palynology, and deposition history of the late Neogene Tauronge Group at Ohinewai, lower Waikato Lowland, South Auckland, New Zealand. *New Zealand Journal of Geology and Geophysics*, 31, 21–40.

Nelson, C.S., Cooke, P.J., Hendy, C.H. and Cuthbertson, A.M. 1993. Oceanographic and climatic changes over the past 160,000 years at Deep Sea Drilling Project site 594 off southeastern New Zealand, southwest Pacific Ocean. *Paleoceanography*, 8, 435–458.

Nelson, R.H. 1993. Environmental Calvinism: the Judeo-Christian roots of eco-theology. *In:* Meiners, R.E. and Yandle, B. (editors). *Taking the Environment Seriously*. Lanham: Rowman and Littlefield, 233–255.

Neumann, A.C. and MacIntyre, I. 1985. Reef response to sea-level rise: keep-up, catch-up or give-up. *In: Proceedings of the 5th International Coral Reef Congress*, 3, 105–110.

Newbrough, J.B. 1932. *Oahspe: A Kosmon Revelation in the Words of Jehovih & His Angel Ambassadors*. Los Angeles: Kosmon, 16 p.

Newton, C.R. 1988. Significance of 'Tethyan' fossils in the American Cordillera. *Science*, 242, 385–391.

Nichols, R.L. 1966. Geomorphology of Antarctica. *In:* Tedrow, J.C.F. (editor). *Antarctic Soils and Soil-Forming Processes*. Washington: American Geophysical Union, 1–46.

Niemi, T. and Hall, N.T. 1992. Late Holocene slip rate and recurrence of great earthquakes on the San

Andreas fault in northern California. *Geology*, 20, 195–198.

Ning, S., Jia-xin, C. and Königsson, L-K. 1993. Late Cenozoic vegetational history and the Pliocene–Pleistocene boundary in the Yushe basin, S.E. Shanxi, China. *Grana*, 32, 260–271.

Noble, D.C., McKee, E.H., Farrar, E. and Peterson, U. 1974. Episodic Cenozoic volcanism and tectonism in the Andes of Peru. *Earth and Planetary Science Letters*, 21, 213–220.

Norton, D.A., Briffa, K.A. and Salinger, M.J. 1989. Reconstruction of New Zealand summer temperatures to 1730 AD using dendroclimatic techniques. *International Journal of Climatology*, 9, 633–644.

Nunn, P.D. 1984. Occurrence and ages of low-level platforms and associated deposits on South Atlantic coasts: appraisal of evidence for regional Holocene high sea level. *Progress in Physical Geography*, 8, 32–60.

Nunn, P.D. 1986. Implications of migrating geoid anomalies for the interpretation of high-level fossil coral reefs. *Geological Society of America, Bulletin*, 97, 946–952.

Nunn, P.D. 1990a. Coastal processes and landforms of Fiji and their bearing on Holocene sea-level changes in the south and west Pacific. *Journal of Coastal Research*, 6, 279–310.

Nunn, P.D. 1990b. Coastal geomorphology of Beqa and Yanuca islands, South Pacific Ocean, and its significance for the geomorphology of the Vatulele-Beqa Ridge. *Pacific Science*, 44, 348–365.

Nunn, P.D. 1991. Keimami sa vakila na liga ni Kalou *(Feeling the hand of God): human and nonhuman impacts on Pacific island environments*. Honolulu: East–West Center, Occasional Paper 13, 69 p.

Nunn, P.D. 1992. Geomorphology of the Astrolabe Islands. *In:* Morrison, R.J. and Naqasima, M.R. (editors). *Fiji's Great Astrolabe Reef and Lagoon: A Baseline Study*. The University of the South Pacific, Institute of Natural Resources, Environmental Studies Report 56, 47–64.

Nunn, P.D. 1994a. *Oceanic Islands*. Oxford, UK, and Cambridge, USA: Blackwell, 418 p.

Nunn, P.D. 1994b. Beyond the naive lands: human history and environmental change in the Pacific Basin. *In:* Waddell, E. and Nunn, P.D. (editors). *The Margin Fades: geographical itineraries in a world of islands*. Suva: Institute of Pacific Studies, The University of the South Pacific, 5–27.

Nunn, P.D. 1994c. *Environmental Change and the Early Settlement of Pacific Islands*. East–West Center, Hawaii, Working Papers, Environmental Series 39, 31 p.

Nunn, P.D. 1995. Holocene sea-level changes in the south and west Pacific. *Journal of Coastal Research,*

Special Issue, 17, 311–319.

Nunn, P.D. 1996. *Emerged Shorelines of the Lau Islands*. Suva: Fiji Mineral Resources Department, Memoir 4, 99 p.

Nunn, P.D. 1997a. Late Quaternary environmental changes on Pacific islands: controversy, certainty and conjecture. *Journal of Quaternary Science*, 12, 443–450.

Nunn, P.D. 1997b. South Pacific climate change: global warning. *The New Internationalist*, 291, 20–21.

Nunn, P.D. 1997c. Keimami sa vakila na liga ni Kalou *(Feeling the hand of God): human and nonhuman impacts on Pacific island environments*. Suva, Fiji: School of Social and Economic Development, The University of the South Pacific, 3rd edition, 72 p.

Nunn, P.D. 1998a. *Pacific Island Landscapes*. Suva: Institute of Pacific Studies, The University of the South Pacific, 332 p.

Nunn, P.D. 1998b. Late Quaternary tectonic change on the islands of the northern Lau-Colville Ridge, southwest Pacific. *In:* Stewart, I. and Vita-Finzi, C. (editors). *Late Quaternary Coastal Tectonics*. Geological Society of London Special Publication, in press.

Nunn, P.D. 1998c. Consequences of sea-level change during the Holocene in the Pacific Basin: introduction. *Journal of Coastal Research*, 14, 1–2.

Nunn, P.D. 1998d. Sea-level changes over the past 1000 years in the Pacific. *Journal of Coastal Research*, 14, 23–30.

Nunn, P.D. 1999a. *Fiji: Beneath the Surface*. Suva: School of Social and Economic Development, The University of the South Pacific, forthcoming.

Nunn, P.D. 1999b. Humans in the Pacific Islands: illuminating the past and future. *Journal of Pacific Studies*, in press.

Nunn, P.D. and Finau, F.T. 1995. Late Holocene emergence history of Tongatapu island, South Pacific. *Zeitschrift für Geomorphologie*, 39, 69–95.

Nunn, P.D. and Omura, A. 1999. Penultimate Interglacial emerged reef around Kadavu Island, Southwest Pacific: implications for late Quaternary island-arc tectonics and sea-level history. *New Zealand Journal of Geology and Geophysics*, submitted.

Nur, A. and Ben-Avraham, Z. 1977. Lost Pacifica continent. *Nature*, 270, 41–43.

Nur, A. and Ben-Avraham, Z. 1989. Ocean plateaus and the Pacific Ocean margins. *In*: Ben-Avraham, Z. (editor). *The Evolution of the Pacific Ocean Margins*. Oxford Monographs in Geology and Geophysics 8, 7–19.

Oba, T., Kato, M., Kitazato, H., Koizumi, I., Omura, A., Sakai, T. and Takayama, T. 1991. Paleo-environmental changes in the Japan Sea during the last 85,000 years. *Paleoceanography*, 6, 499–518.

Ochsenius, C. 1974. Relaciones paleobiostratigráficas y paleoecológicas entre los ambientes lenticos de la Puna de Atacama y Altiplano Boliviano, Tropico de Capricorno. *Boletin de prehistoria de Chile*, 7–8, 99–138.

Ofreneo, R.E. 1993. Japan and the environmental degradation of the Philippines. *In:* Howard, M.C. (editor). *Asia's Environmental Crisis*. Boulder: Westview, 201–219.

Ogden, J., Wilson, A., Hendy, C. and Newnham, R.M. 1992. The late Quaternary history of kauri (*Agathis australis*) in New Zealand and its climatic significance. *Journal of Biogeography*, 19, 611–622.

Ogden, J., Newnham, R.M., Palmer, J.G., Serra, R.G. and Mitchell, N.D. 1993. Climatic implications of macro- and micro-fossil assemblages from late Pleistocene deposits in northern New Zealand. *Quaternary Research*, 39, 107–119.

Oglesby, R.J. 1989. A GCM study of Antarctic glaciation. *Climate Dynamics*, 3, 135–156.

O'Hara, S.L. 1993. Historical evidence of fluctuations in the level of Lake Pátzcuaro, Michoacán, México over the last 600 years. *The Geographical Journal*, 159, 51–62.

O'Hara, S.L., Street-Perrott, F.A. and Burt, T.P. 1993. Accelerated soil erosion around a Mexican highland lake caused by prehispanic agriculture. *Nature*, 362, 48–51.

Ohkouchi, N., Kawamura, K., Nakamura, T. and Taira, A. 1994. Small changes in the sea surface temperature during the last 20,000 years: molecular evidence from the western tropical Pacific. *Geophysical Research Letters*, 21, 2207–2210.

Olivarez, A.M., Owen, R.M. and Rea, D.K. 1991. Geochemistry of eolian dust in Pacific pelagic sediments: implications for paleoclimatic interpretations. *Geochimica et Cosmochimica Acta*, 55, 2147–2158.

Oliver, D. 1991. *Black Islanders: a personal perspective of Bougainville 1937–1991*. Honolulu: University of Hawaii Press, 289 p.

Olivero, E.B., Gasparini, Z., Rinaldi, C.A. and Scasso, R. 1991. First record of fossil dinosaurs in Antarctica (Upper Cretaceous, James Ross Island): palaeogeographical implications. *In:* Thomson, M.R.A., Crame, J.A. and Thomson, J.W. (editors). *Geological Evolution of Antarctica*. Cambridge: Cambridge University Press, 617–622.

Ollier, C.D. 1988. *Volcanoes*. Oxford: Blackwell, 228 p.

Omura, A. 1984. Uranium-series ages of the Riukiu Limestone on Hateruma island, southwestern Ryukyus. *Transactions and Proceedings of the Palaeontological Society of Japan*, 135, 415–426.

Orme, A.R. 1992. Late Quaternary deposits near Point Sal, south-central California: a time frame for coastal-dune emplacement. *In:* Fletcher, C.H. and Wehmiller, J.F. (editors). *Quaternary Coasts of the United States: marine and lacustrine systems*. Tulsa: Society for Sedimentary Geology, 309–315.

Ortiz, J., Mix, A., Hostetler, S. and Kashgarian, M. 1997. The California Current of the last glacial maximum: reconstruction at 42°N based on multiple proxies. *Paleoceanography*, 12, 191–205.

Ortlieb, L., Diaz, A. and Guzman, N. 1996a. A warm interglacial episode during oxygen isotope stage 11 in northern Chile. *Quaternary Science Reviews*, 15, 857–871.

Ortlieb, L., Zazo, C., Goy, J.L., Hillaire-Marcel, C., Ghaleb, B. and Cournoyer, L. 1996b. Coastal deformation and sea-level changes in the northern Chile subduction area (23°S) during the last 330 ky. *Quaternary Science Reviews*, 15, 819–831.

Ortloff, C.R., Moseley, M.E. and Feldman, R.A. 1982. Hydraulic aspects of the Chimu Chicama-Moche intervalley canal. *American Antiquity*, 47, 572–595.

Osborne, P.L., Humphreys, G.S. and Polunin, N.V.C. 1993. Sediment deposition and late Holocene environmental change in a tropical lowland basin: Waigani Lake, Papua New Guinea. *Journal of Biogeography*, 20, 599–613.

Ota, Y. (editor). 1994. *Study on Coral Reef Terraces of the Huon Peninsula, Papua New Guinea: establishment of Quaternary sea level and tectonic history*. Yokohama National University, Japan.

Ota, Y. and Odagiri, S. 1994. Age and deformation of marine terraces on the Ashizuri Peninsula and its vicinity, southwestern Japan. *Journal of Geography*, 103, 243–267 [in Japanese with English abstract].

Ota, Y. and Omura, A. 1991. Late Quaternary shorelines in the Japanese islands. *The Quaternary Research (Japan)*, 30, 175–186.

Ota, Y. and Paskoff, R. 1993. Holocene deposits on the coast of north-central Chile: radiocarbon ages and implications for coastal changes. *Revista Geológica de Chile*, 20, 25–32.

Ota, Y., Hull, A.G., Iso, N., Ikeda, Y., Moriya, I. and Yoshikawa, T. 1992. Holocene marine terraces on the northeast coast of North Island, New Zealand, and their tectonic significance. *New Zealand Journal of Geology and Geophysics*, 35, 273–288.

Ota, Y., Chappell, J., Kelley, R., Yonekura, N., Matsumoto, E., Nishimura, T. and Head, J. 1993. Holocene coral reef terraces and coseismic uplift of Huon Peninsula, Papua New Guinea. *Quaternary Research*, 40, 177–188.

Ota, Y., Miyauchi, T., Paskoff, R. and Koba, M. 1995. Plio-Quaternary marine terraces and their deformation along the Altos de Talinay, north-central Chile. *Revista Geológica de Chile*, 22, 89–102.

Overpeck, J.T., Peterson, L.C., Kipp, N., Imbrie, J. and Rind, D. 1989. Climate change in the circum-North

Atlantic region during the last deglaciation. *Nature*, 338, 553–557.

Pardo, F. and Molnar, P. 1987. Relative motions of the Nazca (Farallon) and South America plates since late Cretaceous time. *Tectonics*, 6, 233–248.

Park, Y.A. 1992. The Holocene marine sediment distribution on the continental shelf of the Korea South Sea and the early Holocene sea-level standing evidence. *Korean Journal of Quaternary Research*, 3, 1–15.

Parrish, J.T. 1987. Global palaeogeography and palaeoclimate of the late Cretaceous and early Tertiary. *In*: Friis, E.M., Chaloner, W.G. and Crane, P.R. (editors). *The Origins of Angiosperms and their Biological Consequences*. Cambridge: Cambridge University Press, 51–73.

Parrish, J.T. 1992. Jurassic climate and oceanography of the Pacific region. *In:* Westerman, G.E.G. (editor). *The Jurassic of the Circum-Pacific*. Cambridge: Cambridge University Press, 365–379.

Parrish, R.R. 1981. Uplift rates of Mt. Logan, Y.T., and British Columbia's Coast Mountains using fission track dating methods (abstract). *EOS, Transactions of the American Geophysical Union*, 62, 59–60.

Parrish, R.R. 1983. Cenozoic thermal evolution and tectonics of the Coast Mountains of British Columbia. 1. Fission track dating, apparent uplift rates, and patterns of uplift. *Tectonics*, 2, 601–631.

Pascual, R., Vucetich, M.G., Scillato-Yané, G.J. and Bond, M. 1985. Main pathways of mammalian diversification in South America. *In:* Stehli, F.G. and Webb, S.D. (editors). *The Great American Biotic Interchange*. New York: Plenum, 219–247.

Paulay, G. 1994. Biodiversity on oceanic islands: its origin and extinction. *American Zoologist*, 34, 134–144.

Pavlides, C. and Gosden, C. 1994. 35,000-year-old sites in the rainforests of west New Britain, Papua New Guinea. *Antiquity*, 68, 604–610.

Pavlov, S.F. 1979. New data on Upper Palaeozoic glacial deposits in the Siberian platform. *Lithology and Mineral Resources* [English translation from Russian *Litologia i Paleznye Iskopaemye*], 15, 254–262.

Pearsall, D.M. 1996. Reconstructing subsistence in the lowland tropics: a case study from the Jama River Valley, Manabí, Ecuador. *In:* Reitz, E.J., Newsom, L.A. and Scudder, S.J. (editors). *Case Studies in Environmental Archaeology*. New York: Plenum, 233–254.

Peart, M. 1991. The Kaiapit landslide: events and mechanisms. *Quarterly Journal of Engineering Geology*, 24, 399–411.

Pedersen, T.F. 1983. Increased productivity in the eastern equatorial Pacific during the last glacial maximum (19,000 to 14,000 yr B.P.). *Geology*, 11, 16–19.

Pellatt, M.G. and Mathewes, R.W. 1997. Holocene tree

line and climate change on the Queen Charlotte Islands, Canada. *Quaternary Research*, 48, 88–99.

Perdue, P.C. 1987. *Exhausting the Earth: state and peasant in Hunan, 1500–1850*. Cambridge, Mass.: Harvard University Press, 331 p.

Petford, N. and Atherton, M.P. 1992. Granitoid emplacement and deformation along a major crustal lineament: the Cordillera Blanca, Peru. *Tectonophysics*, 205, 171–185.

Peucker-Ehrenbrink, B., Ravizza, G. and Hofmann, A.W. 1995. The marine $^{187}Os/^{186}Os$ record of the past 80 million years. *Earth and Planetary Science Letters*, 130, 155–167.

Philander, S.G. 1990. *El Niño, La Niña, and the Southern Oscillation*. San Diego: Academic Press, 293 p.

Philander, S.G. 1998. Who is El Niño? *Eos, Transactions of the American Geophysical Union*, 79, 170.

Pickard, J., Selkirk, P.M. and Selkirk, D.R. 1984. Holocene climates of the Vestfold Hills, Antarctica, and Macquarie Island. *In:* Vogel, J.C. (editor). *Late Cainozoic Palaeoclimates of the Southern Hemisphere*. Rotterdam: Balkema, 173–182.

Pickering, W.H. 1924. The separation of the continents by fission. *Geological Magazine*, 62, 31–34.

Pickup, G., Higgins, R.J. and Warner, R.F. 1980. Erosion and sediment yield in the Fly River drainage basins, Papua New Guinea. *Publications of the International Association of Hydrological Sciences*, 132, 438–456.

Pillans, B. 1983. Upper Quaternary marine terrace chronology and deformation, South Taranaki, New Zealand. *Geology*, 11, 292–297.

Pillans, B, 1986. A late Quaternary uplift map for North Island, New Zealand. *Bulletin of the Royal Society of New Zealand*, 24, 409–417.

Pillans, B. 1991. New Zealand Quaternary stratigraphy: an overview. *Quaternary Science Reviews*, 10, 405–418.

Pillans, B., Holgate, G. and McGlone, M. 1988. Climate and sea level during oxygen isotope stage 7b: on-land evidence from New Zealand. *Quaternary Research*, 29, 176–185.

Pillans, B., Pullar, W.A., Selby, M.J. and Soons, J.M. 1992. The age and development of the New Zealand landscape. *In:* Soons, J.M. and Selby, M.J. (editors). *Landforms of New Zealand*. Auckland: Longman Paul, 31–62.

Pinxian, W., Qiubao, M., Yunhua, B. and Xinrong, C. 1981. Strata of Quaternary transgressions in East China: a preliminary study. *Acta Geologica Sinica*, 55, 1–13.

Piperno, D.R. 1994. Phytolith and charcoal evidence for prehistoric slash-and-burn agriculture in the Darien rain forest of Panama. *The Holocene*, 4, 321–325.

Pirazzoli, P.A. and Delibrias, G. 1983. Late Holocene and recent sea level changes and crustal movements in Kume Island, the Ryukyus, Japan. *Bulletin of the Department of Geography, University of Tokyo*, 15, 63–76.

Pirazzoli, P.A. and Kawana, T. 1986. Détermination de mouvements crustaux quaternaires d'après la déformation des anciens rivages dans les îles Ryukyu, Japon. *Revue de Géologie Dynamique et de Géographie Physique*, 27, 269–278.

Pirazzoli, P.A. and Montaggioni, L.F. 1988. Holocene sea-level changes in French Polynesia. *Palaeogeography, Palaeoclimatology, Palaeoecology*, 68, 153–175.

Pirazzoli, P.A., Koba, T., Montaggioni, L.F. and Person, A. 1987. Anaa (Tuamotus) – an incipient rising atoll? *Marine Geology*, 82, 261–269.

Pisias, N.G. and Rea, D.K. 1988. Late Pleistocene paleoclimatology of the central equatorial Pacific: sea-surface response to the southeast tradewinds. *Paleoceanography*, 3, 21–37.

Pisias, N.G., Martinson, D.G., Moore, T.C., Shackleton, N.J., Prell, W., Hays, J. and Boden, G. 1984. High resolution stratigraphic correlation of benthic oxygen isotopic records spanning the last 300,000 years. *Marine Geology*, 56, 119–136.

Pitulíko, V.V. 1993. An early Holocene site in the Siberian High Arctic. *Arctic Anthropology*, 30, 13–21.

Plafker, G. 1972. The Alaskan earthquake of 1964 and Chilean earthquake of 1960 – implications for arc tectonics. *Journal of Geophysical Research*, 77, 901–925.

Plafker, G. 1981. Late Cenozoic glaciomarine deposits of the Yakataga Formation, Alaska. *In:* Hambrey, M.J. and Harland, W.B. (editors). *Earth's Pre-Pleistocene Glacial Record*. Cambridge: Cambridge University Press, 694–699.

Plafker, G. and Savage, J.C. 1970. Mechanism of the Chilean earthquakes of May 21 and 22, 1960. *Geological Society of America, Bulletin*, 81, 1001–1030.

Pocknall, D.T. 1990. Palynological evidence for the early to middle Eocene vegetation and climate history of New Zealand. *Review of Palaeobotany and Palynology*, 65, 57–69.

Pollak, R. 1971. Introduction. *In:* Brown, T. *Oil on Ice*. San Francisco: Sierra Club, 9–24.

Pope, G.G. 1992. Replacement versus regionally continuous models: the paleobehavioral and fossil evidence from East Asia. *In:* Akazawa, T., Aoki, K. and Kimura, T. (editors). *The Evolution and Dispersal of Modern Humans in Asia*. Japan: Hokusen-sha, 3–14.

Porter, S.C. 1975. Late Quaternary glaciation and tephrochronology of Mauna Kea, Hawaii. *In:*

Suggate, R.P. and Cresswell, M.M. (editors). *Quaternary Studies*. Wellington: Royal Society of New Zealand, 247–251.

Porter, S.C. 1979. Hawaiian glacial ages. *Quaternary Research*, 12, 161–187.

Porter, S.C. 1981. Pleistocene glaciation in the southern Lake District of Chile. *Quaternary Research*, 16, 263–292.

Porter, S.C. 1988. Landscapes of the last ice age in North America. *In:* Carlisle, R.C. (editor). *Americans Before Columbus: Ice-Age Origins*. Pittsburgh: Department of Anthropology, University of Pittsburgh, 1–24.

Porter, S.C. 1997. Late Pleistocene eolian sediments related to pyroclastic eruptions of Mauna Kea Volcano, Hawaii. *Quaternary Research*, 47, 261–276.

Porter, S.C. and An, Z.S. 1995. Correlation between climate events in the North Atlantic and China during the last glaciation. *Nature*, 375, 305–308.

Porter, S.C., Pierce, K.L. and Hamilton, T.D. 1983. Late Wisconsin mountain glaciation in the western United States. *In:* Porter, S.C. (editor). *Late Quaternary Environments of the United States, Vol. 1: The Late Pleistocene*. Minneapolis: University of Minnesota Press, 71–114.

Porter, S.C., Clapperton, C.M. and Sugden, D.E. 1992. Chronology and dynamics of deglaciation along and near the Strait of Magellan, southernmost South America. *Sveriges Geologiska Undersökning*, 81, 233–239.

Prevot, R. and Chatelain, J.L. 1983. *Sismicité et risque sismique à Vanuatu*. Nouméa: ORSTOM, Rapport 5–83.

Prothero, D.R. 1989. Stepwise extinctions and climatic decline during the later Eocene and Oligocene. *In:* Donovan, S.K. (editor). *Mass Extinctions: processes and evidence*. London: Belhaven, 217–234.

Prothero, D.R. 1994. The *Eocene–Oligocene Transition: paradise lost*. New York: Columbia University Press, 291 p.

Quilty, P.G. 1994. The background: 144 million years of Australian palaeoclimate and palaeogeography. *In:* Hill, R.S. (editor). *History of the Australian Vegetation: Cretaceous to Recent*. Cambridge: Cambridge University Press, 14–43.

Quinn, T.M., Taylor, F.W. and Crowley, T.J. 1993. A 173 year stable isotope record from a tropical South Pacific coral. *Quaternary Science Reviews*, 12, 407–418.

Rabassa, J. and Clapperton, C.M. 1990. Quaternary glaciations of the southern Andes. *Quaternary Science Reviews*, 9, 153–174.

Radford, D., Blong, R., d'Aubert, A.M., Kuhnel, I. and Nunn, P.D. 1995. *The Incidence of Tropical Cyclones in the Southwest Pacific Region 1920–*

1994. Townsville, Australia: Greenpeace International, 35 p.

Rampino, M.R. and Self, S. 1993. Climate–volcanism feedback and the Toba eruption of ~74,000 years ago. *Quaternary Research*, 40, 269–280.

Ranere, A.J. and Hansell, P. 1978. Early subsistence patterns along the Pacific coast of central Panama. *In:* Stark, B.L. and Voorhies, B. (editors). *Prehistoric Coastal Adaptations: the economy of maritime middle America*. New York: Academic Press, 43–59.

Raup, D.M. and Sepkoski, J.J. 1982. Mass extinctions in the marine fossil record. *Science*, 215, 1501–1503.

Raymo, M.E. 1994. The initiation of northern hemisphere glaciation. *Annual Review of Earth and Planetary Science*, 22, 353–383.

Raymo, M.E. and Ruddiman, W.F. 1992. Tectonic forcing of late Cenozoic climate. *Nature*, 359, 117–122.

Rea, D.K., Zachos, J.C., Owen, R.M. and Gingerich, P.D. 1990. Global change at the Paleocene–Eocene boundary: climatic and evolutionary consequences of tectonic events. *Palaeogeography, Palaeoclimatology, Palaeoecology*, 79, 117–128.

Rea, D.K., Pisias, N.G. and Newberry, T. 1991. Late Pleistocene paleoclimatology of the central Pacific: flux patterns of biogenic sediments. *Paleoceanography*, 6, 227–244.

Rea, D.K., Basov, I.A., Janecek, T.R and Shipboard Party. 1993. Cenozoic paleoceanography of the North Pacific Ocean: results of ODP leg 145, the North Pacific Transect. *EOS, Transactions of the American Geophysical Union*, 74, 173.

Rees, J.D. 1979. Effects of the eruption of Parícutin Volcano on landforms, vegetation, and human occupancy. *In:* Sheets, P.D. and Grayson, D.K. (editors). *Volcanic Activity and Human Ecology*. New York: Academic Press, 249–292.

Renkin, M.L. and Sclater, J.G. 1988. Depth and age in the North Pacific. *Journal of Geophysical Research*, 93, 2919–2935.

Repenning, C.A. 1993. Global warming debate. *National Geographic Research and Exploration*, 9, 389–390.

Repenning, C.A. 1996. *Allophaiomys* and the age of the Olyor Suite, Krestovka sections, Yakutia. *United States Geological Survey, Bulletin*, 2037.

Repenning, C.A. and Ray, C.E. 1977. The origin of the Hawaiian monk seal. *Proceedings of the Biological Society of Washington*, 89, 667–688.

Repenning, C.A., Brouwers, E.M., Carter, L.D., Marincovich, L. and Ager, T.A. 1987. The Beringian ancestry of *Phenacomys* (Rodentia: Cricetidae) and the beginning of the modern Arctic borderland biota. *United States Geological Survey, Bulletin*, 1687, 29 p.

Repenning, C.A., Fejfar, O. and Heinrich, W-D. 1990. Arvicolid rodent biochronology of the northern hemisphere. *In:* Fejfar, O. and Heinrich, W-D. (editors). *International Symposium on the Evolution, Phylogeny and Biostratigraphy of Arvicolids (Rodentia: Mammalia)*. Prague: Geological Survey, 385–418.

Reynolds, T.E.G. and Kaner, S.C. 1990. Japan and Korea at 18 000 BP. *In:* Soffer, O. and Gamble, C. (editors). *The World at 18 000 BP, Volume 1*. London: Unwin Hyman, 296–311.

Rich, P.V., Rich, T.H., Wagstaff, B.E., Mason, J.M., Douthitt, C.B., Gregory, R.T. and Felton, E.A. 1988. Evidence for low temperature and biologic diversity in Cretaceous high latitudes of Australia. *Science*, 242, 1403–1406.

Rind, D. and Peteet, D. 1985. Terrestrial conditions at the last glacial maximum and CLIMAP sea-surface estimates: are they consistent? *Quaternary Research*, 24, 1–22.

Ringwood, A.E. 1979. *Origin of the Earth and Moon*. New York: Springer-Verlag. 295 p.

Ritchie, J.C. 1984. *Past and Present Vegetation of the far North-west of Canada*. Toronto: University of Toronto Press, 251 p.

Ritchie, J.C. and Harrison, S.P. 1993. Vegetation, lake levels, and climate in western Canada during the Holocene. *In:* Wright, H.E., Kutzbach, J.E., Webb, T., Ruddiman, W.F., Street-Perrott, F.A. and Bartleim, P.J. (editors). *Global Climates since the Last Glacial Maximum*. Minneapolis: University of Minnesota, 401–414.

Rodbell, D.T. 1993. The timing of the last deglaciation in Cordillera Oriental, northern Peru, based on glacial geology and lake sedimentology. *Geological Society of America, Bulletin*, 105, 923–934.

Roe, K.K. and Burnett, W.C. 1985. Uranium geochemistry and dating of Pacific island apatite. *Geochimica et Cosmochimica Acta*, 49, 1581–1592.

Roe, K.K., Burnett, W.C. and Lee, A.I.N. 1983. Uranium disequilibrium dating of phosphate deposits from the Lau group, Fiji. *Nature*, 302, 603–606.

Rogers, W.B. 1896. *Life and Letters of William Barton Rogers*. Cambridge, Mass.: Riverside, 2 volumes.

Roosevelt, A.C., Houseley, R.A., Imazio da Silveira, M., Maranca, S. and Johnson, T. 1991. Eighth millennium pottery from a prehistoric shell midden in the Brazilian Amazon. *Science*, 254, 1621–1624.

Rothé, J.P. 1969. *The Seismicity of the Earth*. Paris: UNESCO, 336 p.

Ruddiman, W.F., Raymo, M. and MacIntyre, A. 1986. Matuyama 41,000-year cycles: North Atlantic Ocean and northern hemisphere ice sheets. *Earth and Planetary Science Letters*, 80, 117–129.

Sabels, B.E. 1966. Climatic variations in the tropical Pacific as evidenced by trace element analysis of soils. *In:* Blumenstock, D.I. (editor). *Pleistocene and Post-Pleistocene Climatic Variations in the Pacific Area.* Honolulu: Bishop Museum Press, 131–151.

Sagan, C. 1996. *The Demon-Haunted World.* New York: Ballantine, 457 p.

Sagan, C., Toon, O.B. and Pollack, J.B. 1979. Anthropogenic albedo changes and the Earth's climate. *Science*, 206, 1363–1368.

Sakaguchi, Y. 1983. Warm and cold stages in the past 7600 years in Japan and their global correlation. *Bulletin of the Department of Geography, University of Tokyo*, 15, 1–31.

Saldarriaga, J.G. and West, D.C. 1986. Holocene fires in the northern Amazon Basin. *Quaternary Research*, 26, 358–366.

Salinger, M.J. 1983. New Zealand climate: the last 5 million years. *In:* Vogel, J.C. (editor). *Late Cainozoic Palaeoclimates of the Southern Hemisphere.* Rotterdam: Balkema, 131–150.

Salinger, M.J. 1988. New Zealand climate: past and present. *In: Climate Change: the New Zealand response.* Wellington: Ministry for the Environment, 17–24.

Sanchez, W.A. and Kutzbach, J.E. 1974. Climate of the American tropics and subtropics in the 1960s and possible comparisons with climatic variations of the last millennium. *Quaternary Research*, 4, 128–135.

Sandweiss, D.H. 1986. The beach ridges at Santa, Peru: El Niño, uplift, and prehistory. *Geoarchaeology*, 1, 17–28.

Sandweiss, D.H. 1996. Environmental change and its consequences for human society on the central Andean coast. *In:* Reitz, E.J., Newsom, L.A. and Scudder, S.J. (editors). *Case Studies in Environmental Archaeology.* New York: Plenum, 127–146.

Santa Luca, A. 1980. *The Ngandong Fossil Hominids.* New Haven: Yale University Press, 175 p.

Savage, D.E. and Russell, D.E. 1983. *Mammalian Paleofaunas of the World.* Reading, Mass.: Addison Wesley.

Savidge, J.A. 1987. Extinction of an island forest avifauna by an introduced snake. *Ecology*, 68, 660–668.

Savin, S.M. and Douglas, R.G. 1985. Sea level, climate, and the central American land bridge. *In:* Stehli, F.G. and Webb, S.D. (editors). *The Great American Biotic Interchange.* New York: Plenum, 303–324.

Scheibner, E. 1985. Suspect terranes in the Tasman Fold Belt system, eastern Australia. *In:* Howell, D.G. (editor). 1985. *Tectonostratigraphic Terranes of the Circum-Pacific Region.* Houston: Circum-Pacific Council for Energy and Mineral Resources, Earth Science Series, No. 1, 493–514.

Schlee, J.S., Karl., H.A. and Torresan, M.E. 1995. Imaging the sea floor. *United States Geological Survey, Bulletin*, 2079, 24 p.

Schmitt, R.C. 1961. *Population trends in Hawaii and French Polynesia.* Honolulu: Romanzo Adams Social Research Laboratory, Report 29.

Schofield, J.C. 1980. Postglacial transgressive maxima and second-order transgressions of the southwest Pacific Ocean. *In:* Mörner, N.-A. (editor). *Earth Rheology, Isostasy and Eustasy.* New York: Wiley, 517–521.

Scholl, D.W. and Creager, J.S. 1973. Geological synthesis of Leg 19 (DSDP) results: far North Pacific, and Aleutian Ridge, and Bering Sea. *In:* Creager, J.S. and Scholl, D.W. (editors). *Initial Reports of the Deep Sea Drilling Project, volume 19.* Washington: United States Government Printing Office, 897–913.

Schweger, C.E., Matthews, J.V., Hopkins, D.M. and Young, S.B. 1982. Paleoecology of Beringia – synthesis. *In:* Hopkins, D.M., Matthews, J.V., Schweger, C.E. and Young, S.B. (editors). *Paleoecology of Beringia.* New York: Academic Press, 425–444.

Scott, D.B., Mudie, P.J., Baki, V., MacKinnon, K.D. and Cole, F.E. 1989. Biostratigraphy and the late Cenozoic paleoceanography of the Arctic Ocean: foraminiferal, lithostratigraphic, and isotopic evidence. *Geological Society of America, Bulletin*, 101, 260–277.

Scott, G.A.J. and Rotondo, G. 1983. A model to explain the differences between Pacific Plate island-atoll types. *Coral Reefs*, 1, 139–149.

Scuderi, L.A. 1987. Glacier variations in the Sierra Nevada, California, as related to a 1200-year tree-ring chronology. *Quaternary Research*, 27, 220–231.

Scuderi, L.A. 1993. A 2000-year tree ring record of annual temperatures in the Sierra Nevada mountains. *Science*, 259, 1433–1436.

Seltzer, G.O. 1993. Late-Quaternary glaciation as a proxy for climate change in the central Andes. *Mountain Research and Development*, 13, 129–138.

Seltzer, G.O. 1994. A lacustrine record of late Pleistocene climatic change in the subtropical Andes. *Boreas*, 23, 105–111.

Sepkoski, J.J. 1993. Ten years in the library: new data confirm paleontological patterns. *Paleobiology*, 19, 43–51.

Serova, M.Y. 1985. Marine Oligocene in stratigraphic section of Paleogene from western Sakhalin. *Izvestiya Academii Nauk. SSSR Seriya Geologicheskaya*, 11, 86–89.

Servant, M. and Fontes, J.-C. 1978. Les lacs quaternaires des hauts plateaux des Andes boliviennes – premieres interprétations paléoclimatiques. *Cahiers ORSTOM, série Géologie*, 10, 9–23.

Shackleton, N.J. and Kennett, J.P. 1975. Paleotemperature history of the Cenozoic and the initiation of Antarctic glaciation: oxygen and carbon isotope analysis in DSDP sites 277, 279 and 281. *Initial Reports of the Deep-Sea Drilling Project*, 29, 743–755.

Shackleton, N.J. and Opdyke, N.D. 1973. Oxygen isotope and palaeomagnetic stratigraphy of equatorial Pacific core V28-238: oxygen isotope temperatures and ice volumes on a 10^5 and 10^6 year scale. *Quaternary Research*, 3, 39–55.

Shackleton, N.J. and Opdyke, N.D. 1976. Oxygen-isotope and paleomagnetic stratigraphy of Pacific core V28-239, late Pliocene to latest Pleistocene. *Geological Society of America, Memoir*, 145, 449–464.

Shackleton, N.J., Berger, A. and Peltier, W.R. 1990. An alternative astronomical calculation of the lower Pleistocene timescale based on ODP Site 677. *Transactions of the Royal Society of Edinburgh, Earth Sciences*, 81, 251–261.

Shackleton, N.J., Crowhurst, S., Hagelberg, T., Pisias, N.G. and Schneider, D.A. 1995. A new late Neogene time scale: application to Leg 138 sites. *In:* Pisias, N.G., Mayer, L.A., Janecek, T.R., Palmer-Julson, A. and van Andel, T.H. (editors). *Proceedings of the Ocean Drilling Program, Scientific Results, Volume 138*. College Station: Ocean Drilling Program, 73–101.

Shane, P., Froggatt, P., Black, T. and Westgate, J. 1995. Chronology of Pliocene and Quaternary bioevents and climatic events from fission-track ages on tephra beds, Wairarapa, New Zealand. *Earth and Planetary Science Letters*, 130, 141–154.

Shemesh, A., Burckle, L.H. and Hays, J.D. 1994. Meltwater input to the Southern Ocean during the Last Glacial maximum. *Science*, 266, 1542–1544.

Sher, A.V. 1997. Late-Quaternary extinction of large mammals in northern Eurasia: a new look at the Siberian contribution. *In:* Huntley, B., Cramer, W., Morgan, A.V., Prentice, H.C. and Allen, J.R.M. (editors). *Past and Rapid Future Environmental Changes: the spatial and evolutionary responses of terrestrial biota*. Berlin: Springer-Verlag, 319–339.

Sherman, C.E., Glenn, C.R., Jones, A.T., Burnett, W.C. and Schwarcz, H.P. 1993. New evidence for two highstands of the sea during the last interglacial, oxygen isotope stage 5e. *Geology*, 21, 1079–1082.

Shi, Y., Kong, Z., Wang, S., Tang, L., Wang, F., Yao, T., Zhao, X., Zhang, P. and Shi, S. 1993. Mid-Holocene climates and environments in China. *Global and Planetary Change*, 7, 219–233.

Shimada, I., Schaaf, C.B., Thompson, L.G. and Mosley-Thompson, E. 1991. Cultural impacts of severe droughts in the prehistoric Andes: application of a 1,500-year ice-core precipitation record. *World Archaeology*, 22, 249–270.

Sibrava, V. 1986. Correlation of European glaciations

and their relation to the deep-sea record. *Quaternary Science Reviews*, 5, 433–441.

Siegel, S.R. and Witham, P. 1991. Case study Colombia. *In:* Kreimer, A. and Munasinghe, M. (editors). *Managing Natural Disaster and the Environment*. Washington, DC: World Bank, 170–171.

Sigé, B. 1972. La faunule de mammifères du Crétacé supérieur de Laguna Umayo (Andes péruviennes). *Museum National d'Histoire Naturelle, Bulletin*, Series 3, 19, 375–409.

Silberling, N.J. 1985. Biogeographic significance of the upper Triassic bivalve Monotis in circum-Pacific accreted terranes. *In:* Howell, D.G. (editor). *Tectonostratigraphic Terranes of the Circum-Pacific Region*. Houston: Circum-Pacific Council for Energy and Mineral Resources, Earth Science Series, No. 1, 63–70.

Simmons, I.G. 1989. *Changing the Face of the Earth: culture, environment, history*. Oxford: Blackwell, 487 p.

Simon, J. 1997. *Endangered Mexico: an environment on the edge*. San Francisco: Sierra Club, 275 p.

Singh, G., Kershaw, A.P. and Clark, R. 1981. Quaternary vegetation and fire history in Australia. *In:* Gill, A.M., Groves, R.H. and Noble, I.R. (editors). *Fire and the Australian Biota*. Canberra: Australian Academy of Science, 23–54.

Sluys, R. 1994. Explanations for biogeographic tracks across the Pacific Ocean: a challenge for paleogeography and historical biogeography. *Progress in Physical Geography*, 18, 42–58.

Smalley, I.J. and Vita-Finzi, C. 1968. The formation of fine particles in sandy deserts and the nature of the desert 'loess'. *Journal of Sedimentary Petrology*, 38, 766–774.

Smart, P.L. and Richards, D.A. 1992. Age estimates for the late Quaternary high sea-stands. *Quaternary Science Reviews*, 11, 687–696.

Smil, V. 1983. Deforestation in China. *Ambio*, 12, 226–231.

Smith, A.G. and Drewry, D.J. 1984. Delayed phase change due to hot asthenosphere causes Transantarctic uplift? *Nature*, 309, 536–538.

Smith, B. 1985. *European Vision and the South Pacific*. New Haven: Yale University Press, 2nd edition, 370 p.

Smith, G.A., Wang, Y., Cerling, T.E. and Geissman, J.W. 1993. Comparison of a paleosol-carbonate isotope record to other records of Pliocene–early Pleistocene climate in the western United States. *Geology*, 21, 691–694.

Smith, G.I. and Street-Perrott, F.A. 1983. Pluvial lakes of the western United States. *In:* Porter, S.C. (editor). *Late Quaternary Environments of the United States, Vol.1: The Late Pleistocene*. Minneapolis: University of Minnesota Press, 190–214.

Smith, G.I., Bischoff, J.L. and Bradbury, J.P. 1997. *Synthesis of the Paleoclimate Record from Owens Lake Core OL-92.* Geological Society of America, Special Paper 317, 143–160.

Smith, P.L. and Westermann, G.E.G. 1990. Paleobiogeography of the ancient Pacific. *Science*, 249, 680.

Smith, W.E. and Smith, A.M. 1975. *Minimata.* New York: Holt, Rinehart and Winston.

Snow, B.E., Shutler, R., Nelson, D.E., Vogel, J.S. and Southon, J.R. 1986. Evidence of early rice cultivation in the Philippines. *Philippine Quarterly of Culture and Society*, 14, 3–11.

Solem, A. 1990. How many Hawaiian land snail species are left, and what can we do for them? B.P. Bishop *Museum (Honolulu), Occasional Paper* 30, 27–40.

Soons, J.M. and Selby, M.J. (editors). 1982. *Landforms of New Zealand.* Auckland: Longman Paul.

Southern, W. 1986. The late Quaternary environmental history of Fiji. *Unpublished PhD thesis,* Australian National University, Canberra.

Spate, O.H.K. 1953. Changing native agriculture in New Guinea. *Geographical Review*, 43, 152–172.

Spaulding, W.G. 1991. A middle Holocene vegetation record from the Mojave Desert of North America and its palaeoclimatic significance. *Quaternary Research*, 35, 427–437.

Speed, R.C. 1979. Collided Palaeozoic microplate in the western United States. *Journal of Geology*, 87, 279–292.

Spörli, K.B. and Ballance, P.F. 1989. Mesozoic ocean floor/continent interaction and terrane configuration, southwest Pacific area around New Zealand. *In:* Ben-Avraham, Z. (editor). *The Evolution of the Pacific Ocean Margins.* New York: Oxford University Press, 176–190.

Sprague de Camp, L. 1970. *Lost Continents: The Atlantis Theme.* New York: Ballantine, 348 p.

Spriggs, M. 1986. Landscape, land use and political transformation in southern Melanesia. *In:* Kirch, P.V. (editor). *Island Societies: archaeological approaches to evolution and transformation.* Cambridge University Press, 6–19.

Spriggs, M. and Anderson, A. 1993. Late colonization of East Polynesia. *Antiquity*, 67, 200–217.

Stafford, R.A. 1988. The long arm of London: Sir Roderick Murchison and imperial science in Australia. *In:* Home, R.W. (editor). *Australian Science in the Making.* Cambridge: Cambridge University Press, 69–101.

Stanley, G.D. and Yancey, T.E. 1990. Paleobiogeography of the ancient Pacific. *Science*, 249, 680–681.

Starratt, S.W. 1993. Late Quaternary paleoceanography of the Pervenets Canyon area of the Bering Sea: evidence from the diatom flora. *Diatom Research*, 8, 159–170.

Steadman, D.W. 1989. Extinction of birds in eastern Polynesia: a review of the record, and comparisons with other Pacific island groups. *Journal of Archaeological Science*, 16, 177–205.

Stearns, H.T. and Macdonald, G.A. 1947. *Geology and ground-water resources of the island of Molokai, Hawaii.* Hawaii Division of Hydrography, Bulletin 7.

Stehli, F.G. and Webb, S.D. (editors). 1985. *The Great American Biotic Interchange.* New York: Plenum Press, 532 p.

Stein, M., Wasserburg, G.J., Aharon, P., Chen, J.H., Zhu, Z.R., Bloom, A. and Chappell, J. 1993. TIMS U-series dating and stable isotopes of the last interglacial event in Papua New Guinea. *Geochimica and Cosmochimica Acta*, 57, 2541–2554.

Stern, T.A. and ten Brink, U.S. 1989. Flexural uplift of the Transantarctic Mountains. *Journal of Geophysical Research*, 94, 10315–10330.

Stets, J., Erben, H.K., Hambach, U., Krumsiek, K. Thein, J. and Wurster, P. 1996. The Cretaceous–Tertiary boundary in the Nanxiong Basin (continental facies, southeast China). *In:* MacLeod, N. and Keller, G. (editors). *Cretaceous–Tertiary Mass Extinctions: biotic and environmental changes.* New York: Norton, 349–371.

Stevens, G.R. 1980. *New Zealand Adrift: the theory of continental drift in a New Zealand setting.* Wellington: Reed, 442 p.

Stevens, G.R. 1989. The nature and timing of biotic links between New Zealand and Antarctica in Mesozoic and early Cenozoic times. *In:* Crame, J.A. (editor). *Origins and Evolution of the Antarctic Biota.* Geological Society of London, Special Publication 47, 141–166.

Stevens, G.R. and Speden, I.G. 1978. New Zealand. *In:* Moullade, M. and Nairn, A.E.M. (editors). *The Phanerozoic Geology of the World II, The Mesozoic A.* Amsterdam: Elsevier, 251–328.

Stevenson, J. and Dodson, J.R. 1995. Palaeoenvironmental evidence for human settlement of New Caledonia. *Archaeology in Oceania*, 30, 36–41.

Stine, S. 1990. Late Holocene fluctuations of Mono Lake, eastern California. *Palaeogeography, Palaeoclimatology, Palaeoecology*, 78, 333–381.

Stonich, S.C. 1995. Development, rural impoverishment, and environmental destruction in Honduras. *In:* Painter, M. and Durham, W.H. (editors). *The Social Causes of Environmental Destruction in Latin America.* Ann Arbor: The University of Michigan Press, 63–99.

Street-Perrott, F.A., Huang, Y., Perrott, R.A., Eglinton, G., Barker, P., Khelifa, L.B., Harkness, D.D. and Olago, D.O. 1997. Impact of lower atmospheric carbon dioxide on tropical mountain ecosystems. *Science*, 278, 1422–1426.

Stringer, C.B. and Andrews, P. 1988. Genetic and fossil

evidence for the origin of modern humans. *Science*, 239, 1263–1268.

Stringer, C. and McKie, R. 1997. *African Exodus: the origins of modern humanity*. London: Pimlico, 267 p.

Stroeven, A.P., Prentice, M.L. and Kleman, J. 1996. On marine microfossil transport and pathways in Antarctica during the late Neogene: evidence from the Sirius Group at Mount Fleming. *Geology*, 24, 727–730.

Stuiver, M. and Pearson, G.W. 1993. High-precision bidecadal calibration of the radiocarbon time scale, AD 1950–500 BC and 2500–6000 BC. *Radiocarbon*, 35, 1–23.

Stuiver, M., Denton, G.H., Hughes, T.J. and Fastook, J.L. 1981. History of the marine ice sheet in West Antarctica during the last glaciation: a working hypothesis. *In:* Denton, G.H. and Hughes, T.J. (editors). *The Last Great Ice Sheets*. New York: Wiley-Interscience, 319–436.

Suggate, R.P. 1990. Late Pliocene and Quaternary glaciations of New Zealand. *Quaternary Science Reviews*, 9, 175–197.

Swadling, P. and Hope, G. 1992. Environmental change in New Guinea since human settlement. *In:* Dodson, J. (editor). *The Naive Lands: prehistory and environmental change in Australia and the southwest Pacific*. Melbourne: Longman Cheshire, 13–42.

Sweet, A.R. and Braman, D.R. 1992. The K–T boundary and contiguous strata in western Canada: interactions between paleoenvironments and palynological assemblages. *Cretaceous Research*, 13, 31–79.

Sylwan, C.A. 1989. Palaeomagnetism, paleoclimate and chronology of late Cenozoic deposits in southern Argentina. *Meddelanden fran Stockholms Universitets Geologiska Instituten*, Nr 277, 110 p.

Szabo, B.J., Ward, W.C., Weidie, A.E. and Brady, M.J. 1979. Age and magnitude of the late Pleistocene sea-level rise on the eastern Yucatan Peninsula. *Geology*, 6, 713–715.

Taylor, F.W. 1978. Quaternary tectonic and sea-level history, Tonga and Fiji, Southwest Pacific. *Unpublished PhD thesis*, Cornell University, Ithaca, New York.

Taylor, F.W., Isacks, B.L., Jouannic, C., Bloom, A.L. and Dubois, J. 1980. Coseismic and Quaternary vertical tectonic movements, Santo and Malekula islands, New Hebrides island arc. *Journal of Geophysical Research*, 85, 5367–5381.

Taylor, F.W., Jouannic, C. and Bloom, A.L. 1985. Quaternary uplift of the Torres islands, northern New Hebrides frontal arc: comparison with Santo and Malekula islands, central New Hebrides frontal arc. *Journal of Geology*, 93, 419–438.

Thiede, J. 1979. Wind regimes over the late Quaternary southwest Pacific Ocean. *Geology*, 7, 259–262.

Thiede, J., Dean, W.E., Rea, D.K., Vallier, T.L. and Adelsack, C.G. 1981. The geologic history of the mid-Pacific mountains in the central North Pacific Ocean: a synthesis of deep sea drilling studies. *In:* Thiede, J. and Vallier, T.L. (editors). *Initial Reports of the Deep-Sea Drilling Project, volume 62*. Washington: United States Government Printing Office, 1073–1120.

Thompson, L.G. and Mosley-Thompson, E. 1987. Evidence of abrupt climatic change during the last 1,500 years recorded in ice cores from the tropical Quelccaya ice cap, Peru. *In:* Berger, W.H. and Labeyrie, L.D. (editors). *Abrupt Climatic Change*. Dordrecht: Reidel, 99–110.

Thompson, L.G. and Mosley-Thompson, E. 1992. *Tropical Ice Core Paleoclimatic Records, Quelccaya Ice Cap, Peru, AD 470–1984*. Byrd Polar Research Center, The Ohio State University, Miscellaneous Publication 321, 106 p.

Thompson, L.G., Mosley-Thompson, E., Davis, M.E., Lin, P.-N., Henderson, K.A., Cole-Dai, J., Bolzan, J.F. and Liu, K.-b. 1995. Late Glacial stage and Holocene tropical ice-core records from Huascarán, Peru. *Science*, 269, 46–50.

Thompson, R.S. 1991. Pliocene environments and climates in the western United States. *Quaternary Science Reviews*, 10, 115–131.

Thompson, R.S., Whitlock, C., Bartlein, P.J., Harrison, S.P. and Spaulding, W.G. 1993. Climatic changes in the western United States since 18,000 yr B.P. *In:* Wright, H.E., Kutzbach, J.E., Webb, T., Ruddiman, W.F., Street-Perrott, F.A. and Bartleim, P.J. (editors). *Global Climates since the Last Glacial Maximum*. Minneapolis: University of Minnesota, 468–513.

Thomson, K.S. 1977. The pattern of diversification among fishes. *In:* Hallam, A. (editor). *Patterns of Evolution as Illustrated by the Fossil Record*. Amsterdam: Elsevier, 377–404.

Thunell, R., Anderson, D., Gellar, D. and Miao, Q. 1994. Sea-surface temperature estimates for the tropical western Pacific during the Last Glaciation and their implications for the Pacific Warm Pool. *Quaternary Research*, 41, 255–264.

Tingey, R.J. 1985. Uplift in Antarctica. *Zeitschrift für Geomorphologie*, Supplementband 54, 85–99.

Tjia, H.D. 1970. Rates of diastrophic movement during the Quaternary in Indonesia. *Geologie en Mijnbouw*, 49, 335–338.

Torgersen, T., Luly, J., De Deckker, P., Jones, M.R., Searle, D.E., Chivas, A.R. and Ullman, W.J. 1988. Late Quaternary environments of the Carpentaria Basin, Australia. *Palaeogeography, Palaeoclimatology, Palaeoecology*, 67, 245–261.

Truswell, E.M. 1991. Antarctica: a history of terrestrial vegetation. *In:* Tingey, R. (editor). *The Geology of Antarctica*. Oxford: Clarendon, 499–537.

Tryon, R. and Tryon, A.F. 1982. *Ferns and Allied Plants with Special Reference to North America*. New York: Springer, 857 p.

Tsuchi, R. 1990. Neogene events in Japan and the Pacific. *Palaeogeography, Palaeoclimatology, Palaeoecology*, 77, 355–365.

Tsuchi, R. 1992. Pacific Neogene climatic optimum and accelerated biotic evolution in time and space. *In:* Tsuchi, R. and Ingle, J.C. (editors). *Pacific Neogene: environment, evolution, and events*. University of Tokyo Press, 237–250.

Tsukada, M. 1967. Vegetation in subtropical Formosa during the Pleistocene glaciations and the Holocene. *Palaeogeography, Palaeoclimatology, Palaeoecology*, 3, 49–64.

Tsukada, M. 1983. Vegetation and climate during the last glacial maximum in Japan. *Quaternary Research*, 19, 212–235.

Tsukada, M. 1986. Vegetation in prehistoric Japan: the last 20,000 years. *In:* Pearson, R.J. (editor). *Windows on the Japanese Past: studies in archaeology and prehistory*. Ann Arbor: Center for Japanese Studies, 11–56.

Turnbull, D. 1994. Comparing knowledge systems: Pacific navigation and western science. *In:* Morrison, J., Geraghty, P. and Crowl, L. (editors). *Science of Pacific Island Peoples, Volume I, Ocean and Coastal Studies*. Suva: Institute of Pacific Studies, The University of the South Pacific, 129–144.

Turner, B.L. 1990. The rise and fall of population and agriculture in the central Maya lowlands: 300 BC to present. *In:* Newman, L.F. (editor). *Hunger in History*. Oxford: Blackwell, 178–211.

Umitsu, M. 1991. Holocene sea-level changes and coastal evolution in Japan. *The Quaternary Research (Japan)*, 30, 187–196.

Upchurch, G.R. and Wolfe, J.A. 1987. Mid-Cretaceous to early Tertiary vegetation and climate: evidence from fossil leaves and woods. *In*: Friis, E.M., Chaloner, W.G. and Crane, P.R. (editors). *The Origins of Angiosperms and their Biological Consequences*. Cambridge: Cambridge University Press, 75–105.

Urich, P.B. 1995. Resource control and environmental change in the Philippines: a case study in the Province of Bohol. *Unpublished PhD thesis*, Australian National University, Canberra, 2 volumes.

Urrutia-Fucugauchi, J., Lozano-García, S., Ortega-Guerrero, B. and Caballero-Miranda, M. 1995. Palaeomagnetic and palaeoenvironmental studies in the southern basin of Mexico – II. Late Pleistocene–Holocene Chalco lacustrine record. *Geofísica Internacional*, 34, 33–53.

Vail, P.R. and Hardenbol, J. 1979. Sea-level changes during the Tertiary. *Oceanus*, 22, 71–79.

Vail, P.R., Mitchum, R.M. and Thompson, S. 1977. Seismic stratigraphy and global changes of sea level, Part 4: global cycles of relative changes of sea level. *In:* Payton, C.E. (editor). *Seismic Stratigraphy – Applications to Hydrocarbon Exploration*. Tulsa: American Association of Petroleum Geologists, Memoir 26, 83–97.

Van Campo, E., Duplessy, J.C., Prell, W.L., Barratt, N. and Sabatier, R. 1990. Comparison of terrestrial and marine temperature estimates for the past 135 kyr off southeast Africa: a test for GCM simulations of palaeoclimate. *Nature*, 348, 209–212.

van der Hammen, T. and Gonzalez, S. 1960. A pollen diagram from 'Laguna de la Herrera' (Sabana de Bogotá). *Leidse Geologische Mededelingen*, 32, 183–191.

Veevers, J.J. 1988. Gondwana facies started when Gondwanaland merged into Pangaea. *Geology*, 16, 732–734.

Veit, H. 1993. Upper Quaternary landscape and climate evolution in the Norto Chico (northern Chile): an overview. *Mountain Research and Development*, 13, 139–144.

Vilas, F., Arche, A., Ferrero, M., Bujalesky, G., Isla, F. and Gonzalez Bonorino, G. 1987. Esquema evolutivo de la sedimentación reciente en la Bahía San Sebastián, Tierra del Fuego, Argentina. *Thalassas*, 5, 33–36.

Villagrán, C. and Varela, J. 1990. Palynological evidence for increased aridity on the central Chilean coast during the Holocene. *Quaternary Research*, 34, 198–207.

Villalba, R. 1990. Climatic fluctuations in northern Patagonia during the last 1000 years as inferred from tree-ring records. *Quaternary Research*, 34, 346–360.

Vincent, E. and Berger, W.H. 1981. Carbon dioxide and polar cooling in the Miocene: the Monterey hypothesis. *In:* Sunquist, E.T. and Broecker, W.S. (editors). *The Carbon Cycle and Atmospheric CO_2: natural variations, Archean to present*. Washington DC: American Geophysical Union, Geophysical Monograph 32, 455–468.

Vogt, P.R. 1979. Global magmatic episodes: new evidence and implications for the steady state mid-oceanic ridge. *Geology*, 7, 93–98.

Waddell, E. 1972. *The Mound Builders: agricultural practices, environment, and society in the central highlands of New Guinea*. Seattle: University of Washington Press, 253 p.

Walker, D. and Chen, Y. 1987. Palynological light on tropical rainforest dynamics. *Quaternary Science Reviews*, 6, 77–92.

Walker, G.P.L. 1990. Geology and volcanology of the Hawaiian islands. *Pacific Science*, 44, 315–347.

Walker, G.P.L. 1995. Plant molds in Hawaiian basalts:

was Oahu a desert, and why? *Journal of Geology*, 103, 85–93.

Wang, H., Ambrose, S.H., Liu. C-L.J. and Follmer, L.R. 1997. Paleosol stable isotope evidence for early hominid occupation of East Asian temperate environments. *Quaternary Research*, 48, 228–238.

Wang, J., Wang, X. and Chen, Z. 1978. Xiachuan culture. *Acta Archaeologica Sinica*, 3, 259–288 [in Chinese].

Wang, L. 1994. Sea surface temperature history of the low latitude western Pacific during the last 5.3 million years. *Palaeogeography, Palaeoclimatology, Palaeoecology*, 108, 379–436.

Wang, P. 1984. Progress in late Cenozoic palaeoclimatology of China: a brief review. *In:* Whyte, R.O. (editor). *The Evolution of the East Asian Environment.* Centre of Asian Studies, University of Hong Kong, 165–187.

Wang, Y. 1998. Sea-level changes, human impacts and coastal responses in China. *Journal of Coastal Research*, 14, 31–36.

Ward, R.G. 1989. Earth's empty quarter? The Pacific islands in the Pacific century. *The Geographical Journal*, 155, 235–246.

Ward, R.G. and Brookfield, M. 1992. The dispersal of the coconut: did it float or was it carried to Panama? *Journal of Biogeography*, 19, 467–480.

Wardle, P. 1973. Variations of the glaciers of Westland National Park and the Hooker Range, New Zealand. *New Zealand Journal of Botany*, 11, 349–388.

Waterhouse, J.B. 1987. Perceptions of the Permian Pacific – the Medusa hypothesis. *In: Pacific Rim Congress 87: the geology, structure, mineralisation and economics of the Pacific Rim.* Parville, Australia: Australian Institute of Mining and Metallurgy, 607–614.

Waythomas, C.F. 1996. Late Quaternary aeolian deposits of the Holitna lowland, interior southwestern Alaska. *In:* West, G.H. (editor). *American Beginnings: the prehistory and palaeoecology of Beringia.* Chicago: The University of Chicago Press, 35–52.

Weaver, S.D., Storey, B.C., Pankhurst, R.J., Mukasa, S.B., DiVenere, V.J. and Bradshaw, J.D. 1994. Antarctica–New Zealand rifting and Marie Byrd Land lithospheric magmatism linked to ridge subduction and mantle plume activity. *Geology*, 22, 811–814.

Webb, P.-N. and Harwood, D.M. 1987. Terrestrial flora of the Sirius Formation: its significance for late Cenozoic glacial history. *Antarctic Journal of the United States*, 22, 7–11.

Webb, P.-N., Harwood, D.M., McKelvey, B.C., Mabin, M.C.G. and Stott, L.D. 1986. Late Cenozoic tectonic and glacial history of the Transantarctic Mountains. *Antarctic Journal of the United States*, 21, 99–100.

Webb, P.-N., Harwood, D.M., Mabin, M.G.C. and McKelvey, B.C. 1996. A marine and terrestrial Sirius Group succession, middle Beardmore Glacier-Queen Alexandra Range, Transantarctic Mountains, Antarctica. *Marine Micropaleontology*, 27, 273–297.

Webb, S.D. 1985. Main pathways of mammalian diversification in North America. *In:* Stehli, F.G. and Webb, S.D. (editors). *The Great American Biotic Interchange.* New York: Plenum Press, 201–247.

Webb, T. 1985. Holocene palynology and climate. *In:* Hecht, A.D. (editor). *Paleoclimate Analysis and Modeling.* Chichester: Wiley, 163–195.

Wegener, A. 1915. *Die Entstehung der Kontinente und Ozeane.* Braunschweig: Vieweg, 231 p.

Wegener, A. 1929. *Die Entstehung der Kontinente und Ozeane [The Origin of Continents and Oceans,* translated by J. Biram, 1967]. London: Methuen, 4th edition (revised), 246 p.

Wells, L.E. 1990. Holocene history of the El Niño phenomenon as recorded in flood sediments of northern coastal Peru. *Geology*, 18, 1134–1137.

Wells, L.E. 1992. Holocene landscape change on the Santa Delta, Peru: impact on archaeological site distributions. *The Holocene*, 2–3, 193–204.

West, R.C. 1994. Aboriginal metallurgy and metalworking in Spanish America: a brief overview. *In:* Craig, A.K. and West, R.C. (editors). *In Quest of Mineral Wealth: aboriginal and colonial mining and metallurgy in Spanish America.* Baton Rouge: Department of Geography and Anthropology, Louisiana State University, 5–20.

Westermann, G.E.G. 1981. Ammonite biochronology and biogeography of the circum-Pacific Middle Jurassic. *In:* House, M.R. and Senior, J.R. (editors). *The Ammonoidea.* London: Academic Press, Systematics Association, Special Volume 18, 459–498.

Westgate, J.A., Stemper, B.A. and Péwé, T.L. 1990. A 3 m.y. record of Pliocene–Pleistocene loess in interior Alaska. *Geology*, 18, 858–861.

White, J.M., Ager, T.A., Adam, D.P., Leopold, E.B., Liu, G., Jetté, H. and Schweger, C.E. 1997. An 18 million year record of vegetation and climate change in northwestern Canada and Alaska: tectonic and global climatic correlates. *Palaeogeography, Palaeoclimatology, Palaeoecology*, 130, 293–306.

White, J.P. 1994. Site 820 and the evidence for early occupation in Australia. *Quaternary Australasia*, 12, 21–23.

White, S.E. 1986. Quaternary glacial stratigraphy and chronology of Mexico. *Quaternary Science Reviews*, 5, 201–205.

Whitney, G.G. 1994. *From Coastal Wilderness to Fruited Plain: a history of environmental change in temperate North America 1500 to the present.* Cambridge: Cambridge University Press, 451 p.

Wickler, S. and Spriggs, M.E. 1988. Pleistocene human occupation of the Solomon Islands, Melanesia. *Antiquity*, 62, 703–706.

Wiles, G.C. and Calkin, P.E. 1994. Late Holocene, high-resolution glacial chronologies and climate, Kenai Mountains, Alaska. *Geological Society of America, Bulletin*, 106, 281–303.

Wilford, G.E. and Brown, P.J. 1994. Maps of late Mesozoic–Cenozoic Gondwana break-up: some palaeogeographical implications. *In:* Hill, R.S. (editor). *History of the Australian Vegetation: Cretaceous to Recent*. Cambridge: Cambridge University Press, 5–13.

Williams, D.F., Thunell, R.C., Tappa, E., Rio, D. and Raffi, I. 1988. Chronology of the Pleistocene oxygen isotope record: 0–1.88 m.y. B.P. *Palaeogeography, Palaeoclimatology, Palaeoecology*, 64, 221–240.

Williams, M.A.J., Dunkerley, D.L., DeDeckker, P., Kershaw, A.P. and Stokes, T. 1993. *Quaternary Environments*. London: Edward Arnold, 329 p.

Willig, J.A. 1996. Environmental context for early human occupation in western North America. *In:* Akasawa, T. and Szathmáry, E.J.E. (editors). *Prehistoric Mongoloid Dispersals*. Oxford: Oxford University Press, 237–272.

Willis, P.M.A. and Molnar, R.E. 1991. A new middle Tertiary crocodile from Lake Palankarinna, South Australia. *South Australian Museum Records*, 2, 39–55.

Wilson, A.T., Hendy, C.H. and Reynolds, C.P. 1979. Short-term climate change and New Zealand temperatures during the last millennium. *Nature*, 279, 315–317.

Wilson, D.J. 1988. *Prehispanic Settlement Patterns in the Lower Santa Valley, Peru*. Washington, DC: Smithsonian Institution Press, 590 p.

Wilson, J.T. 1969. The climatic effects of large-scale surges of ice sheets. *Canadian Journal of Earth Sciences*, 6, 911–918.

Winkler, M.G. and Wang, P.K. 1993. The late-Quaternary vegetation and climate of China. *In:* Wright, H.E., Kutzbach, J.E., Webb, T., Ruddiman, W.F., Street-Perrott, F.A. and Bartleim, P.J. (editors). *Global Climates since the Last Glacial Maximum*. Minneapolis: University of Minnesota, 221–264.

Winograd, I.J., Szabo, B.J., Coplen, T.B., Riggs, A.C. and Kolesar, P.T. 1985. Two-million-year record of deuterium depletion in Great Basin ground waters. *Science*, 227, 519–522.

Wirrmann, D. and Mourguiart, P. 1995. Late Quaternary spatio-temporal limnological variations in the altiplano of Bolivia and Peru. *Quaternary Research*, 43, 344–354.

Wirrmann, D., Mourguiart, P. and de Oliveira Almeida, L.F. 1988. Holocene sedimentology and ostracod repartition in Lake Titicaca – paleohydrological interpretations. *In:* Rabassa, J. (editor). *Quaternary of South America and Antarctic Peninsula, Volume 6*. Rotterdam: Balkema, 89–128.

Wolfe, J.A. 1978. A paleobotanical interpretation of Tertiary climates in the northern hemisphere. *American Scientist*, 66, 694–703.

Wolfe, J.A. 1980. Tertiary climates and floristic relationships at high latitudes in the northern hemisphere. *Palaeogeography, Palaeoclimatology, Palaeoecology*, 30, 313–325.

Wolfe, J.A. 1990. Palaeobotanical evidence for a marked temperature increase following the Cretaceous/Tertiary boundary. *Nature*, 343, 153–156.

Wolfe, J.A. 1994a. Tertiary climate changes at middle latitudes of western North America. *Palaeogeography, Palaeoclimatology, Palaeoecology*, 108, 195–205.

Wolfe, J.A. 1994b. An analysis of Neogene climates in Beringia. *Palaeogeography, Palaeoclimatology, Palaeoecology*, 108, 207–216.

Wolfe, J.A. 1995. Paleoclimatic estimates from Tertiary leaf assemblages. *Annual Review of Earth and Planetary Science*, 23, 119–142.

Woodburne, M.O. and Zinsmeister, W.J. 1982. Fossil land mammal from Antarctica. *Science*, 218, 284–286.

Woodroffe, C.D., Thom, B.G. and Chappell, J. 1985. Development of widespread mangrove swamps in mid-Holocene times in northern Australia. *Nature*, 317, 711–713.

Woodruff, F., Savin, S.M. and Douglas, R.G. 1981. Miocene stable isotope record: a detailed Pacific ocean study and its paleoclimatic implications. *Science*, 212, 665–668.

Worster, D. 1992. *Under Western Skies: nature and history in the American west*. New York: Oxford University Press, 292 p.

Xiao, J., Inouchi, Y., Kumai, H., Yoshikawa, S., Kondo, Y., Liu, T. and An, Z. 1997a. Eolian quartz flux to Lake Biwa, central Japan, over the past 145,000 years. *Quaternary Research*, 48, 48–57.

Xiao, J., Inouchi, Y., Kumai, H., Yoshikawa, S., Kondo, Y., Liu, T. and An, Z. 1997b. Biogenic silica record in Lake Biwa of central Japan over the past 145,000 years. *Quaternary Research*, 47, 277–283.

Xu, D-Y., Zhang, Q-W., Sun, Y-Y., Yan, Z., Chai, Z-F. and He, J-W. 1989. *Astrogeological Events in China*. New York: Van Nostrand Reinhold, 264 p.

Yan, W. 1993. Origins of agriculture and animal husbandry in China. *In:* Aikens, C.M. and Song, N.R. (editors). *Pacific Northeast Asia in Prehistory*. Pullman: Washington State University Press, 113–123.

Yang, Z., Emery, K.O. and Xui, Y. 1989. Historical development and use of thousand-year-old tide prediction tables. *Limnology and Oceanography*, 34, 953–957.

Yasuda, Y. 1976. Early historic forest clearance around the ancient castle site of Tagajo, Miyagi Prefecture, Japan. *Asian Perspectives*, 19, 42–58.

Yates, R.D.S. 1990. War, food shortages, and relief measures in early China. *In:* Newman, L.F. (editor). *Hunger in History*. Oxford: Blackwell, 147–177.

Yinyun, Z. 1991. Human fossils from Anhui, southeast China: coexistence of *Homo erectus* and *Homo sapiens*. *Bulletin of the Indo-Pacific Prehistory Association*, 10, 79–82.

Yokoyama, Y., Nakada, M., Maeda, Y., Nagaoka, S., Okuno, J., Matsumoto, E., Sato, H. and Matsushima, Y. 1996. Holocene sea-level change and hydro-isostasy along the west coast of Kyushu, Japan. *Palaeogeography, Palaeoclimatology, Palaeo-ecology*, 123, 29–47.

Yole, R.W. and Irving, E. 1980. Displacement of Vancouver island: paleomagnetic evidence from the Karmutsen Formation. *Canadian Journal of Earth Sciences*, 17, 1210–1228.

Yonekura, N. 1983. Late Quaternary vertical crustal movements in and around the Pacific as deduced from former shoreline data. *In: Geodynamics of the Western Pacific: Indonesian Region*. Washington: American Geophysical Union, 41–50.

Yonekura, N., Ishii, T., Saito, Y., Maeda, Y., Matsushima, Y., Matsumoto, E. and Kayanne, H. 1988. Holocene fringing reefs and sea-level change in Mangaia island, southern Cook Islands. *Palaeogeography, Palaeoclimatology, Palaeo-ecology*, 68, 177–188.

Young, L.D. 1994. Quaternary lithostratigraphy of the Korean Peninsula. *In: Quaternary Stratigraphy of Asia and the Pacific IGCP 296*. ESCAP, Mineral Resources Development Series 63, 123–136.

Young, R.W. and Bryant, E.A. 1993. Coastal rock platforms and ramps of Pleistocene and Tertiary age in southern New South Wales, Australia. *Zeitschrift für Geomorphologie*, 37, 257–272.

Young, S.B. 1982. The vegetation of land-bridge Beringia. *In:* Hopkins, D.M., Matthews, J.V., Schweger, C.E. and Young, S.B. (editors). *Paleoecology of Beringia*. New York: Academic Press, 179–191.

Zhang, D. 1994. Evidence for the existence of the Medieval Warm Period in China. *Climatic Change*, 26, 289–297.

Zhang, J. (editor). 1988. *The Reconstruction of Climate in China for Historical Times*. Beijing: Science Press. 174 p.

Zhang, J. and Crowley, T.J. 1989. Historical climate records in China and reconstruction of past climates. *Journal of Climate*, 2, 833–849.

Zhang, Z.M. 1985. Tectonostratigraphic terranes of Japan that bear on the tectonics of mainland Asia. *In:* Howell, D.G. (editor). *Tectonostratigraphic Terranes of the Circum-Pacific Region*. Houston: Circum-Pacific Council for Energy and Mineral Resources, Earth Science Series, No. 1, 409–420.

Zhang, Z.M., Liou, J.G. and Coleman, R.G. 1989. The Mesozoic and Cenozoic tectonism in eastern China. *In:* Ben-Avraham, Z. (editor). *The Evolution of the Pacific Ocean Margins*. New York: Oxford University Press, 124–139.

Zhao, X., Tang, L., Wang, S. and Shen, C. 1995. Holocene climatic and sea-level changes in Qingfeng section, Jianhu County, Jiangsu Province: a typical example along the coastal areas in China. *Journal of Coastal Research, Special Issue*, 17, 155–162.

Zheng, X., Zhang, W., Yu, L. and Kunihiko, E. 1994. Paleoenvironmental changes in southern Yangtze delta over the last 20,000 years. *The Quaternary Research (Japan)*, 33, 379–384.

Zhimin, A. 1989. Prehistoric agriculture in China. *In:* Harris, D.R. and Hillman, G.C. (editors). *Foraging and Farming: the evolution of plant exploitation*. London: Unwin Hyman, 643–649.

Ziegler, A.M., Scotese, C.R. and Barrett, S.F. 1983. Mesozoic and Cenozoic paleogeographic maps. *In:* Brosche J. and Sündermann, J. (editors). *Tidal Friction and the Earth's Rotation, Volume II*. Berlin: Springer, 240–252.

Zinsmeister, W.J. 1982. Late Cretaceous–Early Tertiary molluscan biogeography of the southern circum-Pacific. *Journal of Paleontology*, 56, 84–102.

Zinsmeister, W.J. and Feldmann, R.M. 1996. Late Cretaceous faunal changes in the high southern latitudes: a harbinger of global biotic catastrophe? *In:* MacLeod, N. and Keller, G. (editors). *Cretaceous–Tertiary Mass Extinctions: biotic and environmental changes*. New York: Norton, 303–325.

Zubakov, V.A. and Borzhenkova, I.I. 1983. *Paleoclimates of Late Cenozoic*. Leningrad: Gidrometeoizdat, 214 p.

Geographical Index

Subject Index